# Advanced
# mathematics
# 2

## C W Celia
*formerly Principal Lecturer in Mathematics,*
*City of London Polytechnic*

## A T F Nice
*Mathematics Department, Lady Eleanor Holles*
*School, Hampton; formerly Principal Lecturer*
*in Mathematics, Middlesex Polytechnic*

## K F Elliott
*formerly Head of the Division of Mathematics*
*Education, Derby Lonsdale College of Higher*
*Education*

*Consultant Editor: Dr C. Plumpton, Moderator*
*in Mathematics, University of London Schools*
*Examination Department: formerly Reader in*
*Engineering Mathematics, Queen Mary*
*College, London*

MACMILLAN

First published 1982
Reprinted 1983, 1984, 1985

Published by
MACMILLAN EDUCATION LTD
Houndmills, Basingstoke, Hampshire RG21 2XS
and London
Companies and representatives
throughout the world

Printed in Hong Kong

# Contents

# Preface

This is the second of a series of books written for students preparing for A level Mathematics. Books 1 and 2 of the series cover the work required for a single subject A level in Mathematics or in Pure and Applied Mathematics. Book 1 covers the essential core of sixth-form mathematics now accepted by the GCE Boards, whilst Book 2 covers the applied mathematics, i.e. the numerical methods, mechanics and probability, contained in most single-subject syllabuses. Book 3 covers the additional pure mathematics needed by students taking the double subject Mathematics and Further Mathematics, and by those taking Pure Mathematics as a single subject.

Vector notation and vector techniques are used wherever appropriate, particularly where these methods illuminate or simplify the work, but their use is avoided whenever they appear likely to confuse the student. Set language is employed wherever it is considered helpful, but is not introduced at all times as a matter of principle.

The material is arranged under well-known headings and is organised so that the teacher is free to follow his or her own preferred order of treatment. The chapter contents are listed and an index is also provided to make it easy for both the teacher and the student to refer back rapidly to any particular topic. For ease of reference, a list of the notation used is given at the back of the book together with a list of formulae.

The approach in Book 2 is the same as in Book 1, each topic being developed mainly through worked examples. There is a brief introduction to each new piece of work followed by worked examples and numerous simple exercises to build up the student's technical skills and to reinforce his or her understanding. It is hoped that this approach will enable the individual student working alone to make effective use of the books and the teacher to use them with mixed ability groups. At the end of each chapter there are many miscellaneous examples, taken largely from past A level examination papers. In addition to their value as examination preparation, these miscellaneous examples are intended to give the student the opportunity to apply the techniques acquired from the exercises throughout the chapter to a considerable range of problems of the appropriate standard.

We are most grateful to the University of London University Entrance and School Examinations Council (L), the Associated Examining Board (AEB), the University of Cambridge Local Examination Syndicate (C) and the Joint Matriculation Board (JMB) for giving us permission to use questions from their past examination papers.

We are also grateful to the staff of Macmillan, particularly Mrs Janet Hawkins, Mrs Anne Russell and Mr Tony Feldman for the patience they have shown and the help they have given us in the preparations of these books.

C. W. Celia
A. T. F. Nice
K. F. Elliott

# The International System of Units (SI)

| Physical quantity | Name of unit | Unit symbol |
|---|---|---|
| *Basic units* | | |
| mass | kilogram | kg |
| length | metre | m |
| time | second | s |
| *Derived units* | | |
| force | newton | N |
| work, energy | joule | J |
| power | watt | W |
| velocity | metre per second | $m\,s^{-1}$ |
| acceleration | metre per second per second | $m\,s^{-2}$ |
| moment | newton metre | $N\,m$ |
| impulse | newton second | $N\,s$ |
| momentum | kilogram metre per second | $kg\,m\,s^{-1}$ |

| Multiples and submultiples | | |
|---|---|---|
| kilometre | 1000 metres | km |
| centimetre | 0.01 metre | cm |
| millimetre | 0.001 metre | mm |
| kilowatt | 1000 watts | kW |
| tonne | 1000 kilograms | t |

*Notation*
Vectors are denoted by bold face letters, and their magnitudes by italic letters.
Where $\mathbf{F}$ denotes a force, $F$ will denote its magnitude $|\mathbf{F}|$.
Where $\mathbf{v}$ denotes a velocity, $v$ will denote its magnitude $|\mathbf{v}|$.

# 1 Graphical and numerical methods

## 1.1 Linear relations

When the variables $x$ and $y$ satisfy an equation of the form $y = mx + c$, it can be said that there is a linear relation between them. The graph of $y$ against $x$ will then be a straight line with gradient $m$. The constant $c$ gives the value of $y$ at the point at which the line crosses the $y$-axis, i.e. $c$ is the intercept of the line on the $y$-axis.

Suppose that several pairs of values of $x$ and $y$ have been found by measurement, so that they are subject to experimental error. The values of $y$ can be plotted against the values of $x$ on graph paper. When a linear relation is known to exist between $x$ and $y$, we expect the points to lie close to a straight line. Conversely, if the points are found to lie approximately in a straight line, it is possible (but not certain) that such a linear relation exists. To find the linear relation, the straight line that appears to fit the points best should be drawn on the graph. Then choose two points $P(x_1, y_1)$ and $Q(x_2, y_2)$ on this line, well apart as in Fig. 1.1. From the right-angled triangle PQR, the slope of PQ is given by

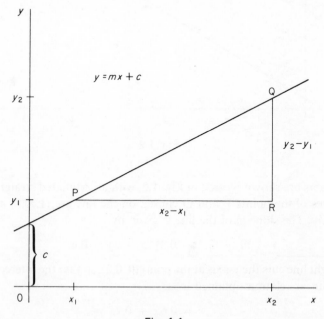

Fig. 1.1

$$m = \frac{QR}{PR} = \frac{y_2 - y_1}{x_2 - x_1}.$$

If $(x_2 - x_1)$ is small, the value found for $m$ may prove to be inaccurate. It is an advantage to choose $x_1$ and $x_2$ so that division by $(x_2 - x_1)$ is simple. If the straight line, produced if necessary, meets the $y$-axis on the graph paper, the value of $c$ can be found directly. Otherwise the intercept $c$ can be found by evaluating $y_1 - mx_1$. The accuracy of the values obtained for $m$ and $c$ often depends on the scales chosen on the two axes. These scales should be taken as large as the graph paper permits, and they should be such that it is easy to make intermediate readings from the axes.

*Example*
The table gives pairs of values of the variables $x$ and $y$ found by experiment. Assuming that a linear relation $y = mx + c$ exists, estimate the values of the constants $m$ and $c$ to one decimal place.

| $x$ | 1.7 | 2.3 | 3.4 | 4.4 | 5.3 |
|---|---|---|---|---|---|
| $y$ | 1.1 | 1.7 | 2.1 | 2.7 | 3.5 |

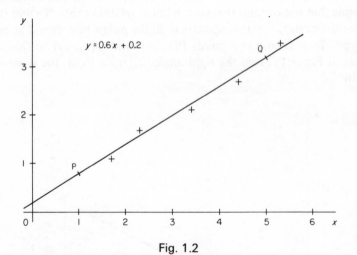

Fig. 1.2

The points are shown plotted in Fig. 1.2, with an estimated straight line. The coordinates of the points P and Q chosen on the line are $(1, 0.8)$ and $(5, 3.2)$ respectively. The slope $m$ of the line is given by

$$m = (3.2 - 0.8)/(5 - 1) = 0.6.$$

The straight line cuts the $y$-axis at the point $(0, 0.2)$, so that the intercept $c$ equals 0.2. The linear relation obtained is

$$y = 0.6x + 0.2.$$

**Exercises 1.1**

1 The coordinates of the points A, B, C, D are respectively (2.1, 1.3), (3.3, 1.9), (1.2, 8.5), (−1.5, −4.1). Find the gradients of the lines AB, AC and AD.

2 In each of the following cases a linear relation exists between the variables. Use the data to find the relation.
   (a) When $t = 2$, $s = 7$, and when $t = 3$, $s = 10$.
   (b) When $u = 1$, $v = 4$, and when $u = 5$, $v = 10$.
   (c) When $p = 2$, $q = -1$, and when $p = 6$, $q = 2$.
   (d) When $x = 1$, $y = 4$, and when $x = 2$, $y = -1$.

3 The length $L$ of a spiral spring is related by a linear law to the tension $T$ in the spring. Use the measurements in the table to obtain a linear relation between $L$ and $T$.

| $L$ (cm) | 23.5 | 28.4 | 32.3 | 35.7 |
|---|---|---|---|---|
| $T$ (newton) | 5 | 10 | 15 | 20 |

4 The variables $v$ and $t$ satisfy a linear relation $v = u + ft$, where $u$ and $f$ are constants. Use the table below to estimate the values of $u$ and $f$ to one decimal place.

| $v$ | 4.1 | 3.7 | 2.9 | 2.2 | 1.6 | 0.9 |
|---|---|---|---|---|---|---|
| $t$ | 1.2 | 1.6 | 2.1 | 2.5 | 2.9 | 3.4 |

5 The variables $x$ and $y$ satisfy a linear relation. The table gives five pairs of measured values, and it is suspected that an error was made in one measurement of $y$. Locate the error and estimate the correct value.

| $x$ | 2.1 | 3.4 | 4.2 | 5.1 | 6.5 |
|---|---|---|---|---|---|
| $y$ | 10.4 | 14.1 | 16.5 | 20.3 | 23.6 |

## 1.2 Determination of laws

When two variables satisfy a law which is not linear, it may be possible to obtain a linear relation by a change of variables. The method of the previous section can then be applied to find the values of the constants.

For a point moving with constant acceleration $f$, the distance $s$ travelled in time $t$ satisfies the law

$$s = ut + \tfrac{1}{2}ft^2.$$

This is not a linear relation between $s$ and $t$, but it can be rewritten in the form

$$s/t = u + \tfrac{1}{2}ft.$$

The substitution $X = t$, $Y = s/t$ produces the linear relation

$$Y = u + \tfrac{1}{2}fX.$$

If values of $Y$ are plotted against values of $X$, the gradient of the straight line obtained will give the value of $\frac{1}{2}f$ and the intercept on the $Y$-axis will give the value of $u$.

Similarly if $x$ and $y$ satisfy the law $ax + by = xy$, the law can be rewritten in the form

$$a/y + b/x = 1$$

Then the substitution $X = 1/x$, $Y = 1/y$ gives

$$aY + bX = 1,$$

i.e. $$Y = (-b/a)X + 1/a.$$

By plotting values of $Y$ against values of $X$, a straight line will be obtained. Its gradient will give the value of $(-b/a)$ and the intercept made on the $Y$-axis will give the value of $1/a$. The values of $a$ and $b$ can then be deduced.

A law of the type $y = ax^n$ or of the type $y = ca^x$ can be converted to a linear relation by means of logarithms. The equation $y = ax^n$ is equivalent to the equation

$$\log y = \log a + n \log x.$$

The substitution $X = \log x$ and $Y = \log y$ gives the linear relation

$$Y = \log a + nX.$$

The graph of $Y$ against $X$ will be a straight line with gradient $n$ making an intercept equal to $\log a$ on the $Y$-axis. Alternatively the values of $y$ against $x$ can be plotted on log-log graph paper, i.e. paper on which each axis is marked with a logarithmic scale. Then the gradient of the straight line obtained will give the value of $n$, and the intercept made on the line $x = 1$ will give the value of $a$.

The equation $y = ca^x$ is equivalent to the equation

$$\log y = \log c + x \log a.$$

The substitution $X = x$ and $Y = \log y$ gives the linear relation

$$Y = \log c + X \log a.$$

In this case the gradient of the straight line will give the value of $\log a$, and the intercept on the $Y$-axis will give the value of $\log c$. Alternatively the constants in the law $y = ca^x$ can be evaluated by plotting values of $x$ and $y$ on semi-logarithmic graph paper, only the $y$-axis having a logarithmic scale. The gradient of the line will give the value of $\log a$, and the intercept made on the $y$-axis will give the value of $c$.

In dealing with the equation $y = ax^n$ or with the equation $y = ca^x$ the choice must be made between logarithms to the base 10 and natural logarithms. If tables are to be used, logarithms to the base 10 may be preferable, but if a calculator is used natural logarithms may be easier to handle.

*Example*

Find estimates of the values of the constants $a$ and $k$ in the law $y = ax^k$ satisfied by $x$ and $y$, given the pairs of measured values in the following table.

| $x$ | 1.32 | 1.78 | 2.19 | 2.51 | 3.02 |
|-----|------|------|------|------|------|
| $y$ | 1.06 | 1.26 | 1.41 | 1.55 | 1.74 |

Taking logarithms, the relation $y = ax^k$ becomes

$$\log y = \log a + \log(x^k)$$
$$= \log a + k \log x.$$

By the change of variables $X = \log x$, $Y = \log y$ this becomes

$$Y = kX + \log a.$$

The graph of $Y$ against $X$ will be a straight line of gradient $k$ making an intercept $\log a$ on the $Y$-axis. Using logarithms to the base 10 the table of values of $X$ and $Y$ is

| $X$ | 0.12 | 0.25 | 0.34 | 0.40 | 0.48 |
|-----|------|------|------|------|------|
| $Y$ | 0.025 | 0.10 | 0.15 | 0.19 | 0.24 |

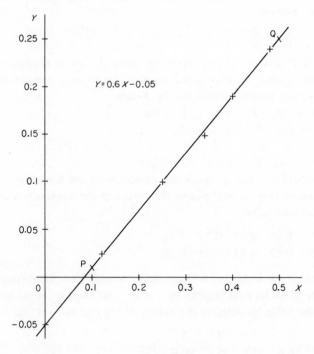

Fig. 1.3

The values of $Y$ are plotted against the values of $X$ in Fig. 1.3 and a suitable straight line is drawn. In order to find the gradient, two points P and Q on this line some distance apart are chosen.

At $P(X_1, Y_1)$, $X_1 = 0.1$, $Y_1 = 0.01$.

At $Q(X_2, Y_2)$, $X_2 = 0.5$, $Y_2 = 0.25$.

Then the gradient $k$ is given by

$$k = \frac{Y_2 - Y_1}{X_2 - X_1} = \frac{0.25 - 0.01}{0.5 - 0.1} = 0.6.$$

It can be seen from the graph that the intercept made by the line on the $Y$-axis is $-0.05$, i.e. $\log_{10} a = -0.05$. Now

$$-0.05 = -1 + 0.95,$$

and so $\log_{10} a$ equals $\bar{1}.95$. From tables, the value of $a$ correct to two decimal places is $0.89$. Therefore our estimate of the law satisfied by $x$ and $y$ is

$$y = 0.89x^{0.6}.$$

**Exercises 1.2**

1  For each of the following laws find a change of variables which will produce a linear relation:

(a) $y = ax^3 + bx$                       (b) $v = ae^{cu}$,

(c) $s = a(1 + t)^n$                   (d) $y = (a + b\sqrt{x})^2$.

2  Show that in each of the following cases the given change of variables produces a linear relation. Find the gradient of the corresponding straight line and the intercept made on the $Y$-axis.

(a) $xy = ay + b$;    $X = y$, $Y = xy$

(b) $a^x = by$;    $X = x$, $Y = \log y$

(c) $y^2 = ax^2 + 2y + b$;    $X = x^2$, $Y = (y - 1)^2$

(d) $x^k y = a$;    $X = \log x$, $Y = \log y$.

3  The variables $x$ and $y$ satisfy an equation of the form $xy = ay + bx$. Estimate to two decimal places the values of the constants $a$ and $b$, given the following table.

| $x$ | 0.25 | 0.33 | 0.47 | 0.71 | 1.96 |
|---|---|---|---|---|---|
| $y$ | 1.28 | 0.62 | 0.43 | 0.35 | 0.28 |

4  A point moves from rest with constant acceleration. Its distance $s$ from a fixed point on its path is given by $s = a + bt^2$, where $t$ is the time. Use the following table to estimate the values of the constants $a$ and $b$.

| $t$ | 1 | 2 | 3 | 4 | 5 |
|---|---|---|---|---|---|
| $s$ | 6.0 | 16.4 | 33.8 | 58.4 | 90.0 |

5  The table gives pairs of values of $x$ and $y$ found by experiment. The two variables satisfy a law of the form $y = ax^n$, where $a$ and $n$ are integers. Evaluate $a$ and $n$.

| $x$ | 0.6 | 0.7 | 0.75 | 0.8 | 0.9 |
|-----|-----|-----|------|-----|-----|
| $y$ | 13 | 24 | 32 | 41 | 66 |

6   The table gives corresponding values of $x$ and $y$. It is suspected that a relation exists of the form $y = a + b/x$. Find suitable values for the constants $a$ and $b$.

| $x$ | 1.00 | 0.40 | 0.25 | 0.17 | 0.12 | 0.10 |
|-----|------|------|------|------|------|------|
| $y$ | 29.2 | 26.1 | 23.5 | 21.0 | 14.9 | 12.5 |

7   The variables $x$ and $y$ satisfy a relation $y = kx^n$ where $k$ and $n$ are integers. Use the following table to find graphically the values of $k$ and $n$.

| $x$ | 1.34 | 3.58 | 7.60 | 12.1 | 14.8 |
|-----|------|------|------|------|------|
| $y$ | 208 | 10.9 | 1.14 | 0.282 | 0.154 |

8   It is known that $x$ and $y$ satisfy a relation of the form $y = ax^2 + bx$ where $a$ and $b$ are integers. The table gives pairs of measured values of $x$ and $y$. By drawing a straight-line graph estimate the values of $a$ and $b$.

| $x$ | 1 | 2 | 3 | 4 | 5 |
|-----|---|---|---|---|---|
| $y$ | 17.1 | 28.0 | 32.9 | 31.9 | 25.0 |

9   Given that $x$ and $y$ satisfy a law of the form $y = ab^x$, use the data in the table to estimate the values of the constants $a$ and $b$ to one decimal place.

| $x$ | 0.2 | 0.5 | 0.9 | 1.2 | 1.3 | 1.5 |
|-----|-----|-----|-----|-----|-----|-----|
| $y$ | 2.07 | 2.19 | 2.36 | 2.49 | 2.54 | 2.63 |

10   The variables $s$ and $t$ satisfy the relation $s = ut + \frac{1}{2}ft^2$. Use the data in the table to draw a straight-line graph from which to estimate $u$ and $f$. Give your answers to two significant figures.

| $t$ | 1 | 3 | 4 | 7 | 8 |
|-----|---|---|---|---|---|
| $s$ | 2.0 | 6.8 | 9.5 | 19.2 | 22.9 |

11   The table gives measured values of $x$ and $y$. Given that a relation of the form $y = ax^k$ exists, estimate the values of the constants $a$ and $k$.

| $x$ | 1.2 | 1.4 | 1.8 | 2.1 | 2.5 |
|-----|-----|-----|-----|-----|-----|
| $y$ | 1.60 | 1.45 | 1.31 | 1.20 | 1.09 |

12   A law of the form $y = ae^{nx}$ is satisfied by the variables $x$ and $y$, where $a$ and $n$ are integers. Use the following table to obtain a straight-line graph from which to evaluate $a$ and $n$.

| $x$ | 0.2 | 0.25 | 0.4 | 0.44 | 0.5 |
|-----|-----|------|-----|------|-----|
| $y$ | 2.68 | 2.43 | 1.80 | 1.66 | 1.47 |

## 1.3   Solution of equations

The real roots of an equation $f(x) = 0$ can be located by sketching the curve

$y = f(x)$ and noting the values of $x$ at the points where the curve meets the $x$-axis. If the values of $f(x)$ for a number of values of $x$ are tabulated, changes in the sign of $f(x)$ can be seen by inspection. Provided that the curve $y = f(x)$ is continuous, the curve will cross the $x$-axis between two values of $x$ for which $f(x)$ has opposite signs.

Let $f(x_1)$ be positive and $f(x_2)$ be negative. Then there is a root of the equation $f(x) = 0$ between $x_1$ and $x_2$. Now let $x_3 = \frac{1}{2}(x_1 + x_2)$. If $f(x_3)$ is positive, the root must lie between $x_3$ and $x_2$. If $f(x_3)$ is negative, the root must lie between $x_3$ and $x_1$. This is the basis of the method of repeated bisection, which is illustrated by the following example.

*Example 1*
Show that a root $\alpha$ of the equation

$$e^x - 3x + 0.2 = 0$$

lies between 1.32 and 1.36. Find $\alpha$ correct to two decimal places by repeated bisection.

Let $f(x) = e^x - 3x + 0.2$.

Then $\quad f(1.32) = 3.743 - 3.96 + 0.2 < 0$
$\qquad f(1.36) = 3.896 - 4.08 + 0.2 > 0.$

Since $f(x)$ is positive when $x = 1.36$ and negative when $x = 1.32$, there is a root between 1.32 and 1.36, i.e.

$$1.32 < \alpha < 1.36.$$

The mid-point of this interval is given by $x = 1.34$.

$$f(1.34) = 3.819 - 4.02 + 0.2 < 0.$$

Since $f(x)$ is positive when $x = 1.36$ and negative when $x = 1.34$,

$$1.34 < \alpha < 1.36.$$

The mid-point of this interval is given by $x = 1.35$.

$$f(1.35) = 3.857 - 4.05 + 0.2 > 0.$$

This gives

$$1.34 < \alpha < 1.35.$$

Finally,

$$f(1.345) = 3.838 - 4.035 + 0.2 > 0.$$

Hence

$$1.34 < \alpha < 1.345.$$

Therefore to two decimal places, $\alpha = 1.34$.

It is sometimes more convenient to find where a suitable curve crosses a line other than the x-axis. In *Advanced mathematics 1*, Section 4.8, the positive root of the equation $x - 2 \sin x = 0$ was found by drawing the curve $y = \sin x$ and the line $y = \frac{1}{2}x$. In Section 2.6, the real root of the equation $x^3 - 7x - 12 = 0$ was found by drawing the curve $y = x^3$ and the line $y = 7x + 12$. It was shown that greater accuracy could be obtained by drawing, on a larger scale, an arc of the curve containing the point at which it crosses the line.

## Linear interpolation

In Fig. 1.4 the arc PQ of the curve $y = f(x)$ crosses the x-axis at point A $(\alpha, 0)$, where $\alpha$ is a root of the equation $f(x) = 0$. The chord PQ crosses the x-axis at B, and the value of $x$ at B gives an approximation to the root $\alpha$.

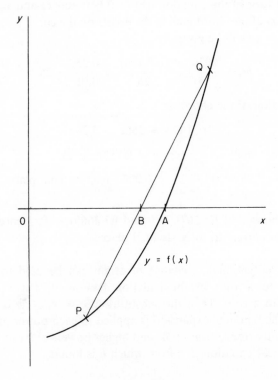

Fig. 1.4

Let P and Q be the points $(x_1, y_1)$ and $(x_2, y_2)$ respectively. Then the equation of the chord PQ is

$$y - y_1 = m(x - x_1),$$

where $m$ is the gradient of PQ, i.e.

$$m = (y_2 - y_1)/(x_2 - x_1).$$

The value of $x$ at B is found by substituting $y = 0$ in the equation of the chord. This gives

$$x = x_1 - y_1/m.$$

This value of $x$ will be a close approximation to the root if the gradient of the curve changes slowly on the arc PQ.

*Example 2*
Show that a root of the equation $x^3 - 7x - 12 = 0$ lies between 3.26 and 3.27. Evaluate this root to three decimal places by linear interpolation.

Put $f(x) = x^3 - 7x - 12$, and let $x_1 = 3.26$, $x_2 = 3.27$. Then to three decimal places, $f(x_1) = -0.174$, $f(x_2) = 0.076$. Since these two values have opposite signs, there is a root of the equation $f(x) = 0$ between $x_1$ and $x_2$.

The gradient $m$ of the chord joining the points on the curve $y = f(x)$ at which $x = x_1$ and $x = x_2$ will be given by

$$m = \frac{0.076 - (-0.174)}{3.27 - 3.26} = \frac{0.25}{0.01} = 25.$$

Hence the equation of the chord is

$$y + 0.174 = 25(x - 3.26).$$

This chord crosses the $x$-axis where $y = 0$. This gives

$$x - 3.26 = 0.174/25 = 0.007 \quad \text{to 3 decimal places.}$$

i.e. $\qquad\qquad x = 3.267.$

A calculation shows that $f(3.267) > 0$ and $f(3.2665) < 0$. Hence 3.267 is the value of the root correct to three decimal places.

When $f(x)$ is a polynomial, a change of origin can be used to improve an approximation to a root of the equation $f(x) = 0$. Let $x_1$ be a close approximation to a root. Then the substitution $x = x_1 + h$ is made in the equation, and the binomial expansion is applied to each power of $x$. Provided that $h$ is sufficiently small, terms in $h^2$ and higher powers of $h$ can be neglected. This leaves a linear equation in $h$ from which $h$ is found.

*Example 3*
The equation $x^3 + 8x - 28 = 0$ has a root near 2. Obtain a better approximation to this root.

When the substitution $x = 2 + h$ is made, the equation becomes

$$(2 + h)^3 + 8(2 + h) - 28 = 0$$

i.e. $\qquad (8 + 12h + 6h^2 + h^3) + 8h - 12 = 0$

Provided that $h$ is small, the terms $6h^2$ and $h^3$ can safely be neglected. Then

$$8 + 12h + 8h - 12 = 0$$
$$20h - 4 = 0$$
$$h = 0.2$$

The new approximation to the root is therefore 2.2.

The process can be repeated with the substitution $x = 2.2 + h$. This gives

$$(2.2 + h)^3 + 8(2.2 + h) - 28 = 0$$
$$(2.2)^3 + 3(2.2)^2 h + \ldots + 17.6 + 8h - 28 = 0$$
$$10.648 + 14.52h + 17.6 + 8h - 28 = 0$$
$$22.52h + 0.248 = 0$$
$$h = -0.01 \quad \text{to two decimal places.}$$

The third approximation to the root is therefore 2.19.

The substitution $x = 2.19 + h$ gives

$$(2.19)^3 + 3(2.19)^2 h + \ldots + 8(2.19 + h) - 28 = 0$$
$$10.50 + 14.39h + 8h - 10.48 = 0$$
$$22.39h + 0.02 = 0$$
$$h = -0.001 \quad \text{to three decimal places.}$$

This gives 2.189 as a close approximation to the root.

**Exercises 1.3**

1  Locate the roots of the equation $x^3 - 5x^2 - 17x + 20 = 0$. Evaluate each root to two decimal places.

2  By drawing the curve $y = \sin x$ and the line $2y = 2x - 1$, show that the equation $2 \sin x = 2x - 1$ has only one root and that its value is approximately 1.5.

3  Draw the curve $y = x^3$ for values of $x$ from $-3$ to 3. By drawing suitable straight lines, locate to one decimal place the roots of the equations
(a) $x^3 - 20 = 0$ (b) $x^3 - 5x = 0$ (c) $x^3 + 10x - 20 = 0$
(d) $x^3 + 5x - 10 = 0$ (e) $x^3 - 5x - 10 = 0$ (f) $x^3 + 5x + 10 = 0$.

4  Sketch the curve $y = e^x$ and use it to find the number of real roots of the equation $e^x = kx$ (a) when $k = 3$ (b) when $k = -3$. Evaluate each root to two decimal places.

5  The chord PQ of a circle of radius $a$ cuts off a segment of area $ka^2$. If the angle between the radii OP and OQ is $x$ radians, show that

$$x - \sin x = 2k.$$

Solve this equation (a) when $k = 1$ (b) when $k = \frac{1}{8}$.

6  Tangents PQ and PR are drawn to a circle with centre O and radius $a$. The area of the region bounded by PQ, PR and the arc QR is $3a^2/2$. If the angle QOR is $2\theta$ radians, show that

$$\tan \theta - \theta = 1.5.$$

Find the value of $\theta$ to two decimal places.

7   Find to the nearest integer the real root of the equation $x^3 - 2x - 5 = 0$. Use linear interpolation to evaluate the root to two decimal places.

8   Locate the roots of the equation $x^3 - 3x + 1 = 0$. Evaluate the negative root to two decimal places.

9   Show that the equation $x^2 + \sin x = 1$ has two roots. Find the positive root to two decimal places.

10  Show that the equation $x^3 + 2x^2 - 4 = 0$ has only one real root. Obtain an approximation to this root correct to two decimal places.

11  Show that the equation $x^4 + 4x^3 + 5x^2 = 9$ has only one positive root. Find this root to two decimal places.

12  Given that $\cos (0.7) = 0.7648$ and $\cos (0.8) = 0.6967$, find by linear interpolation an approximation to a root of the equation $x - \cos x = 0$.

13  Given that $e^2 = 7.3891$ and $e^{2.1} = 8.1662$, find to two decimal places the root of the equation $e^x + 4x = 16$.

## 1.4   The Newton-Raphson method

If the curve $y = f(x)$ crosses the $x$-axis at the point A$(a, 0)$, the equation $f(x) = 0$ will have a root equal to $a$. Let P$(x_1, y_1)$ be a point on the curve near A. The gradient of the curve at P is $f'(x_1)$, and the equation of the tangent at P is

$$y - f(x_1) = f'(x_1)(x - x_1).$$

If this tangent meets the $x$-axis at Q$(x_2, 0)$,

$$0 - f(x_1) = f'(x_1)(x_2 - x_1)$$

i.e.
$$x_2 = x_1 - \frac{f(x_1)}{f'(x_1)}.$$

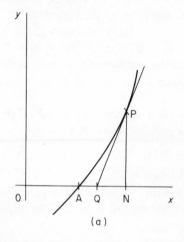

(a)

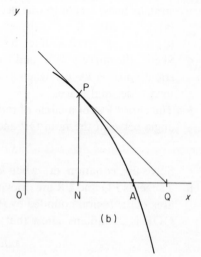

(b)

Fig. 1.5

In Fig. 1.5(a), Q lies between A and N, the foot of the perpendicular from P to the $x$-axis, but in Fig. 1.5(b), A lies between N and Q.

The formula

$$x_2 = x_1 - \frac{f(x_1)}{f'(x_1)}$$

is known as the Newton-Raphson formula. It was published in 1690 by Raphson, though it had already been discovered by Newton. Its purpose is to improve the accuracy of a first approximation $x_1$ to a root of an equation $f(x) = 0$. When Q lies between A and N, as in Fig. 1.5(a), $x_2$ is a better approximation to the root than $x_1$. When N and Q straddle the point A, as in Fig. 1.5(b), it is not always true that $x_2$ is the better approximation.

A great advantage can be gained by choosing the first approximation $x_1$ close to the root. It will then not be necessary to re-apply the Newton-Raphson formula several times.

*Example 1*
The equation $3x^2 - 8x - 2 = 0$ has a root between 2 and 3. Taking the first approximation $x_1$ to be 3, find a second and a third approximation to this root.

When $x_1 = 3$, $f(x_1) = 3x_1^2 - 8x_1 - 2 = 1$
$$f'(x_1) = 6x_1 - 8 = 10.$$

By the Newton-Raphson formula,

$$x_2 = x_1 - \frac{f(x_1)}{f'(x_1)} = 3 - 0.1 = 2.9.$$

The third approximation is given by

$$x_3 = x_2 - \frac{f(x_2)}{f'(x_2)},$$

where $f(x_2) = 25.23 - 23.2 - 2 = 0.03$
$$f'(x_2) = 17.4 - 8 = 9.4.$$

Therefore $x_3 = 2.9 - (0.03)/(9.4)$
$$= 2.8968 \quad \text{to 4 decimal places.}$$

It can be confirmed that this is the correct value of the root to 4 decimal places by showing that $3x^2 - 8x - 2$ is negative when $x = 2.89675$ and positive when $x = 2.89685$.

Note that the gradient of the curve

$$y = 3x^2 - 8x - 2,$$

which is given by $6x - 8$, is positive near $x = 3$, and that the gradient increases as $x$ increases. This means that the curve is similar to that shown in Fig. 1.5(a).

*Example 2*

Taking 2 as the initial approximation, find a second approximation to the cube root of 10 by applying the Newton-Raphson method
(a) to the equation $x^3 - 10 = 0$
(b) to the equation $x^2 - 10/x = 0$.
(Although these equations have the same root, it will be shown that they lead to different approximations.)

(a) If $f(x) = x^3 - 10$, $f(2) = 8 - 10 = -2$
$\quad f'(x) = 3x^2$, $f'(2) = 12$.

The second approximation $x_2$ is given by

$$x_2 = x_1 - \frac{f(x_1)}{f'(x_1)} = 2 - \frac{(-2)}{12} = 2.1\dot{6}.$$

(b) If $f(x) = x^2 - 10/x$, $f(2) = 4 - 5 = -1$
$\quad f'(x) = 2x + 10/x^2$, $f'(2) = 4 + 2.5 = 6.5$.

In this case, $x_2$ is given by

$$x_2 = 2 - (-1)/(6.5)$$
$$= 2.154 \quad \text{to 3 decimal places.}$$

This is the correct value of the cube root of 10 to 3 decimal places.

As the last example shows, a rearrangement of an equation changes the approximation produced by the Newton-Raphson method. In the following exercises it is intended that the method be applied to the given equations as they stand.

**Exercises 1.4**

1　Apply the Newton-Raphson method with the given value for the first approximation $x_1$ to find a second approximation $x_2$ to a root of each of the following equations. Verify by a diagram in each case that $x_2$ lies between $x_1$ and the root. Give your answers to two places of decimals.
(a) $x^4 + 4x - 20 = 0$; $x_1 = 2$
(b) $x^3 + 7 = 0$; $x_1 = -2$
(c) $x^3 - 4x + 2 = 0$; $x_1 = 0.5$
(d) $x^3 - 4x^2 - x + 2 = 0$; $x_1 = 1$.

2　Find the cube root of 3 to two decimal places by applying the Newton-Raphson method to the equation $x^3 - 3 = 0$. Take 1.5 as the initial approximation.

3　Show that the equation $e^x - 5x = 0$ has two real roots. Find each root correct to two decimal places.

4　Show that the equation $x^4 - x - 1 = 0$ has a root between 1.2 and 1.3. Use the Newton-Raphson method to show that to three decimal places this root equals 1.221.

5 Show that the equation $x - \ln x - 4 = 0$ has a root between 5 and 6. Apply the Newton-Raphson method to find a second approximation to two decimal places.
  (a) taking 5 as the first approximation
  (b) taking 6 as the first approximation.

6 Show that the equation $\cos(\pi x) - 3x = 0$ has only one real root. Taking 0.25 as the first approximation, evaluate the root to three decimal places.

7 Taking 2.8 as the first approximation, use the Newton-Raphson method to find a second approximation to a root of the equation $3x^2 - 8x - 1 = 0$. Show that this second approximation gives the root correct to three decimal places.

8 By applying the Newton-Raphson method to the equation $x^3 - 9 = 0$, show that to four decimal places the cube root of 9 is 2.0801.

9 Show that to three decimal places the real root of the equation $xe^x - 1 = 0$ is 0.567.

10 Taking 0.75 as the first approximation, find to four decimal places the real root of the equation $x - \cos x = 0$.

## 1.5  An iterative method for square roots

Consider the equation $x^2 - N = 0$, where $N$ is positive.

Let $f(x) = x^2 - N$, so that $f'(x) = 2x$. If $x_1$ is a first approximation to the positive square root of $N$, the second approximation given by the Newton-Raphson method will be $x_2$ where

$$x_2 = x_1 - \frac{(x_1^2 - N)}{2x_1},$$

i.e.

$$x_2 = \frac{1}{2}\left(x_1 + \frac{N}{x_1}\right).$$

If the same method is applied again, it will give

$$x_3 = \frac{1}{2}\left(x_2 + \frac{N}{x_2}\right).$$

The $n$th application will give

$$x_{n+1} = \frac{1}{2}\left(x_n + \frac{N}{x_n}\right).$$

This is an example of an iterative formula. Once the first approximation $x_1$ has been chosen, the repeated application of the formula produces a sequence of numbers $x_2, x_3, x_4, \ldots$. It can be shown that if $x_1$ is positive the sequence produced by the iterative formula above will converge to the square root of $N$, i.e. the limit of $x_n$ as $n$ tends to infinity will be $\sqrt{N}$. In general the behaviour of a sequence produced by an iterative formula will depend on the choice of $x_1$.

When $N = 2$, the above iteration formula becomes

$$x_{n+1} = \frac{1}{2}\left(x_n + \frac{2}{x_n}\right).$$

If $x_1$ is taken to be 1, this gives

$$x_2 = \tfrac{1}{2}(1 + 2) = 1.5$$

$$x_3 = \frac{1}{2}\left(\frac{3}{2} + \frac{4}{3}\right) = \frac{17}{12} = 1.4166\ldots$$

$$x_4 = \frac{1}{2}\left(\frac{17}{12} + \frac{24}{17}\right) = \frac{577}{408} = 1.414215\ldots$$

$$x_5 = \frac{1}{2}\left(\frac{577}{408} + \frac{816}{577}\right) = 1.414\,213\,562\ldots$$

This value of $x_5$ gives $\sqrt{2}$ correct to nine decimal places.

### Exercises 1.5
1  Use the method above to evaluate $\sqrt{10}$ to two decimal places taking
   (a) $x_1 = 3$  (b) $x_1 = 4$.
2  Show that with $x_1 = 2$, two applications of the iterative method in Section
   1.5 will give $\sqrt{3}$ correct to four decimal places.
3  Taking $x_1 = 2$, show that three applications of the above method give $\sqrt{6}$
   correct to five decimal places.

## 1.6  Equations of the form $x = F(x)$
When an equation $f(x) = 0$ can be put in the form $x = F(x)$, it may be possible
to solve the equation by the iterative formula

$$x_{n+1} = F(x_n).$$

Let $a$ be a root of the equation $x = F(x)$, so that $F(a) = a$. As $x$ tends to $a$, the
value of the expression

$$\frac{F(x) - F(a)}{x - a}$$

tends to the limit $F'(a)$.
Therefore, if $x_n$ is a value of $x$ close to $a$,

$$F(x_n) - F(a) \approx (x_n - a)F'(a).$$

Since $F(x_n) = x_{n+1}$ and $F(a) = a$, this gives

$$(x_{n+1} - a) \approx (x_n - a)F'(a).$$

Two very important deductions can be made.
(i) If $F'(a) > 0$, $x_n$ and $x_{n+1}$ will both be greater than $a$ or both smaller than $a$.

If $F'(a) < 0$, the root will lie between $x_n$ and $x_{n+1}$.

(ii) Since

$$|x_{n+1} - a| \approx |x_n - a| \times |F'(a)|,$$

$|x_{n+1} - a|$ will be less than $|x_n - a|$ if $|F'(a)| < 1$.

This means that when $|F'(a)| < 1$, $x_{n+1}$ will be closer to $a$ than $x_n$, and the sequence of approximations to the root will converge. In practice, $|F'(x)|$ needs to be considerably smaller than 1 for all the values of $x$ involved. Then the sequence will converge rapidly.

*Example 1*

The equation $2x^2 - 4x + 1 = 0$ can be put in the form $x = 2 - 1/(2x)$.

Let $F(x) = 2 - 1/(2x)$. Then the iterative formula $x_{n+1} = F(x_n)$ becomes

$$x_{n+1} = 2 - 1/(2x_n).$$

If $x_1 = 2$, then $x_2 = 1.75$, $x_3 = 1.714 \ldots$, $x_4 = 1.7083 \ldots$. This sequence converges to the root $1 + 1/\sqrt{2}$ of the equation. In this example, $F'(x)$ is $1/(2x^2)$, which is positive and less than 1 for the values of $x$ considered.

*Example 2*

The equation $2x^2 - 4x - 1 = 0$ can be put in the form $x = 2 + 1/(2x)$.

With $F(x) = 2 + 1/(2x)$, this leads to the iterative formula

$$x_{n+1} = 2 + 1/(2x_n).$$

With $x_1 = 2$, this produces the sequence $x_2 = 2.25$, $x_3 = 2.\dot{2}$, $x_4 = 2.225$. It can be shown that this sequence converges to the positive root of the equation $2x^2 - 4x - 1 = 0$, i.e. to $\frac{1}{2}(2 + \sqrt{6})$. Note that $F'(x) = -1/2x^2$, which is negative. The terms in the sequence are alternately greater than and less than the limit to which the sequence tends.

*Example 3*

The equation $x^2 - 4x + 2 = 0$ can be put in the form $x = (x^2 + 2)/4$.

If $F(x) = (x^2 + 2)/4$, $F'(x) = x/2$. Let $x_1 = 1$. Then the iterative formula

$$x_{n+1} = (x_n^2 + 2)/4$$

produces the sequence $x_1 = 1$, $x_2 = 0.75$, $x_3 = 0.64 \ldots$, $x_4 = 0.60 \ldots$. Now the smaller root of the equation $x^2 - 4x + 2 = 0$ is $(2 - \sqrt{2})$, i.e. $0.586 \ldots$. It can be confirmed graphically that the sequence converges to this value by drawing the line $y = x$ and the curve $y = (x^2 + 2)/4$. In Fig. 1.6, P is the point of intersection of the line and the curve, and so the value of $x$ at P is a root of the equation $x = (x^2 + 2)/4$.

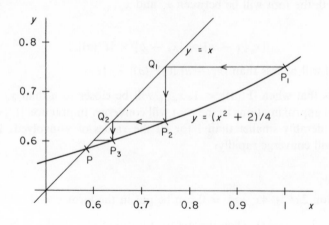

Fig. 1.6

$P_1, P_2, P_3, \ldots$ are the points on the curve at which $x$ equals $x_1, x_2, x_3 \ldots$, and the 'staircase' $P_1Q_1P_2Q_2P_3 \ldots$ represents the steps in the calculation. The graph can be used to investigate the behaviour of the sequence for various choices of $x_1$. It is left to the reader to verify that the iterative formula

$$x_{n+1} = 4 - (2/x_n)$$

can be used to generate a sequence converging to the larger root of the equation $x^2 - 4x + 2 = 0$.

*Example 4*
To solve the equation $x = e^{-x}$, the iterative formula $x_{n+1} = \exp(-x_n)$ can be used.

If $F(x) = e^{-x}$, then $F'(x) = -e^{-x}$ and so $|F'(x)| < 1$ when $x$ is positive. Therefore the sequence will converge if $x_1 > 0$. With $x_1 = 1$ the formula gives (to 3 decimal places)

$$x_2 = 0.368 \quad x_3 = 0.692 \quad x_4 = 0.500$$
$$x_5 = 0.606 \quad x_6 = 0.545 \quad x_7 = 0.580.$$

This sequence converges to $0.567 \ldots$, the root of the equation $x = e^{-x}$. This is illustrated by Fig. 1.7.

Note that, because $F'(x)$ is negative, the members of the sequence are alternately greater than and less than the root.

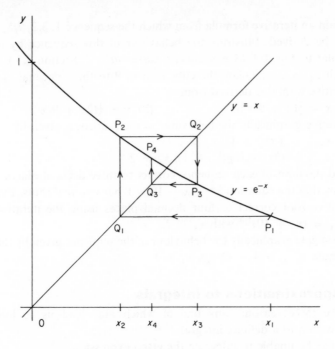

Fig. 1.7

## Exercises 1.6

1 Express $x_n$ in terms of $n$ in the sequences defined by
   (a) $x_{n+1} = x_n + 1$, $x_1 = 1$      (b) $x_{n+1} = x_n - (\frac{1}{2})^n$, $x_1 = 1$
   (c) $x_{n+1} = x_n/(n + 1)$, $x_1 = 1$      (d) $x_{n+1} = x_n + 2n + 1$, $x_1 = 1$.

2 Write down the first five terms of the sequence defined by
   $x_{n+1} = (x_n + 1)/3$, $x_1 = 2$. Show that $(x_{n+1} - \frac{1}{2}) = (x_n - \frac{1}{2})/3$ and deduce
   that as $n$ tends to infinity $x_n$ tends to $\frac{1}{2}$.

3 Using the straight lines $y = x$ and $y = \frac{1}{2}(x + 1)$, determine graphically the
   behaviour of the sequence given by $x_{n+1} = \frac{1}{2}(x_n + 1)$, $x_1 = 4$. State the
   limit of the sequence.

4 Illustrate graphically the sequence defined by $x_{n+1} = \frac{1}{2}(3 - x_n)$, $x_1 = 4$,
   using the straight lines $y = x$ and $x + 2y = 3$. State the limit of the
   sequence.

5 Show that the equation $x^3 - x - 4 = 0$ has only one real root and find the
   integer $k$ nearest to this root. With $x_1 = k$, show that only one of the
   iterative formulae
   (a) $x_{n+1} = (x_n + 4)^{1/3}$    (b) $x_{n+1} = (x_n + 4)/x_n^2$
   gives a sequence converging to this root. Evaluate the root to three decimal
   places.

6 By writing the equation $x^2 = 2$ in the form

$$x = (x + 2)/(x + 1),$$

obtain an iterative formula from which the sequence 1, 3/2, 7/5, 17/12, ... can be derived. Illustrate the behaviour of this sequence by a diagram similar to Fig. 1.7. (See *Advanced mathematics 1*, Section 8.5.)

7  With $x_1 = 2$, calculate the cube root of 9 to three decimal places using iterative formulae derived from

(a) $x = \frac{1}{2}(x + 9/x^2)$                  (b) $x = \frac{1}{3}(2x + 9/x^2)$.

8  Examine graphically the behaviour of the sequences given by

(a) $x_{n+1} = 1/(x_n + 1)$, $x_1 = 2$,

(b) $x_{n+1} = (2x_n + 1)/(x_n + 1)$, $x_1 = 2$.

Find the limit of each sequence correct to three decimal places.

9  Show that the equation $x^4 - x - 10 = 0$ has two real roots. Evaluate the positive root correct to four decimal places using the iterative formula $x_{n+1} = (x_n + 10)^{1/4}$ with $x_1 = 2$.

10  Investigate graphically the behaviour of the sequence given by the iterative formula $x_{n+1} = 4 - (2/x_n)$, $x_1 = 1$.

## 1.7 Approximations to integrals

There are three reasons, any one of which may lead us to look for an approximation to a definite integral:

(a) we may be unable to integrate the given expression,

(b) the amount of computation needed to find the exact value may be excessive,

(c) we may not know the expression to be integrated, but have only a set of its values.

Three of the many possible methods for finding such an approximation will be considered.

### (i) The trapezium rule

Let f$(x)$ be positive for all values of $x$ between $a$ and $b$. Then the definite integral

$$\int_a^b f(x)dx$$

gives the area of the region bounded by the arc of the curve $y = f(x)$, the $x$-axis and the lines $x = a$, $x = b$. If in Fig. 1.8 the arc AB of the curve were replaced by the straight line AB, the region would become a trapezium. The area of this trapezium is given by the product of its width $(b - a)$ and its average height $\frac{1}{2}\{f(a) + f(b)\}$. This gives the approximation

$$\int_a^b f(x)dx \approx \frac{1}{2}(b - a)\{f(a) + f(b)\}.$$

This is the basic trapezium rule. It can be seen in Fig. 1.8(a) that this approximation gives a value which is too large when the gradient of the curve is increasing. In Fig. 1.8(b) the gradient of the curve is decreasing, and the approximation gives a value which is too small.

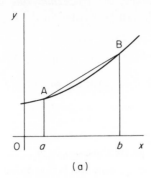 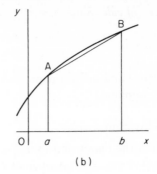

(a)                    (b)

Fig. 1.8

To obtain a more accurate approximation, divide the interval $a \leqslant x \leqslant b$ into $n$ equal parts of length $h$, where $h = (b - a)/n$. The region is then divided into $n$ strips of equal width by lines parallel to the $y$-axis, as in Fig. 1.9.

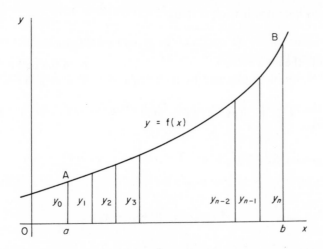

$y = f(x)$

Fig. 1.9

The $(n + 1)$ ordinates $y_0, y_1, \ldots, y_n$ are given by $y_0 = f(a), y_1 = f(a + h), y_2 = f(a + 2h), \ldots, y_n = f(b)$. By the trapezium rule above, the area of the first strip is approximately equal to $\frac{1}{2}h(y_0 + y_1)$, the area of the second strip is approximately equal to $\frac{1}{2}h(y_1 + y_2)$, and so on. The area of the whole region will be approximately equal to

$$\tfrac{1}{2}h\{(y_0 + y_1) + (y_1 + y_2) + (y_2 + y_3) + \ldots + (y_{n-1} + y_n)\},$$

i.e. $$\int_a^b f(x)\,dx \approx h(\tfrac{1}{2}y_0 + y_1 + y_2 + \ldots + y_{n-1} + \tfrac{1}{2}y_n).$$

This extension of the basic rule is often called the *trapezium rule* or the *trapezoidal rule*. Since $h = (b - a)/n$, the rule can be written in the form

$$\int_a^b f(x)dx \approx (b - a) \left\{ \frac{y_0 + 2y_1 + 2y_2 + \ldots + 2y_{n-1} + y_n}{2n} \right\}.$$

The right-hand side is the product of the width $(b - a)$ and a weighted average of the ordinates.

### (ii) Simpson's rule

Consider the region shown in Fig. 1.10 bounded by the arc LMN of the curve $y = f(x)$, the x-axis and the lines $x = -h$, $x = h$. This region is divided into two strips of equal width by the y-axis. Let the parabola given by $y = px^2 + qx + r$ pass through the points $L(-h, y_0)$, $M(0, y_1)$ and $N(h, y_2)$. Then the coefficients $p$, $q$ and $r$ must satisfy the equations

$$\begin{aligned} y_0 &= ph^2 - qh + r, \\ y_1 &= \phantom{ph^2 - qh +} r, \\ y_2 &= ph^2 + qh + r. \end{aligned}$$

From these equations it is clear that

$$y_0 + 4y_1 + y_2 = 2ph^2 + 6r.$$

Let $A$ denote the area of the region bounded by the arc of the parabola through L, M and N, by the x-axis and by the lines $x = -h$, $x = h$. Then we have

$$A = \int_{-h}^h (px^2 + qx + r)dx$$

$$= \left[ px^3/3 + qx^2/2 + rx \right]_{-h}^h = \tfrac{1}{3}h(2ph^2 + 6r).$$

Therefore

$$A = \tfrac{1}{3}h(y_0 + 4y_1 + y_2).$$

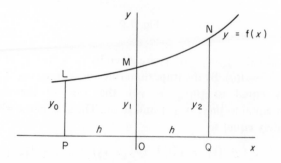

Fig. 1.10

This gives approximately the area of the region shown in Fig. 1.10. If the origin O is moved to a different point on the x-axis, the area of the region will be unchanged. If the vertical sides of the region are now the lines $x = a$, $x = b$, the width $h$ of each strip becomes $(b - a)/2$, and the result can be expressed in the form

$$\int_a^b f(x)dx \approx \tfrac{1}{3}h(y_0 + 4y_1 + y_2),$$

where $h = (b - a)/2$. This is the basic form of Simpson's rule, using two strips of width $h$ (and three ordinates $y_0$, $y_1$ and $y_2$).

If the region is divided into four strips of equal width, the value of $h$ becomes $(b - a)/4$. Let the five ordinates be $y_0$, $y_1$, $y_2$, $y_3$ and $y_4$.

$$\int_a^b f(x)dx = \int_a^{a+2h} f(x)dx + \int_{a+2h}^b f(x)dx$$

$$\approx \tfrac{1}{3}h(y_0 + 4y_1 + y_2) + \tfrac{1}{3}h(y_2 + 4y_3 + y_4),$$

i.e. $\quad\displaystyle\int_a^b f(x)dx \approx \tfrac{1}{3}h(y_0 + 4y_1 + 2y_2 + 4y_3 + y_4).$

In the general case, with the region divided into an *even* number $n$ of strips (and $n + 1$ ordinates), Simpson's rule gives the approximation

$$\int_a^b f(x)dx \approx \tfrac{1}{3}h(y_0 + 4y_1 + 2y_2 + 4y_3 + 2y_4 + \ldots + 4y_{n-1} + y_n),$$

where $h = (b - a)/n$. Note that each of these approximations is in the form of the product of $(b - a)$ and a weighted average of the ordinates.

## (iii) Expansion in series

It is sometimes possible to find an approximation to an integral by replacing the integrand by the first few terms of its expansion in a series. The error in the approximation will depend on the magnitude of the terms neglected, and in order to determine the accuracy of the result it is necessary to estimate the size of the integral of the terms neglected.

*Example 1*

Find an approximation to the value of $\displaystyle\int_1^5 x^4 dx$, (a) by the trapezium rule (b) by Simpson's rule, using four strips.

The five ordinates are $y_0 = 1$, $y_1 = 16$, $y_2 = 81$, $y_3 = 256$, $y_4 = 625$. The width of each strip is 1, i.e. $h = 1$.

(a) The trapezium rule gives the approximation

$$h(\tfrac{1}{2}y_0 + y_1 + y_2 + y_3 + \tfrac{1}{2}y_4),$$

which equals $(\frac{1}{2} + 16 + 81 + 256 + 312\frac{1}{2})$, i.e. 666.

(b) Simpson's rule gives the approximation

$$\tfrac{1}{3}h(y_0 + 4y_1 + 2y_2 + 4y_3 + y_4),$$

which equals $\frac{1}{3}(1 + 64 + 162 + 1024 + 625)$, i.e. $625\frac{1}{3}$.
Compare these with the exact value.

$$\int_1^5 x^4 dx = \left[ x^5/5 \right]_1^5 = 624\frac{4}{5}.$$

*Example 2*

Using Simpson's rule with six strips, estimate the value of the integral $\displaystyle\int_0^{30} y\,dx$
from the data in the table.

| x | 0 | 5 | 10 | 15 | 20 | 25 | 30 |
|---|---|---|----|----|----|----|----|
| y | 41 | 34 | 29 | 25 | 22 | 21 | 20 |

The width of each strip is 5, i.e. $h = 5$. Let $S_1 = y_0 + y_6$, $S_2 = y_1 + y_3 + y_5$
and $S_3 = y_2 + y_4$. Then

$$\tfrac{1}{3}h(y_0 + 4y_1 + 2y_2 + 4y_3 + 2y_4 + 4y_5 + y_6) = (5/3)(S_1 + 4S_2 + 2S_3).$$

| | | |
|---|---|---|
| $y_0 = 41$ | $y_1 = 34$ | $y_2 = 29$ |
| $y_6 = 20$ | $y_3 = 25$ | $y_4 = 22$ |
| — — | $y_5 = 21$ | — — |
| $S_1 = 61$ | — — | $S_3 = 51$ |
| | $S_2 = 80$ | |

$$\int_0^{30} y\,dx \approx (5/3)(61 + 4 \times 80 + 2 \times 51) = 805.$$

*Example 3*

Use Simpson's rule with ten strips to estimate the value of $\displaystyle\int_0^1 \sqrt{(1 + x^2)}dx$.

The approximation is given by $\frac{1}{3}h(S_1 + 4S_2 + 2S_3)$, where $h = 0.1$,
$S_1 = y_0 + y_{10}$, $S_2 = (y_1 + y_3 + y_5 + y_7 + y_9)$, $S_3 = (y_2 + y_4 + y_6 + y_8)$.

| | | |
|---|---|---|
| $y_0 = 1$ | $y_1 =$ 1.0050 | $y_2 = 1.0198$ |
| $y_{10} = 1.4142$ | $y_3 =$ 1.0440 | $y_4 = 1.0770$ |
| — | $y_5 =$ 1.1180 | $y_6 = 1.1662$ |
| $S_1 = 2.4142$ | $y_7 =$ 1.2207 | $y_8 = 1.2806$ |
| | $y_9 =$ 1.3454 | — |
| | | $S_3 = 4.5436$ |
| | $S_2 =$ 5.7331 | $2S_3 = 9.0872$ |
| | $4S_2 =$ 22.9324 | |

With these values,

$$\tfrac{1}{3}h(S_1 + 4S_2 + 2S_3) = \tfrac{1}{3}(0.1)(2.4142 + 22.9324 + 9.0872)$$

$$= 1.14779.$$

To four decimal places the estimated value of the integral is 1.1478. This result can be confirmed by direct integration using the method of substitution.

*Example 4*

Evaluate the integral $\displaystyle\int_0^{0.2} \frac{1}{1 - x^3}dx$ to four decimal places.

By the binomial expansion, valid when $|x| < 1$,

$$(1 - x^3)^{-1} = 1 + x^3 + x^6 + x^9 + \dots.$$

If we propose to use $(1 + x^3)$ as an approximation to the integrand, the sum of the terms neglected must be considered. These terms form the geometric series
$x^6 + x^9 + x^{12} + \dots.$
When $x$ takes its largest value 0.2, the sum of this series is $(0.2)^6/\{1 - (0.2)^3\}$, which is less than $10^{-4}$. Therefore the difference between the integrals

$$\int_0^{0.2} \frac{1}{1 - x^3}dx \quad \text{and} \quad \int_0^{0.2} (1 + x^3)dx \text{ is less than } \int_0^{0.2} 10^{-4}dx,$$

i.e. less than $2 \times 10^{-5}$.

Now $\displaystyle\int_0^{0.2} (1 + x^3)dx = \left[ x + x^4/4 \right]_0^{0.2} = 0.2004.$

This is the correct value of the given integral to four decimal places, since the error in the approximation is too small to affect the fourth decimal place.

**Exercises 1.7**

1   Use the trapezium rule with four strips to estimate the area of the region bounded by the parabola $y = 1 + 4x - x^2$, the axes and the line $x = 4$. Show that Simpson's rule using two strips (i.e. three ordinates) gives the area exactly.

2   Use the trapezium rule with ten strips to estimate the value of $\displaystyle\int_1^2 \log_{10} x \, dx$. Work to four decimal places and give your answer to three decimal places.

3   Estimate the value of $\displaystyle\int_1^2 \frac{1}{x}dx$, using Simpson's rule with four strips.

**4** Show that Simpson's rule gives the exact value of $\int_a^b x^3 dx$.

**5** Use Simpson's rule with nine ordinates to evaluate $\int_9^{25} \sqrt{x}\, dx$. Work to three decimal places, and compare your answer with the exact value.

**6** Observed values of $y$ for given values of $x$ are given in the table. Draw the graph of $y$ against $x$, and estimate the area beneath the curve (a) by the trapezium rule (b) by Simpson's rule.

| $x$ | 0 | 0.5 | 1.0 | 1.5 | 2.0 | 2.5 | 3.0 |
|---|---|---|---|---|---|---|---|
| $y$ | 4.20 | 4.46 | 4.80 | 5.29 | 6.00 | 7.01 | 7.40 |

**7** The graph of $y$ against $x$ is drawn for the values given in the table. The area beneath the curve for $0 \leqslant x \leqslant 8$ is rotated once about the x-axis. Find, to two decimal places, the volume swept out.

| $x$ | 0 | 2 | 4 | 6 | 8 |
|---|---|---|---|---|---|
| $y$ | 1 | 0.578 | 0.445 | 0.374 | 0.338 |

**8** A railway cutting is 160 m long. The table gives the values of $y$, the cross-sectional area, against $x$, the distance from one end. Estimate the volume of soil excavated.

| $x$ (m) | 0 | 20 | 40 | 60 | 80 | 100 | 120 | 140 | 160 |
|---|---|---|---|---|---|---|---|---|---|
| $y$ (m²) | 13 | 42 | 82 | 122 | 130 | 130 | 105 | 62 | 23 |

**9** Use a series expansion to show that

(a) $\int_0^{\frac{1}{4}} \dfrac{1}{(1-x^4)}dx \approx 0.2502$ 　　 (b) $\int_0^{\frac{1}{2}} \dfrac{1}{\sqrt{(1-x^5)}}dx \approx 0.5015$.

**10** Evaluate $\int_0^{\pi/2} \sqrt{(1 - \frac{1}{4}\sin^2 \theta)}d\theta$ to two decimal places

(a) by Simpson's rule using seven ordinates
(b) by means of a binomial expansion.

**Miscellaneous exercises 1**

**1** The table below gives measured values of the variables $u$ and $v$, which are related by an equation of the form $v = au + b/u$. Find approximately the values of the constants $a$ and $b$ by plotting $uv$ against $u^2$.

| $u$ | 1 | 2 | 3 | 4 | 5 |
|---|---|---|---|---|---|
| $v$ | 12.5 | 7.0 | 5.5 | 5.0 | 4.9 |

[L]

**2** The values of $x$ and $y$ given in the table below satisfy approximately the relationship $e^y = kx^a$. By drawing a suitable linear graph estimate the values of the constants $a$ and $k$.

| $x$ | 1 | 2 | 3 | 4 | 5 | 6 | 7 |
|---|---|---|---|---|---|---|---|
| $y$ | 2 | 8.9 | 13.0 | 15.9 | 18.1 | 19.9 | 21.5 |

[L]

3   The variables $s$ and $t$ satisfy a relation of the form

$$(s/t)^a = be^{-t},$$

where the constants $a$ and $b$ are positive integers. Given the table below, draw a linear graph and find the values of $a$ and $b$.

| $t$ | 0.2 | 0.4 | 0.6 | 0.8 | 1.0 |
|---|---|---|---|---|---|
| $s$ | 1.09 | 1.96 | 2.67 | 3.22 | 3.64 |

4   The variables $x$ and $y$ are connected by the relation $y = \log_{10}(a + bx)$, where $a$ and $b$ are constants. Use the table below to plot $10^y$ against $x$, and hence obtain to two significant figures estimates of the values of $a$ and $b$.

| $x$ | 1 | 2 | 3 | 4 | 5 | 6 |
|---|---|---|---|---|---|---|
| $y$ | 0.857 | 0.924 | 0.982 | 1.033 | 1.079 | 1.121 |

5   The temperature, $T$ degrees, of a body while cooling is given in terms of time, $t$ minutes, by the law $T = T_0 e^{-kt}$, where $T_0$ degrees is the temperature at zero time and $k$ is a constant. Measured values of $T$ at various times are given in the table below. By drawing a suitable linear graph estimate the values of $T_0$ and $k$.

| $t$ | 20 | 40 | 60 | 80 | 100 |
|---|---|---|---|---|---|
| $T$ | 75.5 | 59.9 | 47.6 | 37.8 | 30.0 |

6   The table shows approximate values of a variable $y$ corresponding to certain values of another variable $x$. By drawing a suitable linear graph, verify that these values of $x$ and $y$ satisfy approximately a relationship of the form $y = ax^k$. Use your graph to find approximate values of the constants $a$ and $k$.

| $x$ | 5 | 10 | 15 | 20 | 25 | 30 |
|---|---|---|---|---|---|---|
| $y$ | 45 | 63 | 77 | 89 | 100 | 110 |

[L]

7   The variables $x$ and $y$ below are believed to be related by a law of the form $y = \log_e(ax^2 + bx)$, where $a$ and $b$ are constants. Plot a suitable graph to show that this is so and determine the probable values of $a$ and $b$.

| $x$ | 1 | 2 | 3 | 4 | 5 | 6 |
|---|---|---|---|---|---|---|
| $y$ | $-1.897$ | 0.588 | 1.599 | 2.262 | 2.757 | 3.153 |

[AEB]

8   The following values of $x$ and $y$ are believed to obey a law of the form

$$y = \frac{a}{bx + c},$$ where $a$, $b$ and $c$ are constants. Show that they do obey this

law and hence estimate the value of the ratios $a:b:c$.

| $x$ | 0 | 1 | 2 | 3 | 4 |
|---|---|---|---|---|---|
| $y$ | 1.00 | 0.67 | 0.50 | 0.40 | 0.33 |

[AEB]

9  Show that the equation $e^x + e^{-x} - 4 = 0$ has a root between 1.3 and 1.4. Find this root correct to two decimal places by linear interpolation.

10  The following values of $x$ and $y$ are believed to obey a law of the form $ax^2 - by^2 = 4$, where $a$ and $b$ are constants. Show graphically that they do obey this law and determine approximate values of $a$ and $b$.

| $x$ | 0.50 | 1.00 | 1.50 | 2.00 | 2.50 |
|---|---|---|---|---|---|
| $y$ | 2.80 | 2.70 | 2.54 | 2.30 | 1.96 |

[AEB]

11  In an experiment, sets of values of the related variables $(x, y)$ are obtained. State how you would determine whether $x$ and $y$ are connected by a law of the form
(i) $y = a^{x+b}$
(ii) $ay^2 = (x + b) \log_e x$,
where in each case $a$ and $b$ are unknown constants.

   State briefly how you would be able to determine the values of $a$ and $b$ for each law.

[AEB]

12  It is believed that the variables $x$, $y$, $z$ are connected by a relationship of the form $z = Ax^m y^n$, where $A$, $m$, $n$ are constants. Show graphically that the values given in the following tables support this belief and find values for $A$, $m$, $n$.

| | | | | | |
|---|---|---|---|---|---|
| $x = 0.45$ | $y$ | 175 | 603 | 1130 | 1850 |
| | $z$ | 65.5 | 139 | 204 | 276 |
| $x = 5.6$ | $y$ | 175 | 603 | 1130 | 1850 |
| | $z$ | 21 600 | 45 960 | 67 420 | 91 070 |

13  Show that the equation $x^3 - x^2 - 1 = 0$ has only one real root. Evaluate this root to three decimal places (a) by the Newton-Raphson method (b) by any other method.

14  Taking 2 as the first approximation to the positive root of the equation $x^3 + 3x^2 - 9x - 3 = 0$, use the Newton-Raphson method to show that to four decimal places this root is 2.0642.

15  Show that the equation $x^3 - 3x - 3 = 0$ has only one real root. Evaluate this root to three decimal places by the Newton-Raphson method.

16  By applying the Newton-Raphson method to the equation $f(x) = 0$, where $f(x) = 1/x - a$, obtain the relation

$$x_2 = x_1(2 - ax_1).$$

If $a = 0.143$ and $x_1 = 7$, show that $x_2$ gives the reciprocal of 0.143 correct to four decimal places.

17 Show that the equation $8x^4 - 8x - 3 = 0$ has only two real roots. By substituting $x = 1 + h$ and neglecting $h^2$ and higher powers of $h$, show that 1.125 is an approximation to the positive root. By repeating this method evaluate this root to four decimal places. Find the negative root correct to two decimal places.

18 By using Newton's method, or any other suitable iterative method, solve the equation $x + \log_e x = 3$, given that the root is close to 2. Obtain the root correct to two decimal places. [JMB]

19 Show that there is precisely one real root of the equation $x^3 + 2x - 1 = 0$ and that it lies in the interval $0 < x < 1$. Use Newton's method to find the root to three decimal places. [JMB]

20 Show that the equation $x^4 - 6x + 4 = 0$ has two real roots. Evaluate each root to three decimal places by means of an iterative formula.

21 Show that the gradient of the curve $y = 2x^3 + 4x - 5$ is always positive, and deduce that the equation

$$2x^3 + 4x - 5 = 0$$

has only one real root. Find a suitable iterative formula of the form

$$x_{n+1} = (a + bx_n + cx_n^3)/10,$$

and use it to evaluate the root to three decimal places.

22 Locate by means of a graph the roots of the equation

$$x^3 - 2x^2 - 2x + 2 = 0.$$

Evaluate these roots to three decimal places by means of the following iterative formulae:
(a) $x_{n+1} = 2(x_n^2 + x_n - 1)/x_n^2$
(b) $x_{n+1} = (x_n^3 + 2)/(2x_n + 2)$
(c) $x_{n+1} = (2x_n^2 + 10x_n - 2 - x_n^3)/8$.

23 Find each root of the equation $x^3 - 10x + 1 = 0$ correct to three places of decimals.

24 Sketch the curve $y = (x + 1)(x - 2)^2$, and use it to locate the real roots of the equations
(a) $x^3 - 3x^2 + 5 = 0$
(b) $x^3 - 3x^2 + 3 = 0$.
Evaluate each root to one decimal place.

25 Show that the equation $x^3 - x^2 + x - 2 = 0$ has only one real root. Evaluate this root to two decimal places.

26 Find the number of real roots of each of the equations
(a) $x + \cos x = 0$
(b) $x^2 = \cos x$
(c) $2 \sin x + x = \pi$.

27 By drawing the line $y = x + 7$ and the curve $8y = x^4$, show that the equation $x^4 = 8x + 56$ has two real roots. Evaluate the positive root to

two decimal places.

**28** Find to one decimal place the smallest positive root of the equation
$x + \tan(2x) = 1$.

**29** The integral $\displaystyle\int_a^b (x^2 + px + q)dx$ is to be evaluated by the trapezium rule

Show that if only two ordinates are used, the error in the result will be
$(b - a)^3/6$.

Find the error if three ordinates are used.

**30** Find graphically an approximate value of the root of the equation

$$f(x) \equiv xe^{-x} - \cos x = 0$$

in the neighbourhood of $x = 1$, By Newton's iterative method, or
otherwise, evaluate the root to five decimal places. [AEB]

**31** Establish the rule that if $f(x) \equiv a + bx + cx^2 + dx^3$, where $a, b, c$ and
$d$ are constants, then

$$\int_0^1 f(x)dx = \tfrac{1}{6}[f(0) + f(1) + 4f(\tfrac{1}{2})].$$

Prove also that if $f(x)$ contains an additional term $kx^4$, where $k$ is a
constant, the error in still using the rule is $k/120$. [AEB]

**32** Use the Maclaurin series for $\cos x$ to show that

$$\int_0^1 (\sqrt{x}) \cos x \, dx \approx 0.53.$$

**33** Draw the graph of $y = \sin x - \cos x$ for values of $x$ from 0 to $\pi$ radians.
Use the graph to solve the equation

$$x + \sin x = \cos x.$$

**34** Sketch the curve $y = \log_{10} x$ for values of $x$ between 0 and 1. Use the graph
to show that the equation

$$10x + \log_{10} x = 4$$

has only one real root. Evaluate this root correct to three decimal places by
means of the iterative formula

$$x_{n+1} = (4 - \log_{10} x_n)/10.$$

**35** Sketch the curves $x^2 + y = 1$ and $y^2 + 2x = 3$. From your graph deduce
the number of real roots of the equation

$$x^4 - 2x^2 + 2x - 2 = 0.$$

Evaluate each root to two decimal places by an iterative method.

**36** The integral of $x^2$ from $x = 0$ to $x = 1$ is calculated by the trapezium rule
using the values of $x^2$ at the points $x = 0, 1/n, 2/n, \ldots, 1$. Show that the
error in the result is inversely proportional to $n^2$.

**37** The table below gives the coordinates of seven points on a certain curve. Use Simpson's rule to estimate the volume generated by rotating the region under this curve, between $x = 2.4$ and $x = 4.8$, about the $x$-axis.

| $x$ | 2.4 | 2.8 | 3.2 | 3.6 | 4.0 | 4.4 | 4.8 |
|---|---|---|---|---|---|---|---|
| $y$ | 5 | 5 | 3 | 6 | 7 | 7 | 4 |

[L]

**38** A chord of a circle subtends an angle of $\theta$ radians ($\theta < \pi$) at the centre. If the chord divides the circle into two segments whose areas are in the ratio 3:1, prove that

$$\sin \theta = \theta - \pi/2.$$

Obtain graphically an approximate solution of this equation. Give your answer to one decimal place.

**39** Given the values in the table below estimate the value of $\displaystyle\int_0^1 \cos \sqrt{x}\ dx$

(a) by Simpson's rule
(b) by the trapezium rule
(c) by means of the Maclaurin series for $\cos \sqrt{x}$.

| $x$ | 0 | 0.25 | 0.5 | 0.75 | 1 |
|---|---|---|---|---|---|
| $\cos \sqrt{x}$ | 1 | 0.878 | 0.760 | 0.648 | 0.540 |

Give your answers to three decimal places.
Evaluate the integral by the substitution $x = t^2$.

**40** The following table gives to two decimal places the speed $v$ in $m\,s^{-1}$ at intervals of 0.2 s of a particle moving in a straight line. At time $t = 0.2$, the particle passes through the point A, and at time $t = 1$ it passes through the point B. Use Simpson's rule to estimate the distance AB.

| $v$ | 6.03 | 4.04 | 2.71 | 1.82 | 1.22 |
|---|---|---|---|---|---|
| $t$ | 0.2 | 0.4 | 0.6 | 0.8 | 1 |

The speed $v$ is known to be related to the time $t$ by a law of the form $v = ae^{bt}$, where $a$ and $b$ are integers. By plotting a graph of $\ln v$ against $t$, find $a$ and $b$. Hence calculate the distance AB to two decimal places. [L]

**41** Evaluate the integral $\displaystyle\int_0^4 \exp(\sqrt{x})dx$ (a) by Simpson's rule (b) by the trapezium rule, using the values below. Give your answers to three decimal places.

| $x$ | 0 | 1 | 2 | 3 | 4 |
|---|---|---|---|---|---|
| $\exp(\sqrt{x})$ | 1 | 2.7183 | 4.1132 | 5.6522 | 7.3891 |

By using a suitable substitution to evaluate the integral, determine which of these numerical answers is nearer to the exact value. [L]

42 By finding the points of intersection of the graph of the function where $f(x) = 5 - \dfrac{5}{x^2}$ with a suitable straight line, find to two significant figures the positive roots of the equation $x = 5 - \dfrac{5}{x^2}$.

The iterative method based on the relation $x_{n+1} = f(x_n)$, using an initial approximation $x_0 > 2$, is to be employed to evaluate one of these roots. By inspecting your graph, or otherwise, find to which root the iteration will converge. Perform one iteration to improve the approximation you have found for this root. [L]

43 Tabulate values of $f(x) = \sqrt{\{27 + (x - 3)^3\}}$ for integral values of $x$ from 0 to 6 inclusive and sketch the graph of $y = f(x)$ for the interval $0 \leqslant x \leqslant 6$.

Given that $F(t) \equiv \displaystyle\int_0^t f(x)dx$, use Simpson's rule and the calculated values of $f(x)$ to estimate $F(2)$ and $F(6)$. [L]

44 The following pairs of values of $x$ and $y$ satisfy approximately a relation of the form $y = ax^n$, where $a$ and $n$ are integers. By plotting the graph of lg $y$ against lg $x$ find the values of the integers $a$ and $n$. [lg $N$ denotes $\log_{10} N$.]

| $x$ | 0.7 | 0.9 | 1.1 | 1.3 | 1.5 |
|-----|------|------|------|------|-------|
| $y$ | 1.37 | 2.92 | 5.32 | 8.80 | 13.50 |

Estimate the value of the integral $\displaystyle\int_{0.7}^{1.5} y \, dx$

(a) by Simpson's rule, using five ordinates and clearly indicating your method

(b) by using the relation $y = ax^n$ with the values found for $a$ and $n$. [L]

45 Show that the equation $x(x^2 + 2) - 4 = 0$ has only one root, and that this root lies between 1 and 1.5. Taking 1.2 as a first approximation to this root, use Newton's method to obtain a second approximation, giving your result to two places of decimals. [JMB]

46 Use Simpson's rule with 5 ordinates (4 strips) to find an approximate value of

$$\int_0^4 xe^{-x}dx.$$

Give your answer correct to two decimal places. [JMB]

47 Show that $x^3 + 6x - 16 = 0$ has only one real root $\alpha$, and that $1 < \alpha < 2$. Use linear interpolation to determine the number $k$, expressed to one decimal place, such that $k < \alpha < k + 0.1$. The arithmetical calculations involved in this question must be shown in the solution. [JMB]

48 Show that the equation $x^3 - x - 2 = 0$ has a root between 1 and 2. Using Newton's approximation with starting point 1.5 (and showing all relevant

working) determine by means of two iterations an approximation to this root, giving your answer to two decimal places. [JMB]

49 Draw the graph of $y = \log_{10} x$ for values of $x$ in the range $1 \leqslant x \leqslant 10$. By drawing a second graph solve the equation $x^{10} = 10^{x+1}$. [AEB]

50 Show that a root of the equation $\tan \theta = 1 + \sin \theta$ occurs in the vicinity of $\theta = 1.08$ rad. Use this value as a first approximation to calculate the value of this root correct to three decimal places. [AEB]

51 A solution of the equation $x^4 + x^3 + x^2 + x = 5$ is known to be $x = 1 + h$ where $h$ is small. Neglecting powers of $h$ above the first and using the binomial theorem, show that the solution of the equation is $x = 1.1$ approximately. [AEB]

52 The values of $x$ and $y$ given in the table below were obtained in an experiment. It is believed that $x$ and $y$ are connected by a relationship of the form $x(y - b) = a$, where $a$ and $b$ are constants. Show, by drawing a straight line graph, that this is so. Use your graph to find probable values of $a$ and $b$.

| $x$ | 2.31 | 3.12 | 4.41 | 5.23 | 6.54 |
|---|---|---|---|---|---|
| $y$ | 5.87 | 5.64 | 5.45 | 5.38 | 5.31 |

[AEB]

53 An equation can be written in the form $x = F(x)$ and it is known that the equation has only one real root and that this root is near $x_1$. Explain, with the aid of a diagram, how, if $|F'(x)| < 1$, the iterative formula

$$x_{r+1} = F(x_r), \quad (r = 1, 2, \ldots)$$

will give the root to whatever accuracy is required.

Show that the cubic equation $x^3 + 3x - 15 = 0$ has only one real root and that this root is near 2. This cubic equation can be written in any one of the forms

(a) $x = \frac{1}{3}(15 - x^3)$
(b) $x = 15/(x^2 + 3)$
(c) $x = (15 - 3x)^{1/3}$.

Determine which of these forms would be suitable for the use of the previous iterative formula. [L]

54 Show graphically that the equation $x^3 - x - 1 = 0$ has only one real root and find the integer $n$ such that the root $\alpha$ satisfies $n < \alpha < n + 1$.

An iterative process for finding this root is defined by

$$x_1 = 1, x_{m+1} = (x_m + 1)^{1/3} \text{ for all } m \in \mathbb{N}.$$

Obtain, to three places of decimals, the values of $x_2$ and $x_3$.

Show, on a sketch graph, the line $y = x$ and the curve $y = (x + 1)^{1/3}$, indicating on this graph the relation between $x_1, x_2, x_3$ and the root $\alpha$. [L]

**55** If $x_1$ is an approximation to the root of the equation $x^3 - N = 0$, show that the second approximation obtained by the Newton-Raphson method is given by $x_2$, where

$$x_2 = \frac{2x_1}{3} + \frac{N}{3x_1^2}.$$

By taking $x_1 = 2.7$, find the cube root of 20 to three places of decimals. Find also to three decimal places the cube root of 200.

# 2 Geometry in three dimensions

## 2.1 The straight line

Consider the straight line which is parallel to the vector $\mathbf{b}$ and which passes through the point A whose position vector is $\mathbf{a}$. Any vector parallel to the vector $\mathbf{b}$ can be expressed in the form $t\mathbf{b}$, where $t$ is a scalar, i.e. a real number. If P is any point on this straight line, the vector $\overrightarrow{AP}$ is parallel to $\mathbf{b}$ and so $\overrightarrow{AP}$ is a scalar multiple of $\mathbf{b}$. Hence the position vector $\mathbf{r}$ of the point P is given by

$$\mathbf{r} = \mathbf{a} + t\mathbf{b}.$$

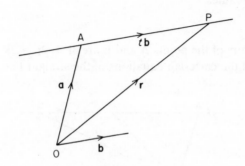

Fig. 2.1

For all values of the parameter $t$, the point with position vector $\mathbf{a} + t\mathbf{b}$ will lie on the given line, and so the vector equation of this line is

$$\mathbf{r} = \mathbf{a} + t\mathbf{b}.$$

The unit vectors $\mathbf{i}$, $\mathbf{j}$ and $\mathbf{k}$ are in the directions of the axes $Ox$, $Oy$ and $Oz$ respectively. If P is the point $(x, y, z)$ its position vector $\mathbf{r}$ is given by

$$\mathbf{r} = x\mathbf{i} + y\mathbf{j} + z\mathbf{k}.$$

Let $\mathbf{a} = a_1\mathbf{i} + a_2\mathbf{j} + a_3\mathbf{k}$ and $\mathbf{b} = b_1\mathbf{i} + b_2\mathbf{j} + b_3\mathbf{k}$. The equation of the line can be put in terms of row vectors,

$$(x, y, z) = (a_1, a_2, a_3) + t(b_1, b_2, b_3)$$

or in terms of column vectors

$$\begin{pmatrix} x \\ y \\ z \end{pmatrix} = \begin{pmatrix} a_1 \\ a_2 \\ a_3 \end{pmatrix} + t\begin{pmatrix} b_1 \\ b_2 \\ b_3 \end{pmatrix}.$$

By equating components we obtain the three equations

$$x = a_1 + tb_1, \quad y = a_2 + tb_2, \quad z = a_3 + tb_3.$$

This is the parametric form of the cartesian equations of the straight line. It follows that

$$\frac{x - a_1}{b_1} = \frac{y - a_2}{b_2} = \frac{z - a_3}{b_3},$$

since each fraction is equal to $t$. These are the cartesian equations of the straight line which passes through the point $A(a_1, a_2, a_3)$ and which is parallel to the vector $\mathbf{b}$. The ratios $b_1:b_2:b_3$ are known as the direction ratios of the line.

Let the line pass through the point $C(c_1, c_2, c_3)$. Then the equations

$$\frac{x - c_1}{b_1} = \frac{y - c_2}{b_2} = \frac{z - c_3}{b_3}$$

represent the same line.

*Example 1*
The position vectors of the points A and B are $3\mathbf{i} + 2\mathbf{j} - 2\mathbf{k}$ and $5\mathbf{i} - 2\mathbf{j} + \mathbf{k}$ respectively. Find the cartesian equations of the straight line AB.

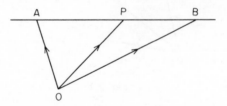

Fig. 2.2

The direction of the line is given by the vector $\overrightarrow{AB}$.

$$\overrightarrow{AB} = (5\mathbf{i} - 2\mathbf{j} + \mathbf{k}) - (3\mathbf{i} + 2\mathbf{j} - 2\mathbf{k})$$
$$= 2\mathbf{i} - 4\mathbf{j} + 3\mathbf{k}.$$

Therefore the direction ratios of the line are $2: -4:3$. Since the point $A(3, 2, -2)$ lies on the line, its cartesian equations are

$$\frac{x - 3}{2} = \frac{y - 2}{-4} = \frac{z + 2}{3}.$$

An alternative form using the coordinates of B is

$$\frac{x - 5}{2} = \frac{y + 2}{-4} = \frac{z - 1}{3}.$$

Note that, for each value of $t$, the equations

$$x = 3 + 2t, \quad y = 2 - 4t, \quad z = -2 + 3t$$

give the coordinates of a point P on the line AB.

## Example 2
A straight line is parallel to the vector $4\mathbf{i} - \mathbf{j} + 3\mathbf{k}$. Find its vector equation if (a) it passes through the origin O (b) it passes through the point A with position vector $2\mathbf{i} - 3\mathbf{j} + \mathbf{k}$.

(a) If the line passes through the origin, the position vector $\mathbf{r}$ of any point on the line will be a scalar multiple of the vector $4\mathbf{i} - \mathbf{j} + 3\mathbf{k}$. Hence the vector equation of the line is

$$\mathbf{r} = t(4\mathbf{i} - \mathbf{j} + 3\mathbf{k}).$$

(b) Let $\mathbf{r}$ be the position vector of a point P on the line through A parallel to the vector $4\mathbf{i} - \mathbf{j} + 3\mathbf{k}$. Then

$$\mathbf{r} = \overrightarrow{OP} = \overrightarrow{OA} + \overrightarrow{AP}.$$

Now the vector $\overrightarrow{AP}$ is a scalar multiple of the vector $4\mathbf{i} - \mathbf{j} + 3\mathbf{k}$. Therefore the vector equation of the line is

$$\mathbf{r} = (2\mathbf{i} - 3\mathbf{j} + \mathbf{k}) + t(4\mathbf{i} - \mathbf{j} + 3\mathbf{k}).$$

Each value of the parameter $t$ corresponds to a point on the line.

## Example 3
Show that the lines given by the vector equations

$$\mathbf{r} = (5\mathbf{i} + 2\mathbf{j} + 3\mathbf{k}) + s(2\mathbf{i} + 3\mathbf{j} + \mathbf{k}),$$
$$\mathbf{r} = (3\mathbf{i} + \mathbf{j} - 2\mathbf{k}) + t(2\mathbf{i} + 2\mathbf{j} + 3\mathbf{k}),$$

intersect and find the position vector of their common point.

If the lines meet, there will be values of $s$ and $t$ such that

$$(5\mathbf{i} + 2\mathbf{j} + 3\mathbf{k}) + s(2\mathbf{i} + 3\mathbf{j} + \mathbf{k}) = (3\mathbf{i} + \mathbf{j} - 2\mathbf{k}) + t(2\mathbf{i} + 2\mathbf{j} + 3\mathbf{k}).$$

i.e. $\quad (2 + 2s - 2t)\mathbf{i} + (1 + 3s - 2t)\mathbf{j} + (5 + s - 3t)\mathbf{k} = 0.$

Since $\mathbf{i}, \mathbf{j}$ and $\mathbf{k}$ are independent vectors the coefficient of each one must be zero. This gives three equations

$$2 + 2s - 2t = 0,$$
$$1 + 3s - 2t = 0,$$
$$5 + s - 3t = 0.$$

If these equations have no solution the lines cannot meet. From the first two equations, $s = 1$ and $t = 2$. These values satisfy the third equation, showing that the three equations are consistent.

The substitution $s = 1$ in the equation for the first line or the substitution $t = 2$ in the equation for the second line gives the position vector $7\mathbf{i} + 5\mathbf{j} + 4\mathbf{k}$ of the point at which the lines meet.

*Example 4*

Show that the lines given by the equations

$$\frac{x - 1}{3} = \frac{y + 3}{-1} = \frac{z - 4}{2},$$

$$\frac{x + 5}{2} = \frac{y - 2}{4} = \frac{z + 1}{-1}$$

are at right angles. Find the direction ratios of a straight line which is at right angles to both these lines.

The direction ratios of the first line are $3: -1:2$, so that this line is parallel to the vector $3\mathbf{i} - \mathbf{j} + 2\mathbf{k}$. The direction ratios of the second line are $2:4: -1$, so that this line is parallel to the vector $2\mathbf{i} + 4\mathbf{j} - \mathbf{k}$. These two vectors will be at right angles if their scalar product is zero.

$$(3\mathbf{i} - \mathbf{j} + 2\mathbf{k})\cdot(2\mathbf{i} + 4\mathbf{j} - \mathbf{k}) = 3 \times 2 - 1 \times 4 + 2 \times (-1) = 0.$$

This shows that the two lines are at right angles.

Let $a\mathbf{i} + b\mathbf{j} + c\mathbf{k}$ be a vector which is at right angles to each of the vectors $3\mathbf{i} - \mathbf{j} + 2\mathbf{k}$ and $2\mathbf{i} + 4\mathbf{j} - \mathbf{k}$. The scalar product of the vector $a\mathbf{i} + b\mathbf{j} + c\mathbf{k}$ with each of these vectors will be zero. This gives the equations

$$3a - b + 2c = 0,$$
$$2a + 4b - c = 0.$$

Elimination of $c$ gives $7a + 7b = 0$, i.e. $b = -a$. Substitution in either equation gives $c = -2a$. Hence the ratios $a:b:c$ are $1: -1: -2$, and these are the direction ratios of a straight line at right angles to both of the given lines.

## Direction cosines

Let the vector $\mathbf{a}$ make angles $\alpha$, $\beta$ and $\gamma$ with the positive direction of the $x$, $y$ and $z$ axes respectively. Then $\cos \alpha$, $\cos \beta$ and $\cos \gamma$ are given by the scalar products

$$\mathbf{a}\cdot\mathbf{i} = a \cos \alpha, \quad \mathbf{a}\cdot\mathbf{j} = a \cos \beta, \quad \mathbf{a}\cdot\mathbf{k} = a \cos \gamma$$

where $a = |\mathbf{a}|$. These three cosines are known as the direction cosines of the vector $\mathbf{a}$ and are denoted by the letters $l$, $m$ and $n$ respectively.

Let $\mathbf{a} = a_1\mathbf{i} + a_2\mathbf{j} + a_3\mathbf{k}$. Then

$$l = \cos \alpha = (\mathbf{a}\cdot\mathbf{i})/a = a_1/a,$$
$$m = \cos \beta = (\mathbf{a}\cdot\mathbf{j})/a = a_2/a,$$
$$n = \cos \gamma = (\mathbf{a}\cdot\mathbf{k})/a = a_3/a.$$

Hence $l^2 + m^2 + n^2 = (a_1^2 + a_2^2 + a_3^2)/a^2 = 1$.

The direction cosines of the vector $-\mathbf{a}$ will be $-l$, $-m$ and $-n$.

The direction cosines of any line parallel to the vector $\mathbf{a}$ may be taken to be $l$, $m$, $n$ or $-l$, $-m$, $-n$.

The direction cosines of a line can be found directly from its direction ratios. If the line is parallel to the vector $\mathbf{a}$, its direction ratios are $a_1:a_2:a_3$. Then its direction cosines are $a_1/a$, $a_2/a$, $a_3/a$ where $a^2 = a_1^2 + a_2^2 + a_3^2$.

*Example 5*

Find the direction cosines of the line

$$\frac{x - 2}{6} = \frac{y - 4}{-6} = \frac{z + 3}{7}.$$

The direction ratios of the line are $6: -6: 7$ and

$$6^2 + (-6)^2 + 7^2 = 121 = 11^2.$$

Therefore

$$l = \cos \alpha = 6/11, \ m = \cos \beta = -6/11, \ n = \cos \gamma = 7/11.$$

This shows that the vector $6\mathbf{i} - 6\mathbf{j} + 7\mathbf{k}$ makes an acute angle with $Ox$ and with $Oz$ but an obtuse angle with $Oy$. The vector $-6\mathbf{i} + 6\mathbf{j} - 7\mathbf{k}$ will make an acute angle with $Oy$ but an obtuse angle with $Ox$ and with $Oz$.

*Example 6*

The position vectors $\mathbf{a}$, $\mathbf{b}$, $\mathbf{c}$ and $\mathbf{d}$ of the points A, B, C and D are given by $\mathbf{a} = \mathbf{i} + \mathbf{k}$, $\mathbf{b} = 3\mathbf{i} - \mathbf{j} + 3\mathbf{k}$, $\mathbf{c} = -4\mathbf{i} + \mathbf{j} + 2\mathbf{k}$, $\mathbf{d} = 2\mathbf{i} + 4\mathbf{j}$. Find vector equations for the straight lines AB and CD, and find the acute angle between them.

$$\overrightarrow{AB} = \mathbf{b} - \mathbf{a} = (3\mathbf{i} - \mathbf{j} + 3\mathbf{k}) - (\mathbf{i} + \mathbf{k})$$
$$= 2\mathbf{i} - \mathbf{j} + 2\mathbf{k}.$$

The equation of the line AB is

$$\mathbf{r} = \mathbf{a} + t(\mathbf{b} - \mathbf{a})$$

i.e.
$$\mathbf{r} = (\mathbf{i} + \mathbf{k}) + t(2\mathbf{i} - \mathbf{j} + 2\mathbf{k})$$

$$\overrightarrow{CD} = \mathbf{d} - \mathbf{c} = (2\mathbf{i} + 4\mathbf{j}) - (-4\mathbf{i} + \mathbf{j} + 2\mathbf{k})$$
$$= 6\mathbf{i} + 3\mathbf{j} - 2\mathbf{k}.$$

The equation of the line CD is

$$\mathbf{r} = \mathbf{c} + s(\mathbf{d} - \mathbf{c})$$

i.e.
$$\mathbf{r} = (-4\mathbf{i} + \mathbf{j} + 2\mathbf{k}) + s(6\mathbf{i} + 3\mathbf{j} - 2\mathbf{k}).$$

Let $\theta$ be the angle between the vectors $2\mathbf{i} - \mathbf{j} + 2\mathbf{k}$ and $6\mathbf{i} + 3\mathbf{j} - 2\mathbf{k}$. Since $|2\mathbf{i} - \mathbf{j} + 2\mathbf{k}| = 3$ and $|6\mathbf{i} + 3\mathbf{j} - 2\mathbf{k}| = 7$,

$$(3 \times 7) \cos \theta = (2\mathbf{i} - \mathbf{j} + 2\mathbf{k}) \cdot (6\mathbf{i} + 3\mathbf{j} - 2\mathbf{k}) = 5$$
$$\cos \theta = 5/21.$$

Hence $\theta$ equals $76°$ to the nearest degree, and this is the acute angle between the lines.

*Example 7*

Find the shortest distance from the point A with position vector $2\mathbf{i} + 3\mathbf{j} - 2\mathbf{k}$ to the line with equation

$$\mathbf{r} = 3\mathbf{i} + \mathbf{k} + t(2\mathbf{i} - \mathbf{j}).$$

Let P be the point on the line such that AP is at right angles to the line. Let the value of $t$ at P be $t_0$. The vector $\overrightarrow{AP}$ is given by

$$\overrightarrow{AP} = 3\mathbf{i} + \mathbf{k} + t_0(2\mathbf{i} - \mathbf{j}) - (2\mathbf{i} + 3\mathbf{j} - 2\mathbf{k})$$
$$= (1 + 2t_0)\mathbf{i} - (3 + t_0)\mathbf{j} + 3\mathbf{k}.$$

Since this vector is perpendicular to the vector $2\mathbf{i} - \mathbf{j}$, their scalar product is zero.

$$(2 + 4t_0) + (3 + t_0) = 0$$
$$t_0 = -1$$

Hence $\overrightarrow{AP}$ equals $-\mathbf{i} - 2\mathbf{j} + 3\mathbf{k}$, and so the length of AP is $\sqrt{14}$.

### Exercises 2.1

1  A straight line is drawn through the point with position vector $\mathbf{a}$ parallel to the vector $\mathbf{b}$. Find a vector equation for the line in the following cases:
   (a) $\mathbf{a} = \mathbf{i}, \mathbf{b} = \mathbf{j}$
   (b) $\mathbf{a} = \mathbf{i} - \mathbf{j}, \mathbf{b} = 2\mathbf{i} + \mathbf{j} + \mathbf{k}$
   (c) $\mathbf{a} = 4\mathbf{i} + 3\mathbf{j} - 2\mathbf{k}, \mathbf{b} = \mathbf{j} + \mathbf{k}$
   (d) $\mathbf{a} = 2\mathbf{i} + 3\mathbf{k}, \mathbf{b} = 2\mathbf{i} + 3\mathbf{k}.$

2  The points P and Q have position vectors $\mathbf{p}$ and $\mathbf{q}$ respectively. Find a vector equation for the straight line PQ in the following cases:
   (a) $\mathbf{p} = \mathbf{i} + \mathbf{k}, \mathbf{q} = \mathbf{j} + \mathbf{k}$
   (b) $\mathbf{p} = \mathbf{i} + \mathbf{j} + \mathbf{k}, \mathbf{q} = \mathbf{j} + \mathbf{k}$
   (c) $\mathbf{p} = 2\mathbf{i} - \mathbf{j} + 3\mathbf{k}, \mathbf{q} = 2\mathbf{i} + \mathbf{j} + \mathbf{k}$
   (d) $\mathbf{p} = \mathbf{j} - \mathbf{k}, \mathbf{q} = 0.$

3  Find the direction ratios of the straight line through the points $(2, 3, 4)$ and $(4, 7, 5)$ and write down cartesian equations for this line.

4  Find the cartesian equations of the straight lines through the points which have the following position vectors:
   (a) $\mathbf{i} - \mathbf{j} + 2\mathbf{k}, 4\mathbf{i} + \mathbf{j} + 4\mathbf{k}$
   (b) $\mathbf{j} - 2\mathbf{k}, 8\mathbf{i} + 7\mathbf{j}$
   (c) $2\mathbf{i} - 3\mathbf{j}, -4\mathbf{j} + \mathbf{k}$
   (d) $-2\mathbf{i} + \mathbf{k}, \mathbf{i} + 4\mathbf{j} + 3\mathbf{k}.$

**5** Show that each of the following pairs of lines is a pair of intersecting lines, and in each case find the position vector of the common point.

(a) $\mathbf{r} = \mathbf{i} + \mathbf{k} + s\mathbf{j}, \mathbf{r} = 2\mathbf{i} + 2\mathbf{j} + \mathbf{k} + t\mathbf{i}$

(b) $\mathbf{r} = 2\mathbf{i} + 4\mathbf{j} + s\mathbf{k}, \mathbf{r} = \mathbf{i} + 4\mathbf{j} + \mathbf{k} + t(\mathbf{i} - \mathbf{k})$

(c) $\mathbf{r} = \mathbf{i} + 5\mathbf{j} + 3\mathbf{k} + s(\mathbf{j} + \mathbf{k}), \mathbf{r} = \mathbf{i} + t\mathbf{j}$

(d) $\mathbf{r} = \mathbf{i} + \mathbf{j} - \mathbf{k} + s(\mathbf{j} - \mathbf{k}), \mathbf{r} = \mathbf{i} + t(3\mathbf{i} + \mathbf{j} + \mathbf{k})$.

**6** Find the direction cosines of the vector $\overrightarrow{AB}$ when

(a) A is the point $(2, 2, 3)$ and B is the point $(3, 1, 2)$

(b) A is the point $(4, 0, 4)$ and B is the point $(6, 1, 2)$

(c) A is the point $(3, -4, -1)$ and B is the point $(-3, 3, 5)$

(d) A is the point $(4, 1, 0)$ and B is the point $(-2, 4, 2)$.

**7** Find the cartesian equations of the straight line through the origin which is perpendicular to each of the lines

$$(x - 1)/3 = (y + 3)/(-2) = z,$$
$$x + 2 = y = (z - 1)/2.$$

**8** Show that the triangle with vertices at the points $(4, 2, 1)$, $(2, 3, 2)$ and $(1, 0, -3)$ is right-angled.

**9** Find in parametric form the cartesian equations of the straight line through the points $(2, 1, 3)$ and $(4, 7, -1)$. Show that this line passes through the points $(5, 10, -3)$ and $(0, -5, 7)$.

**10** Show that the two lines given by the equations

$$x = 3 + t, \quad y = 4 + 2t, \quad z = -2t,$$
$$x = 8 - 2s, \quad y = 5 - s, \quad z = -4 + 2s$$

intersect. Find the coordinates of their common point. Find also the acute angle between these lines.

## 2.2 The vector equation of a plane

Consider the plane which is perpendicular to the vector $\mathbf{a}$ and which passes through the point B with position vector $\mathbf{b}$. Let $\mathbf{r}$ be the position vector of any point P in this plane, and let N be the foot of the perpendicular to the plane from the origin O (Fig. 2.3). The vector $\overrightarrow{BP}$, which equals $\mathbf{r} - \mathbf{b}$, is at right angles to the vector $\mathbf{a}$. Therefore the scalar product of $\mathbf{r} - \mathbf{b}$ and $\mathbf{a}$ is zero, i.e.

$$(\mathbf{r} - \mathbf{b}) \cdot \mathbf{a} = 0$$
$$\mathbf{r} \cdot \mathbf{a} = \mathbf{b} \cdot \mathbf{a}$$

This is the vector equation of the plane, since it is satisfied if and only if the point with position vector $\mathbf{r}$ lies in the plane.

It follows that any equation of the form $\mathbf{r} \cdot \mathbf{a} = d$ represents a plane at right angles to the vector $\mathbf{a}$.

Let the perpendicular distance from O to the plane $\mathbf{r} \cdot \mathbf{a} = \mathbf{b} \cdot \mathbf{a}$ be $p$. Then

$$p = ON = OB \cos BON.$$

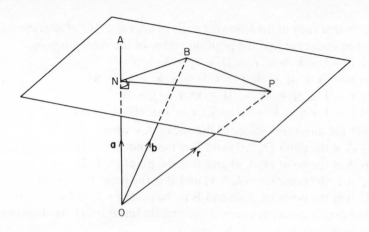

Fig. 2.3

Now if $OA = a = |\mathbf{a}|$ and $OB = b = |\mathbf{b}|$,

$$\mathbf{b} \cdot \mathbf{a} = ba \cos \text{BON} = a \times OB \cos \text{BON}$$

i.e. $\qquad\qquad \mathbf{b} \cdot \mathbf{a} = ap.$

Therefore the equation $\mathbf{r} \cdot \mathbf{a} = \mathbf{b} \cdot \mathbf{a}$ is equivalent to the equation $\mathbf{r} \cdot \mathbf{a} = ap$.
Let $\mathbf{n}$ be the unit vector in the direction of the vector $\mathbf{a}$, i.e. $\mathbf{n} = \mathbf{a}/a$. Then the equation $\mathbf{r} \cdot \mathbf{a} = ap$ can be written in the form

$$\mathbf{r} \cdot \mathbf{n} = p.$$

This equation represents a plane which is perpendicular to the unit vector $\mathbf{n}$ and which is at a distance $p$ from the origin. The vector $\mathbf{n}$ is the unit normal to the plane. Since $\mathbf{a} = a\mathbf{n}$, the equation $\mathbf{r} \cdot \mathbf{a} = d$ is equivalent to the equation $\mathbf{r} \cdot \mathbf{n} = d/a$. Therefore the perpendicular distance from the origin to the plane $\mathbf{r} \cdot \mathbf{a} = d$ is $d/a$.

There is an alternative form for the vector equation of a plane. In Fig. 2.4, $\overrightarrow{OA} = \mathbf{a}$, $\overrightarrow{AB} = \mathbf{b}$ and $\overrightarrow{AC} = \mathbf{c}$ where $\mathbf{a}, \mathbf{b}$ and $\mathbf{c}$ are independent vectors. The points A, B and C define a plane. For any point P in this plane the vector $\overrightarrow{AP}$ can be expressed in the form

$$\overrightarrow{AP} = s\overrightarrow{AB} + t\overrightarrow{AC},$$

where $s$ and $t$ are parameters.

Let $\mathbf{r}$ be the position vector of P. Then

$$\mathbf{r} = \overrightarrow{OP} = \overrightarrow{OA} + \overrightarrow{AP}$$
$$= \mathbf{a} + s\overrightarrow{AB} + t\overrightarrow{AC}$$

i.e. $\qquad\qquad \mathbf{r} = \mathbf{a} + s\mathbf{b} + t\mathbf{c}.$

This is the equation of the plane which passes through the point with position vector $\mathbf{a}$ and which contains lines parallel to the vectors $\mathbf{b}$ and $\mathbf{c}$.

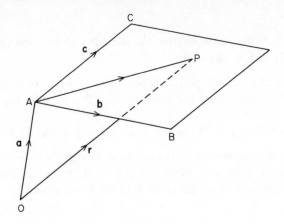

Fig. 2.4

## Example 1

Find in vector form the equation of the plane which is perpendicular to the vector $3\mathbf{i} - 6\mathbf{j} - 2\mathbf{k}$ and which passes through the point with position vector $2\mathbf{i} - \mathbf{j} - \mathbf{k}$. Find the perpendicular distance from the origin to this plane.

The plane is given by the equation

$$\mathbf{r}\cdot(3\mathbf{i} - 6\mathbf{j} - 2\mathbf{k}) = (2\mathbf{i} - \mathbf{j} - \mathbf{k})\cdot(3\mathbf{i} - 6\mathbf{j} - 2\mathbf{k}),$$

i.e. $\qquad \mathbf{r}\cdot(3\mathbf{i} - 6\mathbf{j} - 2\mathbf{k}) = 6 + 6 + 2 = 14.$

This is the equation of the plane in the form $\mathbf{r}\cdot\mathbf{a} = d$.

Now $|3\mathbf{i} - 6\mathbf{j} - 2\mathbf{k}| = \sqrt{(9 + 36 + 4)} = 7$. Therefore the equation of the plane in the form $\mathbf{r}\cdot\mathbf{n} = p$ is

$$\mathbf{r}\cdot(3\mathbf{i} - 6\mathbf{j} - 2\mathbf{k})/7 = 2.$$

Hence the required distance is 2.

## Example 2

Find the acute angle between the planes with vector equations

$$\mathbf{r}\cdot(\mathbf{i} + 2\mathbf{j} - \mathbf{k}) = 5 \quad \text{and} \quad \mathbf{r}\cdot(\mathbf{j} - \mathbf{k}) = 7.$$

The angle between the planes equals the angle between the normals to the planes. The unit normal to the plane $\mathbf{r}\cdot(\mathbf{i} + 2\mathbf{j} - \mathbf{k}) = 5$ is $(\mathbf{i} + 2\mathbf{j} - \mathbf{k})/\sqrt{6}$. The unit normal to the plane $\mathbf{r}\cdot(\mathbf{j} - \mathbf{k}) = 7$ is $(\mathbf{j} - \mathbf{k})/\sqrt{2}$.

Let $\theta$ be the angle between the two unit normals. Then their scalar product equals $\cos\theta$.

$$\cos\theta = (\mathbf{i} + 2\mathbf{j} - \mathbf{k})\cdot(\mathbf{j} - \mathbf{k})/\sqrt{12}$$
$$= 3/\sqrt{12} = \sqrt{3}/2.$$

Therefore the acute angle between the planes is $30°$.

*Example 3*

Find the perpendicular distance from the point P with position vector $4\mathbf{i} + \mathbf{k}$ to the plane $\mathbf{r} \cdot (2\mathbf{i} + \mathbf{j} + 2\mathbf{k}) = 4$.

Since $|2\mathbf{i} + \mathbf{j} + 2\mathbf{k}| = 3$, the perpendicular distance from the origin to the given plane is 4/3. The equation of the plane through P parallel to the given plane is

$$\mathbf{r} \cdot (2\mathbf{i} + \mathbf{j} + 2\mathbf{k}) = (4\mathbf{i} + \mathbf{k}) \cdot (2\mathbf{i} + \mathbf{j} + 2\mathbf{k}) = 10.$$

The perpendicular distance from the origin to this plane is 10/3. Therefore the distance between these two planes is $(10 - 4)/3$, i.e. 2. This is also the perpendicular distance from the point P to the given plane.

*Example 4*

Show that the line with equation $\mathbf{r} = \mathbf{j} + t(\mathbf{i} + \mathbf{k})$ makes an angle of 30° with the plane $\mathbf{r} \cdot (\mathbf{j} + \mathbf{k}) = 2$.

The line is parallel to the vector $\mathbf{i} + \mathbf{k}$, and the vector $\mathbf{j} + \mathbf{k}$ is at right angles to the plane. Let $\theta$ be the angle between these two vectors. Since $|\mathbf{i} + \mathbf{k}| = \sqrt{2}$ and $|\mathbf{j} + \mathbf{k}| = \sqrt{2}$,

$$(\mathbf{i} + \mathbf{k}) \cdot (\mathbf{j} + \mathbf{k}) = 2 \cos \theta$$

This gives $2 \cos \theta = 1$, i.e. $\cos \theta = 1/2$, $\theta = 60°$. This is the angle between the line and the normal to the plane. Hence the angle between the line and the plane is 30°.

*Example 5*

Find a unit vector which is at right angles to the plane through the points A, B and C which have position vectors $\mathbf{a}$, $\mathbf{b}$ and $\mathbf{c}$ where $\mathbf{a} = 2\mathbf{i} + \mathbf{j} + \mathbf{k}$, $\mathbf{b} = 4\mathbf{i} + 5\mathbf{j} + 4\mathbf{k}$ and $\mathbf{c} = 5\mathbf{i} + 2\mathbf{j} + 3\mathbf{k}$. Obtain the equation of this plane in vector form.

Let the unit vector $l\mathbf{i} + m\mathbf{j} + n\mathbf{k}$ be perpendicular to the plane ABC. Then this vector will be at right angles to the vectors $\overrightarrow{AB}$ and $\overrightarrow{AC}$.

$\overrightarrow{AB} = \mathbf{b} - \mathbf{a} = 2\mathbf{i} + 4\mathbf{j} + 3\mathbf{k}$.

This vector and the vector $l\mathbf{i} + m\mathbf{j} + n\mathbf{k}$ are at right angles.

$$(2\mathbf{i} + 4\mathbf{j} + 3\mathbf{k}) \cdot (l\mathbf{i} + m\mathbf{j} + n\mathbf{k}) = 0$$
$$2l + 4m + 3n = 0$$

$\overrightarrow{AC} = \mathbf{c} - \mathbf{a} = 3\mathbf{i} + \mathbf{j} + 2\mathbf{k}$.

This vector and the vector $l\mathbf{i} + m\mathbf{j} + n\mathbf{k}$ are at right angles.

$$(3\mathbf{i} + \mathbf{j} + 2\mathbf{k}) \cdot (l\mathbf{i} + m\mathbf{j} + n\mathbf{k}) = 0$$
$$3l + m + 2n = 0$$

From these two equations in $l$, $m$ and $n$,

$$4(3l + m + 2n) - (2l + 4m + 3n) = 0$$
$$10l + 5n = 0$$
$$n = -2l$$

$$3l + m - 4l = 0$$
$$m = l$$

Therefore $\quad\quad\quad li + mj + nk = l(i + j - 2k).$

Since this is a unit vector its modulus equals 1. It follows that $l = 1/\sqrt{6}$, so that the required vector is

$$(i + j - 2k)/\sqrt{6}.$$

The equation of the plane ABC will be of the form

$$r \cdot (i + j - 2k) = d$$

where $d$ is a constant. Since the point A lies on the plane,

$$a \cdot (i + j - 2k) = d.$$

Therefore $\quad\quad d = (2i + j + k) \cdot (i + j - 2k) = 1.$

Hence the equation of the plane ABC is

$$r \cdot (i + j - 2k) = 1.$$

*Example 6*

Write down a vector equation for the plane which passes through the point with position vector $i + k$ and which contains lines parallel to the vectors $2i + j$ and $j + k$. Find the perpendicular distance from the origin to this plane.

A vector equation for this plane is

$$r = i + k + s(2i + j) + t(j + k).$$

Let the vector $ai + bj + ck$ be at right angles to each of the vectors $2i + j$ and $j + k$. Then

$$(ai + bj + ck) \cdot (2i + j) = 0$$
$$2a + b = 0$$
$$(ai + bj + ck) \cdot (j + k) = 0$$
$$b + c = 0$$

This gives $b = -2a$ and $c = 2a$, so that

$$ai + bj + ck = a(i - 2j + 2k).$$

It can be verified by inspection that this vector is at right angles to the vectors $2i + j$ and $j + k$.

Since $|\mathbf{i} - 2\mathbf{j} + 2\mathbf{k}| = 3$, the vector $a(\mathbf{i} - 2\mathbf{j} + 2\mathbf{k})$ will be a unit vector when $a = 1/3$. Hence the unit vector normal to the plane is $(\mathbf{i} - 2\mathbf{j} + 2\mathbf{k})/3$. The plane with equation

$$\mathbf{r} \cdot (\mathbf{i} - 2\mathbf{j} + 2\mathbf{k})/3 = p$$

will pass through the point with position vector $\mathbf{i} + \mathbf{k}$ if

$$p = (\mathbf{i} + \mathbf{k}) \cdot (\mathbf{i} - 2\mathbf{j} + 2\mathbf{k})/3$$

i.e. if $p = 1$. Therefore the vector equation of the same plane in the form $\mathbf{r} \cdot \mathbf{n} = p$ is

$$\mathbf{r} \cdot (\mathbf{i} - 2\mathbf{j} + 2\mathbf{k})/3 = 1,$$

showing that the distance from the origin to the plane is 1.

**Exercises 2.2**

1  Find in the form $\mathbf{r} \cdot \mathbf{n} = p$, where $\mathbf{n}$ is a unit vector, the equation of the plane which passes through the point with position vector $\mathbf{i} + 2\mathbf{j}$ and which is
   (a) perpendicular to the vector $\mathbf{i}$
   (b) perpendicular to the vector $\mathbf{i} + 2\mathbf{j}$
   (c) parallel to the plane $\mathbf{r} \cdot \mathbf{k} = 1$
   (d) parallel to the plane $\mathbf{r} \cdot (2\mathbf{i} + 2\mathbf{j} + \mathbf{k}) = 1$.

2  Find the perpendicular distance from the origin to the plane with vector equation
   (a) $\mathbf{r} \cdot (\mathbf{i} + \mathbf{j}) = 2$
   (b) $\mathbf{r} \cdot (4\mathbf{i} + 4\mathbf{j} + 7\mathbf{k}) = 9$
   (c) $\mathbf{r} \cdot (2\mathbf{i} - \mathbf{j} - 2\mathbf{k}) = 6$
   (d) $\mathbf{r} \cdot (2\mathbf{i} + 6\mathbf{j} - 3\mathbf{k}) = 21$.

3  Find the acute angle between the planes
   (a) $\mathbf{r} \cdot (3\mathbf{i} + 4\mathbf{j} + 5\mathbf{k}) = 4, \quad \mathbf{r} \cdot (6\mathbf{i} + 8\mathbf{j}) = 5$
   (b) $\mathbf{r} \cdot (\mathbf{j} - \mathbf{k}) = 3, \quad \mathbf{r} \cdot (\mathbf{i} + 2\mathbf{j} - \mathbf{k}) = 2$
   (c) $\mathbf{r} = \mathbf{i} + s\mathbf{j} + t\mathbf{k}, \quad \mathbf{r} = \mathbf{k} + s(\mathbf{i} - \mathbf{k}) + t\mathbf{j}$
   (d) $\mathbf{r} \cdot (2\mathbf{i} + \mathbf{j} - \mathbf{k}) = 1, \quad \mathbf{r} = s(\mathbf{i} + \mathbf{j} + \mathbf{k}) + t(2\mathbf{i} + 3\mathbf{j} + 2\mathbf{k})$.

4  Find the acute angle between the line and the plane given by the equations
   (a) $\mathbf{r} = t(3\mathbf{i} + 4\mathbf{j} + 4\mathbf{k}), \quad \mathbf{r} \cdot (4\mathbf{i} - 5\mathbf{j} + 2\mathbf{k}) = 6$
   (b) $\mathbf{r} = \mathbf{i} + t(\mathbf{j} + \mathbf{k}), \quad \mathbf{r} \cdot (2\mathbf{i} - 2\mathbf{j} - \mathbf{k}) = 4$
   (c) $\mathbf{r} = 2\mathbf{i} + \mathbf{k} + t(2\mathbf{i} + \mathbf{j} + \mathbf{k}), \quad \mathbf{r} \cdot (\mathbf{i} + 2\mathbf{j} - \mathbf{k}) = 1$
   (d) $\mathbf{r} = \mathbf{j} + t(2\mathbf{i} - \mathbf{j} + \mathbf{k}), \quad \mathbf{r} \cdot (\mathbf{i} - \mathbf{j}) = 2$.

5  Find the perpendicular distance from the point with position vector $5\mathbf{i} + 4\mathbf{j} + \mathbf{k}$ to the plane with equation
   (a) $\mathbf{r} \cdot (2\mathbf{i} + 5\mathbf{j} - 14\mathbf{k}) = 1$
   (b) $\mathbf{r} \cdot (\mathbf{i} + 4\mathbf{j} + 8\mathbf{k}) = 2$
   (c) $\mathbf{r} \cdot (2\mathbf{i} + 6\mathbf{j} - 9\mathbf{k}) = 3$
   (d) $\mathbf{r} = \mathbf{i} + s\mathbf{j} + t\mathbf{k}$.

6  Show that the two planes $\mathbf{r} \cdot (\mathbf{i} + 2\mathbf{j} + 2\mathbf{k}) = 0$ and $\mathbf{r} \cdot (2\mathbf{i} - \mathbf{j}) = 0$ are at

right angles. Find the equation of a third plane through the origin which is at right angles to each of these planes.

7  A plane passes through the point with position vector $2\mathbf{i} + \mathbf{j}$ and is parallel to the vectors $\mathbf{i} + \mathbf{j}$ and $2\mathbf{j} + 3\mathbf{k}$. Show that this plane passes through the points with position vectors $\mathbf{i} + 2\mathbf{j} + 3\mathbf{k}$ and $6\mathbf{i} - \mathbf{j} - 9\mathbf{k}$. Find the distance from the origin to this plane.

8  The position vectors of the points A, B and C are $\mathbf{i}$, $\mathbf{j}$ and $\mathbf{k}$ respectively. Find the vector equation of
   (a) the plane which bisects the line AB at right angles
   (b) the plane through the points A, B and C
   (c) the line in which these two planes meet.

9  Find the vector equation of the plane which passes through the origin and through the points which have position vectors $\mathbf{i} + 3\mathbf{j} - 2\mathbf{k}$ and $\mathbf{i} + 2\mathbf{j} - \mathbf{k}$.

10  Find the vector equation of the plane which passes through the points A, B and C with position vectors $\mathbf{a}$, $\mathbf{b}$ and $\mathbf{c}$, where $\mathbf{a} = 2\mathbf{i} - \mathbf{j} - \mathbf{k}$, $\mathbf{b} = 4\mathbf{i} + \mathbf{j} + 2\mathbf{k}$ and $\mathbf{c} = \mathbf{i} - 2\mathbf{j} + \mathbf{k}$.

## 2.3  The cartesian equation of a plane

Let $\mathbf{r} = x\mathbf{i} + y\mathbf{j} + z\mathbf{k}$ and $\mathbf{a} = a_1\mathbf{i} + a_2\mathbf{j} + a_3\mathbf{k}$. Then the scalar product $\mathbf{r} \cdot \mathbf{a}$ equals $a_1x + a_2y + a_3z$, and the vector equation $\mathbf{r} \cdot \mathbf{a} = d$ of a plane becomes

$$a_1x + a_2y + a_3z = d.$$

This is the cartesian equation of a plane at right angles to the vector $\mathbf{a}$. The distance from the origin to this plane is $d/a$, where

$$a = \sqrt{(a_1^2 + a_2^2 + a_3^2)}$$

The vector equation of a plane at right angles to the unit vector $l\mathbf{i} + m\mathbf{j} + n\mathbf{k}$ and at a distance $p$ from the origin is

$$\mathbf{r} \cdot (l\mathbf{i} + m\mathbf{j} + n\mathbf{k}) = p.$$

When $\mathbf{r}$ is replaced by $x\mathbf{i} + y\mathbf{j} + z\mathbf{k}$, this equation becomes

$$lx + my + nz = p.$$

This is the cartesian equation of a plane which is at right angles to the unit vector $l\mathbf{i} + m\mathbf{j} + n\mathbf{k}$ and which is at a distance $p$ from the origin.

*Example 1*
The coordinates of the points A and B are $(1, 1, 2)$ and $(3, -1, 8)$ respectively. Find the cartesian equation of the plane at right angles to AB which passes through the mid-point of AB.

The vector $\overrightarrow{AB}$ is the normal to the plane.

$$\overrightarrow{AB} = (3\mathbf{i} - \mathbf{j} + 8\mathbf{k}) - (\mathbf{i} + \mathbf{j} + 2\mathbf{k}) = 2\mathbf{i} - 2\mathbf{j} + 6\mathbf{k}.$$

The direction ratios of $\overrightarrow{AB}$ are $2:-2:6$, i.e. $1:-1:3$.

The equation of any plane perpendicular to AB is of the form

$$x - y + 3z = d.$$

This plane will pass through the point $(2, 0, 5)$, which is the mid-point of AB, if $d = 2 + 15 = 17$. Therefore the required equation is

$$x - y + 3z = 17.$$

This equation can also be found by taking any point $P(x, y, z)$ on the plane and equating $PA^2$ to $PB^2$.

*Example 2*

Find the cartesian equations of the two planes which are at right angles to the vector $4\mathbf{i} - 4\mathbf{j} + 7\mathbf{k}$ and which are at unit distance from the origin.

The equations will be of the form $4x - 4y + 7z = d$. Since $4^2 + 4^2 + 7^2 = 9^2$, the length of the vector $4\mathbf{i} - 4\mathbf{j} + 7\mathbf{k}$ is 9. The unit vector in the same direction is $(4\mathbf{i} - 4\mathbf{j} + 7\mathbf{k})/9$.

The plane $4x - 4y + 7z = d$ will pass through the point $\left(\dfrac{4}{9}, -\dfrac{4}{9}, \dfrac{7}{9}\right)$ if $d = (16 + 16 + 49)/9 = 9$.

The plane $4x - 4y + 7z = d$ will pass through the point $\left(-\dfrac{4}{9}, \dfrac{4}{9}, -\dfrac{7}{9}\right)$ if $d = (-16 - 16 - 49)/9 = -9$.

Hence the equations of the two planes are $4x - 4y + 7z = \pm 9$.

*Example 3*

Find the cartesian equation of the plane which passes through the points $A(1, 1, 1)$, $B(3, 4, 3)$ and $C(2, 5, 3)$.

The normal to the plane is at right angles to $\overrightarrow{AB}$ and to $\overrightarrow{AC}$.

$$\overrightarrow{AB} = (3\mathbf{i} + 4\mathbf{j} + 3\mathbf{k}) - (\mathbf{i} + \mathbf{j} + \mathbf{k}) = 2\mathbf{i} + 3\mathbf{j} + 2\mathbf{k}.$$
$$\overrightarrow{AC} = (2\mathbf{i} + 5\mathbf{j} + 3\mathbf{k}) - (\mathbf{i} + \mathbf{j} + \mathbf{k}) = \mathbf{i} + 4\mathbf{j} + 2\mathbf{k}.$$

Let the vector $a\mathbf{i} + b\mathbf{j} + c\mathbf{k}$ be normal to the plane. Then

$$(a\mathbf{i} + b\mathbf{j} + c\mathbf{k}) \cdot (2\mathbf{i} + 3\mathbf{j} + 2\mathbf{k}) = 0$$
$$2a + 3b + 2c = 0.$$
$$(a\mathbf{i} + b\mathbf{j} + c\mathbf{k}) \cdot (\mathbf{i} + 4\mathbf{j} + 2\mathbf{k}) = 0$$
$$a + 4b + 2c = 0.$$

These equations give $b = a$ and $c = -5a/2$, so that

$$a\mathbf{i} + b\mathbf{j} + c\mathbf{k} = (a/2)(2\mathbf{i} + 2\mathbf{j} - 5\mathbf{k}).$$

Hence the equation of the plane will take the form

$$2x + 2y - 5z = d.$$

Since the plane passes through the point $(1, 1, 1)$, the constant $d$ equals $-1$ and the equation of the plane is

$$2x + 2y - 5z = -1.$$

*Example 4*
Find the perpendicular distance from the point $P$ $(x_1, y_1, z_1)$ to the plane $ax + by + cz = d$.

The perpendicular distance from the origin to the plane $ax + by + cz = d$ is $d/\sqrt{(a^2 + b^2 + c^2)}$.
The equation of the parallel plane passing through $P$ is

$$ax + by + cz = ax_1 + by_1 + cz_1.$$

The perpendicular distance from the origin to this plane is

$$(ax_1 + by_1 + cz_1)/\sqrt{(a^2 + b^2 + c^2)}.$$

Therefore the distance between these two planes is

$$(ax_1 + by_1 + cz_1 - d)/\sqrt{(a^2 + b^2 + c^2)}.$$

and this expression gives the perpendicular distance from $P$ to the plane $ax + by + cz = d$.

For points on one side of the plane this expression is positive, and for points on the other side it is negative.

*Example 5*
Find the equation of the plane through the point $(1, 2, 1)$ which passes through the line of intersection of the planes $3x - 5y + z = 6$ and $2x + y - 3z = -2$.

Consider the equation

$$(3x - 5y + z - 6) + \lambda(2x + y - 3z + 2) = 0,$$

where $\lambda$ is a constant. This equation represents a plane, because it is of the form $ax + by + cz + d = 0$.
If $P$ is a point on the line of intersection of the two given planes, its coordinates will satisfy the equation

$$(3x - 5y + z - 6) + \lambda(2x + y - 3z + 2) = 0,$$

since the sum of the terms in each bracket will be zero. Therefore the plane represented by this equation passes through the line in which the given planes intersect. The value of the constant $\lambda$ can be chosen so that the plane passes through the point $(1, 2, 1)$. The substitution $x = 1$, $y = 2$, $z = 1$ gives

$$-12 + 3\lambda = 0, \quad \text{i.e. } \lambda = 4.$$

Hence the required equation is $11x - y - 11z = -2$.

*Example 6*

Find the cartesian equations of the line of intersection of the planes $x + 2y + z = 5$ and $3x - y + 2z = 1$.

Let the direction ratios of the line be $a:b:c$. Since the line will be at right angles to the normal to each plane,

$$a + 2b + c = 0,$$
$$3a - b + 2c = 0.$$

These equations give $a:b:c = 5:1:-7$.

Since the line is not parallel to the $x$–$y$ plane, there will be a point on the line at which $z = 0$. When $z = 0$ the equations of the planes become

$$x + 2y = 5,$$
$$3x - y = 1.$$

Hence $x = 1$, $y = 2$. This shows that the point $(1, 2, 0)$ lies on the line. The cartesian equations of the line which passes through the point $(1, 2, 0)$ and which has direction ratios $5:1:-7$ are

$$\frac{x - 1}{5} = \frac{y - 2}{1} = \frac{z}{-7}.$$

**Exercises 2.3**

1  Find the perpendicular distance from the origin to
   (a) the plane $2x - 14y + 5z = 30$
   (b) the plane $20x + 4y - 5z = 21$.

2  The planes $2x - 2y + nz = 2$, $2x + my + z = 8$ are parallel and are at a distance $d$ apart. Find $m$, $n$ and $d$.

3  Find the angle between the planes $2x - y + z = 6$, $x + y + 2z = 3$.

4  Find the perpendicular distance to the plane $3x + 4y - 12z = 3$
   (a) from the point $(2, -1, 1)$
   (b) from the point $(3, 2, -1)$.

5  Find the equation of the plane which bisects at right angles the line joining the points $(4, 5, -2)$, $(-2, -1, 6)$.

6  Show that the line

$$x - 3 = y - 1 = (z - 2)/2$$

   lies in the plane $x - 5y + 2z = 2$.

7  Find the equation of the plane through the origin which passes through the line of intersection of the planes $x - y - 3z = 2$, $x - 2y - 5z = 3$.

8  Find the cartesian equation of the plane which passes through the points $(1, 1, 1)$, $(2, 2, 4)$ and $(3, 5, 5)$.

9  Find the cartesian equation of the plane which passes through the point $(1, 4, 2)$ and through the line of intersection of the planes $2x + 3y - z = 4$, $x - y + 2z = 5$.

**10** Find the cartesian equations of the line of intersection of the planes $5x - 2y - z = 9, x - 2y + z = 7$.

## Miscellaneous exercises 2

**1** The position vectors of four vertices of a cube are $\mathbf{0}, \mathbf{i}, \mathbf{j}$ and $\mathbf{k}$. Find the vector equations of the six faces of the cube.

Find also the vector equation of the diagonal of the cube which passes through the origin.

**2** A straight line is given by the parametric equations

$$x = 1 - 6t, \ y = 3 + 2t, \ z = 2 + 3t.$$

Find the distance between the points in which this line meets the planes $4x + 3y + z = 0, x - 5y - 2z = 4$.

**3** A straight line makes an angle of $60°$ with the $x$-axis and with the $y$-axis. Find the acute angle which it makes with the $z$-axis.

**4** Find the vector equation of the plane which is perpendicular to the vector $\mathbf{i} + \mathbf{j} + \mathbf{k}$ and which passes through the point with position vector $3\mathbf{i} + 2\mathbf{j} - \mathbf{k}$.

**5** Find the vector equation of the straight line which passes through the point with position vector $\mathbf{i} + \mathbf{k}$ and which is perpendicular to the plane $\mathbf{r} \cdot \mathbf{j} = 2$.

**6** Find the position vector of the point on the line $\mathbf{r} = \mathbf{k} + t(\mathbf{i} + \mathbf{j} - \mathbf{k})$ which is nearest to the origin.

**7** The position vectors of the points A and B are $5\mathbf{i} - 2\mathbf{j} + \mathbf{k}$ and $2\mathbf{i} - 7\mathbf{j} - 4\mathbf{k}$. Find the position vector of the point in which the line AB meets the plane $\mathbf{r} \cdot \mathbf{j} = 8$.

**8** A plane passes through the points $(2, 0, 0)$, $(0, 2, 0)$ and $(0, 0, 1)$. Find the perpendicular distance to the plane from the origin.

**9** Find the angles made with the axes by the line of intersection of the planes $x - y + 2z = 2, 2x - 2y + 3z = 3$.

**10** Find the cartesian equation of the plane through the points $(3, 2, -1)$, $(1, 0, 1)$ and $(4, 3, -2)$.

**11** The cartesian equations of two lines are

$$x = (y + 3)/(-2) = z + 1$$
and
$$(x + 1)/5 = y + 1 = (z + 2)/(-3).$$

Show that the lines intersect at right angles and find the cartesian equations of the line joining their common point to the origin.

**12** Find the position vector of the point in which the straight line $\mathbf{r} = \mathbf{i} + t(\mathbf{j} + 2\mathbf{k})$ meets the plane $\mathbf{r} \cdot (2\mathbf{i} - \mathbf{j} + \mathbf{k}) = 5$.

**13** Find the position vector of the foot of the perpendicular to the plane $\mathbf{r} \cdot (\mathbf{i} - 2\mathbf{j} + \mathbf{k}) = 3$ from the point with position vector $\mathbf{i} - \mathbf{k}$.

**14** The points A, B and C have coordinates $(5, -4, 2)$, $(7, 4, 6)$ and $(9, 6, 8)$ respectively. Show that the line with equation

$$\begin{pmatrix} x \\ y \\ z \end{pmatrix} = \begin{pmatrix} 1 \\ 1 \\ 1 \end{pmatrix} + t \begin{pmatrix} 4 \\ -5 \\ 1 \end{pmatrix}$$

passes through A and intersects BC at right angles. Find the coordinates of the point of intersection.

15 The points A, B and C have coordinates $(2, 1, -1)$, $(1, -7, 3)$ and $(-2, 5, 1)$ respectively. Calculate the cosine of the angle BAC and show that the area of the triangle BAC is $\sqrt{629}$. [JMB]

16 A plane passes through three points A, B and C with position vectors $2\mathbf{i} - \mathbf{j} + \mathbf{k}$, $3\mathbf{i} + 2\mathbf{j} - \mathbf{k}$ and $-\mathbf{i} + 3\mathbf{j} + 2\mathbf{k}$ respectively. Show that a unit vector normal to the plane is $(11\mathbf{i} + 5\mathbf{j} + 13\mathbf{k})/(3\sqrt{35})$. Show that the equation of the plane is $11x + 5y + 13z = 30$. [JMB]

17 Forces $\mathbf{P} = 3\mathbf{i} - \mathbf{j} + 6\mathbf{k}$ and $\mathbf{Q} = \mathbf{i} + 5\mathbf{j} + 6\mathbf{k}$ act through points whose position vectors are $2\mathbf{i} - \mathbf{j} + 3\mathbf{k}$ and $7\mathbf{i} - 4\mathbf{j} + 12\mathbf{k}$ respectively. Show that the lines of action of the forces intersect and find the position vector of their point of intersection. [JMB]

18 The lines $L_1$ and $L_2$ are given by the equations

$$L_1 : \mathbf{r} = \begin{pmatrix} 1 \\ 6 \\ 3 \end{pmatrix} + t \begin{pmatrix} 2 \\ -1 \\ 1 \end{pmatrix}, \quad L_2 : \mathbf{r} = \begin{pmatrix} 3 \\ 3 \\ 8 \end{pmatrix} + s \begin{pmatrix} 1 \\ 0 \\ 1 \end{pmatrix}.$$

(i) Calculate the angle between the directions of $L_1$, $L_2$.
(ii) Show that the lines do not intersect.
(iii) Verify that the vector $\mathbf{a} = \begin{pmatrix} 1 \\ 1 \\ -1 \end{pmatrix}$ is perpendicular to each of the lines.

The point P on $L_1$ is given by $t = p$; the point Q on $L_2$ is given by $s = q$. Write down the column vector representing $\overrightarrow{PQ}$. Hence calculate $p$ and $q$ so that $\overrightarrow{PQ}$ and $\mathbf{a}$ are parallel. [JMB]

19 Let $\mathbf{a} = \mathbf{i} - 2\mathbf{j} + \mathbf{k}$, $\mathbf{b} = 2\mathbf{i} + \mathbf{j} - \mathbf{k}$. Given that $\mathbf{c} = \lambda\mathbf{a} + \mu\mathbf{b}$ and that $\mathbf{c}$ is perpendicular to $\mathbf{a}$, find the ratio of $\lambda$ to $\mu$.

Let A, B be the points with position vectors $\mathbf{a}$, $\mathbf{b}$ respectively with respect to an origin O. Write down in terms of $\mathbf{a}$ and $\mathbf{b}$ a vector equation of the line $l$ through A, in the plane of O, A and B, which is perpendicular to OA. Find the position vector of P, the point of intersection of $l$ and OB. [JMB]

20 Find the vector equation of the plane which passes through the points with position vectors $\mathbf{i} - \mathbf{j} + 3\mathbf{k}$, $4\mathbf{i} - 2\mathbf{j} + \mathbf{k}$ and $6\mathbf{i} - 7\mathbf{j} + 4\mathbf{k}$.

21 The position vector $\mathbf{r}$ of a particle at time $t$ is

$$\mathbf{r} = 2t^2\mathbf{i} + (t^2 - 4t)\mathbf{j} + (3t - 5)\mathbf{k}.$$

Find the velocity and acceleration of the particle at time $t$. Show that when $t = 2/5$ the velocity and acceleration are perpendicular to each other.

The velocity and acceleration are resolved into components along and perpendicular to the vector $\mathbf{i} - 3\mathbf{j} + 2\mathbf{k}$. Find the velocity and acceleration components parallel to this vector when $t = 2/5$. [AEB]

22 The equations of two planes $P_1$ and $P_2$ are $3x + 4y = 7$ and $2x - y + 2z = 1$ respectively. Show that the acute angle between the planes is $\cos^{-1}(2/15)$.

Find

(i) the distances from the origin and from the point $A(4, -3, 0)$ to the plane $P_1$

(ii) the cartesian equations of the set of points whose distances from the plane $P_1$ are equal to the distance of the point A from the plane $P_1$

(iii) a vector equation of the line of intersection of the planes $P_1$ and $P_2$. [AEB]

23 The position vectors of the points A and B are $3\mathbf{i} - 2\mathbf{j} - \mathbf{k}$ and $5\mathbf{i} + 4\mathbf{j} - 5\mathbf{k}$. Find cartesian and vector equations for the plane which bisects AB at right angles.

24 Four points in space A, B, C and D have coordinates $(1, 0, 0)$, $(2, 1, 0)$, $(3, 0, 2)$ and $(4, 1, 1)$ respectively, the distances being measured in metres. If E, F, G and H are the mid-points of AB, BC, CD and DA respectively, prove that EFGH is a parallelogram. Find

(i) the area of the parallelogram EFGH

(ii) the angle between the vectors $\overrightarrow{GH}$ and $\overrightarrow{BA}$, and show that the lines GH and BA do not intersect. [AEB]

25 The vertices of a tetrahedron are at the points $A(0, 2, 1)$, $B(1, 2, 6)$, $C(-2, 3, 2)$ and $D(0, 1, 4)$, the unit of length being the metre. Find the equation of the plane BCD.

Find equations of the line L through A perpendicular to the plane BCD. Find the coordinates of the point E where this line intersects the plane.

Show that the distance AE is $\frac{1}{2}\sqrt{14}\,\text{m}$ and find the coordinates of the point F on L such that AF:FE is 1:2.

Find also the area of the triangle BCD. [AEB]

26 The equations of two planes are

$$\mathbf{r} \cdot (\mathbf{i} + 2\mathbf{j} + \mathbf{k}) = 1,$$
$$\mathbf{r} \cdot (2\mathbf{i} + \mathbf{j} - \mathbf{k}) = 2.$$

Show that a vector equation of L, the line of intersection of the two planes, is

$$\mathbf{r} = \mathbf{i} + t(\mathbf{i} - \mathbf{j} + \mathbf{k}),$$

where $t$ is a scalar.

Find the cartesian equation of the plane which contains the line L and the origin.

Also find the shortest distance from the point with position vector $2\mathbf{i} - 4\mathbf{j} + 4\mathbf{k}$ to the plane $\mathbf{r} \cdot (2\mathbf{i} + \mathbf{j} - \mathbf{k}) = 2$. [AEB]

**27** Show that the line which passes through the points with position vectors $3\mathbf{i} + 6\mathbf{k}$ and $-2\mathbf{i} + 2\mathbf{j} + 2\mathbf{k}$ is parallel to the plane with equation $\mathbf{r} \cdot (4\mathbf{i} + 20\mathbf{j} + 5\mathbf{k}) = 0$. Find the distance of the line from the plane.

**28** The position vectors of points A, B, C and D are respectively $\mathbf{i} - 3\mathbf{k}$, $-2\mathbf{i} + 2\mathbf{j} + \mathbf{k}$, $-\mathbf{j} + 4\mathbf{k}$ and $3\mathbf{i} - 3\mathbf{j}$. Find, in vector form, the equations of the lines AC and BD, and prove that these two lines meet in a point P.

Prove that the quadrilateral ABCD is a rectangle, and find the vector equation of the line through P perpendicular to the plane ABCD.     [L]

**29** Show that the straight line with vector equation

$$\mathbf{r} = \mathbf{i} + t\mathbf{j} + t\mathbf{k}$$

passes through the points P and Q which have position vectors $\mathbf{i} - \mathbf{j} - \mathbf{k}$ and $\mathbf{i} + \mathbf{j} + \mathbf{k}$ respectively. Find the vector equation of the straight line through the points R and S which have position vectors $\mathbf{i} + \mathbf{j}$ and $2\mathbf{i} + \mathbf{j} + \mathbf{k}$ respectively. Show that PQ and RS are inclined at $60°$ to one another.

Show that the vector $\mathbf{a}$, where $\mathbf{a} = \mathbf{i} + \mathbf{j} - \mathbf{k}$, is perpendicular to both PQ and RS. Find the scalar product of $\mathbf{a}$ with $\overrightarrow{PR}$, and hence find the shortest distance between PQ and RS.     [L]

**30** A plane passes through the three points A, B, C whose position vectors referred to an origin O are $\mathbf{i} + 3\mathbf{j} + 3\mathbf{k}$, $3\mathbf{i} + \mathbf{j} + 4\mathbf{k}$, $2\mathbf{i} + 4\mathbf{j} + \mathbf{k}$ respectively. Find in the form $l\mathbf{i} + m\mathbf{j} + n\mathbf{k}$ a unit vector normal to this plane.

Find also the cartesian equation of the plane and the perpendicular distance from the origin to this plane.     [L]

**31** Show that the lines $l_1$, $l_2$, with vector equations

$$\mathbf{r} = (5\mathbf{i} - 2\mathbf{j} + 3\mathbf{k}) + s(-3\mathbf{i} + \mathbf{j} - \mathbf{k}),$$
$$\mathbf{r} = (10\mathbf{i} - 3\mathbf{j} + 6\mathbf{k}) + t(4\mathbf{i} - \mathbf{j} + 2\mathbf{k})$$

respectively, intersect and find a vector equation of the plane $\Pi$ containing $l_1$ and $l_2$.

Show that the point Q with position vector $(6\mathbf{i} + 7\mathbf{j} - 2\mathbf{k})$ lies on the line which is perpendicular to $\Pi$ and which passes through the intersection of $l_1$ and $l_2$. Find a vector equation of the plane which passes through Q and is parallel to $\Pi$.     [L]

**32** A tetrahedron OABC with vertex O at the origin is such that $\overrightarrow{OA} = \mathbf{a}$, $\overrightarrow{OB} = \mathbf{b}$ and $\overrightarrow{OC} = \mathbf{c}$. Show that the line segments joining the mid-points of opposite edges bisect one another.

Given that two pairs of opposite edges are perpendicular, prove that $\mathbf{a} \cdot \mathbf{b} = \mathbf{b} \cdot \mathbf{c} = \mathbf{c} \cdot \mathbf{a}$ and show that the third pair of opposite edges is also perpendicular.

Prove also that, in this case, $OA^2 + BC^2 = OB^2 + AC^2$.     [L]

33 Show that the straight lines given by the equations
$$x = 2 + 2t, \quad y = 1 - 3t, \quad z = 5 - 6t$$
and $\qquad x = 3s, \quad y = 2 - 4s, \quad z = 3 - 7s$
intersect. Find the cartesian equation of the plane which contains both lines.

34 Find the shortest distance to the line
$$x = 3 - 2t, \quad y = 1 + t, \quad z = 4 + 2t$$
(a) from the origin (b) from the point (1, 2, 3).

35 Find a vector equation for the line of intersection of the planes with vector equations
$$\mathbf{r} \cdot (2\mathbf{i} + 2\mathbf{j} + \mathbf{k}) = 1, \quad \mathbf{r} \cdot (4\mathbf{i} + 7\mathbf{j} + 4\mathbf{k}) = 1.$$

# 3 Kinematics

## 3.1 Uniformly accelerated motion in a straight line

Consider a particle moving along a straight line BOA (Fig. 3.1). The position of the particle is given by its distance from O, distances to the right of O being taken to be positive and distances to the left of O being taken to be negative. The particle is at the point P, where $OP = s$, at time $t$ and at the point Q, where $OQ = s + \delta s$, at time $t + \delta t$. Thus, in moving from P to Q, the particle travels a distance $\delta s$ in time $\delta t$. The average velocity of the particle over this interval of time $\delta t$ is $\dfrac{\delta s}{\delta t}$. The velocity at time $t$, i.e. when the particle is at P, is defined as

$$\lim_{\delta t \to 0} \frac{\delta s}{\delta t}, \quad \text{i.e.} \quad \frac{ds}{dt}.$$

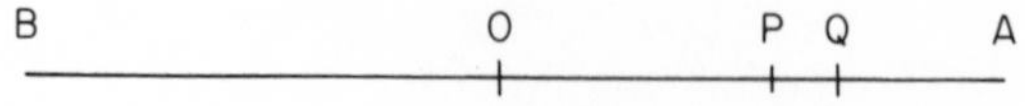

B          O        P Q    A

Fig. 3.1

If the velocity of the particle at time $t$ is $v$ and at time $t + \delta t$ is $v + \delta v$, the velocity of the particle has increased by $\delta v$ in time $\delta t$. The average acceleration of the particle over the interval $\delta t$ is $\dfrac{\delta v}{\delta t}$. The acceleration $a$ at time $t$, i.e. when the particle is at P, is defined as

$$\lim_{\delta t \to 0} \frac{\delta v}{\delta t}, \quad \text{i.e.} \quad \frac{dv}{dt}.$$

By the chain rule for differentiation

$$\frac{dv}{dt} = \frac{dv}{ds} \times \frac{ds}{dt} = v\frac{dv}{ds}.$$

Also

$$\frac{dv}{dt} = \frac{d}{dt}\left(\frac{ds}{dt}\right) = \frac{d^2s}{dt^2}.$$

Thus the velocity    $v = \dfrac{ds}{dt}$

and acceleration    $a = \dfrac{dv}{dt} = v\dfrac{dv}{ds} = \dfrac{d^2s}{dt^2}.$

Note that $s$, $v$ and $a$ are all directed quantities, $v$ and $a$ also being taken to be positive when in the direction OA and negative when in the direction OB.

The special case when the acceleration $a$ is constant will now be considered.

$$\frac{dv}{dt} = a.$$

Integrating $\qquad v = at + C_1$ where $C_1$ is a constant.

If the particle has velocity $u$ when $t = 0$, then $C_1 = u$,

giving $\qquad v = u + at$ $\qquad\qquad$ ...(1)

i.e. $\qquad \dfrac{ds}{dt} = u + at.$

Integrating again $\quad s = ut + \frac{1}{2}at^2 + C_2$ where $C_2$ is a constant.

If the particle is at O when $t = 0$, i.e. $s = 0$ when $t = 0$, then $C_2 = 0$,

giving $\qquad s = ut + \frac{1}{2}at^2.$ $\qquad\qquad$ ...(2)

Also $\qquad v\dfrac{dv}{ds} = a.$

Integrating $\displaystyle\int v\frac{dv}{ds}ds = \int a\,ds,$

i.e. $\displaystyle\int v\,dv = \int a\,ds,$

giving $\qquad \frac{1}{2}v^2 = as + C_3$ where $C_3$ is a constant.

The particle has velocity $u$ when $t = 0$, i.e. $v = u$ when $s = 0$, so $C_3 = \frac{1}{2}u^2$,

giving $\qquad \frac{1}{2}v^2 = as + \frac{1}{2}u^2,$

i.e. $\qquad v^2 = u^2 + 2as.$ $\qquad\qquad$ ...(3)

From (1) $\qquad v - u = at$

From (3) $\quad v^2 - u^2 = 2as.$

Dividing (3) by (1) gives

$$v + u = \frac{2s}{t},$$

i.e. $\qquad s = \left(\dfrac{u + v}{2}\right)t$ $\qquad\qquad$ ...(4)

Equations (1), (2), (3) and (4) are now used to solve problems concerned with particles and other objects moving with constant acceleration. These equations are used so frequently that they are now collected together for ease of reference.

$$v = u + at \qquad \qquad \qquad \ldots(1)$$

$$s = ut + \tfrac{1}{2}at^2 \qquad \qquad \ldots(2)$$

$$v^2 = u^2 + 2as \qquad \qquad \ldots(3)$$

$$s = \left(\frac{u + v}{2}\right)t \qquad \qquad \ldots(4)$$

At this stage only the motion is being considered and not how it is produced. This study of motion without consideration of how that motion is produced is known as *kinematics*.

*Example 1*

A particle moving in a straight line with constant acceleration starts with velocity $10 \text{ m s}^{-1}$ and after 2 seconds its velocity is $20 \text{ m s}^{-1}$. Find the acceleration and the distance moved in this time.

With the notation of equations (1) to (4)
$u = 10$, $v = 20$, $t = 2$ and $a$ and $s$ have to be found.
Using equation (1), $v = u + at$, $\quad 20 = 10 + 2a$, giving $a = 5$.

Using equation (4), $s = \left(\dfrac{u + v}{2}\right)t$, $\quad s = \left(\dfrac{10 + 20}{2}\right)2 = 30$,

i.e. the acceleration is $5 \text{ m s}^{-2}$ and the distance moved in 2 seconds is 30 m.

*Example 2*

A car travelling with constant acceleration along a straight road moves a distance of 160 m as the velocity increases from $10 \text{ m s}^{-1}$ to $30 \text{ m s}^{-1}$. Find the acceleration and the time taken.

With the usual notation
$u = 10$, $v = 30$, $s = 160$ and $a$ and $t$ have to be found.

Using equation (4), $s = \left(\dfrac{u + v}{2}\right)t$, $160 = \left(\dfrac{10 + 30}{2}\right)t$, giving $t = 8$.

Using equation (3), $v^2 = u^2 + 2as$, $30^2 = 10^2 + 2a(160)$, giving $a = 2.5$.
Alternatively, having found $t$, equation (1), $v = u + at$, could have been used to obtain the acceleration ($30 = 10 + 8a$ giving $a = 2.5$). However by this method an error in $t$ would result in an incorrect value also being obtained for the acceleration.

The acceleration of the car is $2.5 \text{ m s}^{-2}$ and the time taken is 8 seconds.

*Example 3*

A train travelling at $162 \text{ km h}^{-1}$ is brought to rest with constant retardation in 1.5 km. Find the retardation and the time taken in coming to rest.

Start by converting $162 \text{ km h}^{-1}$ into metres per second.

$1 \text{ m s}^{-1} = \dfrac{1}{1000} \text{ km s}^{-1} = \dfrac{1}{1000} \times 60 \times 60 \text{ km h}^{-1} = 3.6 \text{ km h}^{-1}.$

(It may be useful to remember that $10 \text{ m s}^{-1} = 36 \text{ km h}^{-1}$.)

Thus $162 \text{ km h}^{-1} = \dfrac{162}{3.6} \text{ m s}^{-1} = 45 \text{ m s}^{-1}$ and $1.5 \text{ km} = 1500 \text{ m}$.

With the usual notation

$u = 45, v = 0, s = 1500$ and $a$ and $t$ have to be found.

Using equation (4), $1500 = \left(\dfrac{45}{2}\right)t$, giving $t = 66\frac{2}{3}$.

Using equation (3), $0 = 45^2 + 2a(1500)$, giving $a = -\dfrac{27}{40} = -0.675$.

The $-$ sign signifies a negative acceleration, i.e. a retardation.

As in Example 2 an alternative method is to use $v = u + at$ after finding $t$.

   The retardation of the train is $0.675 \text{ m s}^{-2}$ and the time taken in bringing it to rest is $66\frac{2}{3}$ seconds.


*Example 4*

A particle moving in a straight line with constant acceleration moves 26 m in the 4th second and 30 m in the 5th second of its motion. Calculate (a) its acceleration (b) its initial velocity (c) the distance travelled in the 6th second.

Choosing any one of the possible intervals of time, e.g. between $t = 3$ and $t = 4$ or between $t = 4$ and $t = 5$, there is not enough information to solve this problem by means of the four equations used so far. However if two of these intervals are chosen, two simple simultaneous equations in two unknowns are obtained and then the problem can be solved.

(a) Let $u_1 =$ the velocity at time $t = 3$ and $a =$ acceleration.

Using equation (2), $s = ut + \frac{1}{2}at^2$, the two intervals $t = 3$ to $t = 4$ and $t = 3$ to $t = 5$, give

$$26 = u_1 + \tfrac{1}{2}a$$

and
$$56 = 2u_1 + \tfrac{1}{2}a(4).$$

Solving these equations gives $u_1 = 24$ and $a = 4$.

Thus the acceleration is $4 \text{ m s}^{-2}$.

(b) Let $u =$ velocity when $t = 0$.

Using the equation $v = u + at$ for the interval $t = 0$ to $t = 3$, $24 = u + 4(3)$, giving $u = 12$, i.e. the initial velocity of the particle is $12 \text{ m s}^{-1}$.

(c) Using $s = ut + \frac{1}{2}at^2$ gives

when $t = 6$, $s = 12(6) + \frac{1}{2}(4)(36) = 144$

when $t = 5$, $s = 12(5) + \frac{1}{2}(4)(25) = 110$.

$\therefore$ the distance moved in the 6th second $= (144 - 110) \text{ m} = 34 \text{ m}$.

An alternative semi-graphical solution to this problem is given in Example 3 of Section 3.3.

**Exercises 3.1**

1  A particle moves in a straight line with constant acceleration $a$ m s$^{-2}$. It starts from the point O with velocity $u$ m s$^{-1}$ and after $t$ seconds it has velocity $v$ m s$^{-1}$ and is at a distance $s$ m from 0.

(a)  Given that $u = \phantom{0}6$, $v = \phantom{0}12$, $t = \phantom{00}4$, find $a$ and $s$.
(b)  Given that $u = 10$, $v = \phantom{00}4$, $t = \phantom{00}2$, find $a$ and $s$.
(c)  Given that $u = \phantom{0}8$, $v = \phantom{0}20$, $s = \phantom{0}60$, find $a$ and $t$.
(d)  Given that $u = 16$, $v = \phantom{00}6$, $s = \phantom{0}40$, find $a$ and $t$.
(e)  Given that $u = \phantom{0}5$, $v = \phantom{0}25$, $a = \phantom{00}2$, find $s$ and $t$.
(f)  Given that $u = 24$, $v = \phantom{0}12$, $a = -3$, find $s$ and $t$.
(g)  Given that $u = 12$, $a = \phantom{0}4$, $t = \phantom{00}2$, find $s$ and $v$.
(h)  Given that $u = \phantom{0}8$, $a = -4$, $t = \phantom{00}4$, find $s$ and $v$.
(i)  Given that $v = 10$, $s = \phantom{0}30$, $t = \phantom{00}4$, find $u$ and $a$.
(j)  Given that $v = -12$, $s = -10$, $t = \phantom{00}6$, find $u$ and $a$.
(k)  Given that $u = -6$, $s = -36$, $t = \phantom{00}3$, find $v$ and $a$.
(l)  Given that $u = \phantom{0}5$, $s = \phantom{0}50$, $t = \phantom{00}4$, find $v$ and $a$.
(m) Given that $s = 80$, $t = \phantom{0}10$, $a = \phantom{00}4$, find $u$ and $v$.
(n)  Given that $s = -60$, $t = \phantom{00}5$, $a = -6$, find $u$ and $v$.
(o)  Given that $u = \phantom{0}9$, $a = \phantom{0}4$, $s = \phantom{0}48$, find $v$ and $t$.
(p)  Given that $u = -4$, $a = \phantom{0}8$, $s = \phantom{0}30$, find $v$ and $t$.
(q)  Given that $t = \phantom{0}4$, $v = \phantom{0}20$, $a = \phantom{00}2$, find $u$ and $s$.
(r)  Given that $t = \phantom{0}6$, $v = -12$, $a = -4$, find $u$ and $s$.
(s)  Given that $s = 60$, $v = \phantom{0}32$, $a = \phantom{00}6$, find $u$ and $t$.
(t)  Given that $s = 100$, $v = \phantom{0}10$, $a = -2$, find $u$ and $t$.

2  A car starts from rest and is uniformly accelerated to a velocity of 24 m s$^{-1}$ in 10 seconds. Find the distance travelled in reaching a velocity of (a) 24 m s$^{-1}$ (b) 12 m s$^{-1}$.

3  A train starts from rest and moves with uniform acceleration until a velocity of 144 km h$^{-1}$ is reached. If this takes 2 minutes, find the distance travelled in reaching (a) 144 km h$^{-1}$ (b) 72 km h$^{-1}$ (c) 36 km h$^{-1}$.

4  A car travelling at 30 m s$^{-1}$ is uniformly retarded to 10 m s$^{-1}$ in a distance of 60 m. Find the time taken. Find also the further distance it will travel in coming to rest if the retardation remains the same.

5  A train is travelling at 108 km h$^{-1}$ when the brakes are applied producing a uniform retardation which brings the train to rest after travelling a distance of 2 km. Find the retardation and the time taken in bringing the train to rest.

6  A particle moving in a straight line with uniform acceleration covers 19 m in the 3rd second and 25 m in the 4th second of its motion. Find the acceleration and the velocity at the start of the 3rd second. Find also the initial velocity of the particle.

## 3.2  Vertical motion under gravity

A special case of uniformly accelerated motion in a straight line is vertical motion under gravity when air resistance is neglected. The acceleration of the particle is then the acceleration due to gravity, approximately 9.8 m s$^{-2}$, which is the same for all bodies but which differs slightly at different places on the earth's surface. This acceleration is represented by the letter $g$.

Consider a particle projected vertically upwards with velocity $u$ from a point O. Distances, velocities and accelerations upwards are taken to be positive and those downwards to be negative.

Equation (1), $v = u + at$   becomes   $v = u - gt$,

equation (2), $s = ut + \frac{1}{2}at^2$   becomes   $s = ut - \frac{1}{2}gt^2$,

equation (3), $v^2 = u^2 + 2as$   becomes   $v^2 = u^2 - 2gs$,

and equation (4), $s = \left(\dfrac{u + v}{2}\right)t$ remains unchanged.

If A is the highest point reached by the particle, then at A the velocity of the particle is zero.

Using $v^2 = u^2 - 2gs$ gives $0 = u^2 - 2gs$, i.e. $s = u^2/(2g) = $ OA.

Thus the greatest height reached above the point of projection is $u^2/(2g)$.

Using $v = u - gt$ gives $0 = u - gt$, i.e. $t = u/g = $ time to reach A.

Thus the time to reach the highest point is $u/g$.

When the particle returns to O, then $s = 0$.

Using $s = ut - \frac{1}{2}gt^2$ gives $0 = ut - \frac{1}{2}gt^2$, giving $t = 0$ or $2u/g$.

Thus the particle is at O at $t = 0$, i.e. the instant of projection, and at $t = 2u/g$ so that the time of flight from O back to O is $2u/g$. This is twice the time taken to reach the greatest height so that the time spent moving upwards is the same as that spent moving downwards.

Using $v = u - gt$ with $t = 2u/g$ gives $v = u - 2u = -u$, so that the magnitude of the velocity of the particle when it returns to O is the same as the magnitude of the velocity of projection. The magnitude of the velocity of the particle is known as its speed. Thus on returning to O the particle has the same speed as it had when it left O.


*Example 1*

A particle is projected from a point O with velocity 40 m s$^{-1}$ vertically upwards. Find

(a) the greatest height reached and the time taken to reach this greatest height

(b) the position of the particle after (i) 2 seconds (ii) 10 seconds

(c) the times when the particle is 75 m (i) above O (ii) below O.

(Take $g$ to be 10 m s$^{-2}$.)

With the usual notation, $u = +40$, $a = -10$.

(a) At the highest point $v = 0$.

Using $v^2 = u^2 + 2as$, $0 = 1600 - 20s$, giving $s = 80$.

Using $v = u + at$, $0 = 40 - 10t$, giving $t = 4$.

The greatest height reached is 80 m above O and the time taken is 4 seconds.

(b) (i) $t = 2$.

Using $s = ut + \frac{1}{2}at^2$ gives $s = 40(2) + \frac{1}{2}(-10)(4) = 60$,

i.e. the particle is 60 m above O after 2 seconds.

(ii) $t = 10$.
Similarly $s = 40(10) + \frac{1}{2}(-10)(100) = -100$,
i.e. the particle is 100 m below O after 10 seconds.
(c) (i) $s = +75$.
Using $s = ut + \frac{1}{2}at^2$ gives $75 = 40t - 5t^2$, i.e. $t^2 - 8t + 15 = 0$,
giving $(t - 3)(t - 5) = 0$ and $t = 3$ or $5$.
In other words the particle is 75 m above O when $t = 3$ and when $t = 5$.
When $t = 3$, $v = 40 - 30 = 10$,
when $t = 5$, $v = 40 - 50 = -10$,
i.e. when $t = 3$ the particle is 75 m above O moving upwards at $10 \text{ m s}^{-1}$ and
when $t = 5$ the particle is 75 m above O moving downwards at $10 \text{ m s}^{-1}$.
(ii) $s = -75$.
Using $s = ut + \frac{1}{2}at^2$ again, $-75 = 40t - 5t^2$, i.e. $t^2 - 8t - 15 = 0$.
Using the quadratic equation formula gives $t = 9.57$ or $-1.57$ approx.
When $t \approx 9.57$, $v \approx 40 - 95.7 = -55.7$ and the particle is moving downwards.
When $t \approx -1.57$, $v \approx 40 + 15.7 = +55.7$.
Since the motion started at $t = 0$, $t = -1.57$ can be rejected as a solution to the problem. However had the motion started before $t = 0$, at $t = -1.57$ the particle would have been 75 m below O moving upwards with speed $55.7 \text{ m s}^{-1}$ and at $t = 0$ it would have been at O moving upwards with speed $40 \text{ m s}^{-1}$.


*Example 2*
A stone falls freely from the top of a tower of height 45 m. Find the time taken to reach the ground and the speed with which the stone hits the ground when the stone is
(a) released from rest at the top of the tower
(b) thrown vertically downwards from the top of the tower with speed $20 \text{ m s}^{-1}$.
(Take $g$ to be $10 \text{ m s}^{-2}$.)

(a) Taking the top of the tower as origin O and downwards as the positive direction gives, with the usual notation,
$u = 0$, $a = +10$, $s = +45$ and $t$ and $v$ have to be found.
Using $v^2 = u^2 + 2as$, gives $v^2 = 2(10)(45) = 900$ giving $v = 30$.
Using $s = ut + \frac{1}{2}at^2$, gives $45 = \frac{1}{2}(10)t^2$, i.e. $t^2 = 9$ giving $t = 3$.
(This time could also be obtained by using $v = u + at$ with $v = 30$.)
Although $v = -30$ and $t = -3$ satisfy these equations these values have to be rejected since they clearly do not fit the data. Thus the stone reaches the ground with speed $30 \text{ m s}^{-1}$ after 3 seconds.

(b) Here $u = +20$, $a = +10$, $s = +45$.
As in (a) $v^2 = 400 + 900 = 1300$, giving $v = 10\sqrt{13} \approx 36$.
Using $v = u + at$, $10\sqrt{13} = 20 + 10t$ giving $t \approx 1.6$, i.e. the particle hits the ground with speed $36 \text{ m s}^{-1}$ after 1.6 seconds approximately.

## Exercises 3.2

In the following exercises take $g$ to be $10 \text{ m s}^{-2}$.

1  A particle is projected vertically with velocity $60 \text{ m s}^{-1}$ upwards from a point on the ground. Find the maximum height reached and the time the particle takes to reach the ground again. Find also its position and velocity 4 seconds after projection.

2  A stone is thrown vertically upwards and reaches a maximum height of 20 m. Find
(a) its speed of projection    (b) when it is at a height of 10 m.

3  A coin falls from the top of a tower of height 60 m. Find
(a) the time it takes to reach the ground
(b) the speed with which it hits the ground.
Find also its speed when it has fallen 30 m.

4  A pebble is thrown vertically downwards with speed $12 \text{ m s}^{-1}$ from the top of a cliff 100 m high. Find how long the pebble takes to reach the beach at the foot of the cliff and the speed with which it hits the beach. Find also the speed of the pebble
(a) when it has fallen 50 m
(b) when it has been falling for half its time of flight.

5  A particle is projected vertically with speed $25 \text{ m s}^{-1}$ upwards from a point 50 m above the ground. Find
(a) the greatest height reached above the ground
(b) the speed with which the particle hits the ground.
Find also its speed when it is (i) 80 m (ii) 30 m, above the ground and the times when it is in these positions.

## 3.3   Motion in a straight line with variable acceleration

In Section 3.1, $v = \dfrac{ds}{dt}$.

Thus to find the velocity at any time $T$, $s$ is differentiated to obtain the value of $\dfrac{ds}{dt}$ when $t = T$. Alternatively the graph of $s$ against $t$ could be drawn and the gradient at the point on the graph where $t = T$ would give $\dfrac{ds}{dt}$ where $t = T$ and thus the velocity when $t = T$.

Also in Section 3.1, $a = \dfrac{dv}{dt}$.

To find the acceleration at any time $T$, $v$ is differentiated to obtain the value of $\dfrac{dv}{dt}$ when $t = T$. Alternatively the graph of $v$ against $t$ could be drawn and the gradient at the point on the graph where $t = T$ would give $\dfrac{dv}{dt}$ when $t = T$ and thus the acceleration when $t = T$.

Also since $\dfrac{ds}{dt} = v$, then $s = \displaystyle\int v\,dt.$

The distance travelled in the interval between $t = t_1$ and $t = t_2$ is given by

$$\left[ s \right]_{s_1}^{s_2} = \int_{t_1}^{t_2} v\,dt,$$

where $s_1$ and $s_2$ are the values of $s$ when $t = t_1$ and $t = t_2$ respectively. This distance can also be obtained graphically by evaluating the area between the $v$–$t$ graph and the $t$-axis between $t = t_1$ and $t = t_2$, i.e. the area 'under' the velocity–time graph.

Since $\dfrac{dv}{dt} = a$, then $v = \displaystyle\int a\,dt$

and

$$\left[ v \right]_{v_1}^{v_2} = \int_{t_1}^{t_2} a\,dt,$$

where $v_1$ and $v_2$ are the values of $v$ when $t = t_1$ and $t = t_2$ respectively. Thus the change in velocity, $v_2 - v_1$, is given by

$$\int_{t_1}^{t_2} a\,dt.$$

This change in velocity can also be obtained graphically by evaluating the area between the $a$–$t$ graph and the $t$-axis between $t = t_1$ and $t = t_2$, i.e. the area 'under' the acceleration–time graph.

The following table summarises the results of this section.

| Graph | Gradient | Area 'under' curve |
|-------|----------|--------------------|
| distance–time | $\dfrac{ds}{dt} = \text{velocity}$ | |
| velocity–time | $\dfrac{dv}{dt} = \text{acceleration}$ | $\displaystyle\int_{t_1}^{t_2} v\,dt = \text{distance moved}$ |
| acceleration–time | | $\displaystyle\int_{t_1}^{t_2} a\,dt = \text{change in velocity}$ |

The velocity–time graph is particularly valuable since, from it, can be obtained both the acceleration, from the gradient, and the distance moved, from the area under the curve.

These methods can be applied to the special case of uniformly accelerated motion considered in Section 3.1.

Figure 3.2 shows the acceleration–time graph. The change in velocity is given by the area under the curve and is $at$. Since the initial velocity is $u$, the velocity at time $t$ is $u + at$, i.e.

$$v = u + at. \qquad \ldots(1)$$

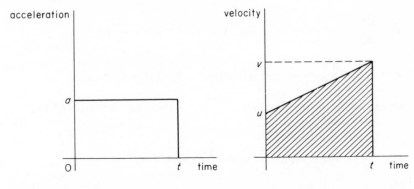

Fig. 3.2                    Fig. 3.3

From this equation it can be seen that the velocity–time graph in this case will be a straight line (Fig. 3.3). The distance travelled in time $t$ is given by the area of the shaded trapezium in Fig. 3.3,

i.e. the distance moved $\quad s = \left(\dfrac{u + v}{2}\right)t.$ $\qquad \ldots(4)$

Putting $v = u + at$ in equation (4) gives

$$s = \left(\frac{u + u + at}{2}\right)t = ut + \tfrac{1}{2}at^2. \qquad \ldots(2)$$

From (1) $\qquad\qquad\qquad\qquad v - u = at.$

From (4) $\qquad\qquad\qquad\qquad v + u = \dfrac{2s}{t}.$

Multiplying gives $\qquad (v - u)(v + u) = (at)\left(\dfrac{2s}{t}\right),$

i.e. $\qquad\qquad\qquad\qquad\qquad v^2 - u^2 = 2as$

or $\qquad\qquad\qquad\qquad\qquad v^2 = u^2 + 2as. \qquad \ldots(3)$

*Example 1*

A particle moves in a straight line and after being in motion for $t$ seconds it has velocity $v$ m s$^{-1}$ and is at a distance $s$ metres from a point O in its path.

(a) If $s = t^3 + 2t^2 + 3t + 4$, find the acceleration and the velocity of the particle when $t = 4$.

(b) If $v = 9t^2 - 8t - 6$, find the acceleration of the particle when $t = 2$ and the distance travelled in the interval between $t = 2$ and $t = 6$.

(c) If $a = 2t + 3$, and $v = 10$ when $t = 2$, find the distance travelled in the interval between $t = 2$ and $t = 4$.

(a) $s = t^3 + 2t^2 + 3t + 4$.

$$\therefore \quad v = \frac{ds}{dt} = 3t^2 + 4t + 3, \text{ giving } v = 67 \text{ when } t = 4.$$

$$a = \frac{dv}{dt} = \frac{d^2s}{dt^2} = 6t + 4, \text{ giving } a = 28 \text{ when } t = 4,$$

i.e. when $t = 4$, the acceleration is 28 m s$^{-2}$ and the velocity is 67 m s$^{-1}$.

(b) $v = 9t^2 - 8t - 6$.

$$\therefore \quad a = \frac{dv}{dt} = 18t - 8, \text{ giving } a = 28 \text{ when } t = 2.$$

$$s = \int_{t_1}^{t_2} v\,dt = \int_{2}^{6} (9t^2 - 8t - 6)dt = \left[ 3t^3 - 4t^2 - 6t \right]_{2}^{6} = 472,$$

i.e. the distance travelled in the interval between $t = 2$ and $t = 6$ is 472 m.

(c) $a = \frac{dv}{dt} = 2t + 3$.

Integrating $v = \int (2t + 3)dt = t^2 + 3t + C$   where $C$ is a constant.

When $t = 2$, $v = 10$, $\therefore$ $10 = 4 + 6 + C$ giving $C = 0$.

$$\therefore \qquad\qquad v = \frac{ds}{dt} = t^2 + 3t.$$

The distance travelled in the interval between $t = 2$ and $t = 4$ is

$$\int_{2}^{4} v\,dt = \int_{2}^{4} (t^2 + 3t)dt = \left[ \tfrac{1}{3}t^3 + \frac{3t^2}{2} \right]_{2}^{4} = 36\tfrac{2}{3} \text{ m}.$$

*Example 2*

The speeds of a car at one-second intervals are given in the following table:

| Time (seconds) | 0 | 1 | 2 | 3 | 4 | 5 | 6 |
|---|---|---|---|---|---|---|---|
| Speed (m s$^{-1}$) | | 3.0 | 3.6 | 4.7 | 6.1 | 8.1 | 10.8 | 15 |

Estimate (a) the acceleration of the car when $t = 3$
         (b) the distance travelled in the six seconds.

Figure 3.4 shows the $v$–$t$ graph.
(a) The gradient at $t = 3$ gives the required acceleration. This gradient is

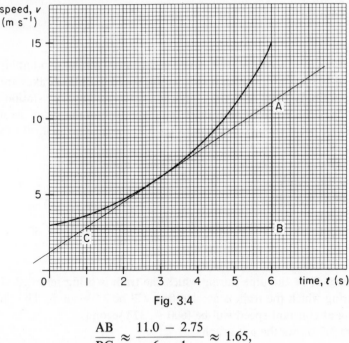

Fig. 3.4

$$\frac{AB}{BC} \approx \frac{11.0 - 2.75}{6 - 1} \approx 1.65,$$

i.e. the acceleration when $t = 3$ is approximately $1.65 \text{ m s}^{-2}$.

(b) The distance travelled in the six seconds is given by the area under the curve between $t = 0$ and $t = 6$. This area can be evaluated in a variety of ways such as counting squares, using the trapezium rule or using Simpson's rule.

Evaluating by Simpson's rule with 7 ordinates gives

$$\text{distance moved} = \int_0^6 v\,dt \approx \tfrac{1}{3}(1)\bigg[ (v_0 + v_6) + 4(v_1 + v_3 + v_5) + 2(v_2 + v_4) \bigg],$$

where $v_0, v_1, v_2, v_3, v_4, v_5, v_6$ are the speeds when $t = 0, 1, 2, 3, 4, 5, 6$. The evaluation by Simpson's rule is tabulated below

| $t$ | $v$ | | |
|---|---|---|---|
| 0 | 3.0 | | |
| 1 | | 3.6 | |
| 2 | | | 4.7 |
| 3 | | 6.1 | |
| 4 | | | 8.1 |
| 5 | | 10.8 | |
| 6 | 15.0 | | |
| | 18.0 | 20.5 | 12.8 |
| | 82.0 | ×4 | |
| | 25.6 | | ×2 |
| | 125.6 | | |

Distance moved $\approx \frac{1}{3}(1)(125.6) \approx 42$ m to the nearest metre.

*Example 3*

A train starts from rest at station **A** and is uniformly accelerated until it reaches a speed of 108 km h$^{-1}$. It then travels at this speed until the brakes are applied and the train is then uniformly retarded until it stops at station **B**. The magnitude of this retardation is twice the magnitude of the initial acceleration. The distance between the stations is 12 km and the time taken for the journey is 10 minutes. Find

(a) the time spent on each of the three stages of the journey

(b) the initial acceleration

(c) the final retardation.

Converting to SI units, $108 \text{ km h}^{-1} = \dfrac{108 \times 1000}{60 \times 60} = \dfrac{108}{3.6} = 30 \text{ m s}^{-1}$,

12 km = 12 000 m and 10 minutes = 600 seconds.

Let $T$ seconds be the time during which the train is being retarded. Then the time during which the train is accelerating will be $2T$ seconds. The time spent travelling at constant speed will be $(600 - 3T)$ seconds.

Figure 3.5 shows the $v$–$t$ graph.

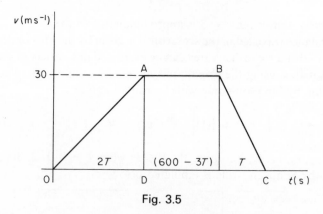

Fig. 3.5

The area of OABC, the trapezium formed by the $v$–$t$ graph and the $t$-axis

$= \frac{1}{2}(\text{AB} + \text{OC})\text{AD} = \frac{1}{2}(600 - 3T + 600)(30)$

$= 15(1200 - 3T)$.

This area gives the total distance travelled which is 12 000 m.

$\therefore \qquad\qquad\qquad 15(1200 - 3T) = 12\,000$

giving $\qquad\qquad\qquad\qquad T = 400/3 = 133\frac{1}{3}$.

Thus the first stage (during acceleration) occupies $266\frac{2}{3}$ seconds and the third stage (during retardation) occupies $133\frac{1}{3}$ seconds, so that the second stage (of constant speed) occupies 200 seconds.

The initial acceleration $= \dfrac{30 \text{ m s}^{-1}}{266\frac{2}{3} \text{ s}} = 0.1125 \text{ m s}^{-2}$ and

the final retardation $= \dfrac{30 \text{ m s}^{-1}}{133\frac{1}{3} \text{ s}} = 0.225 \text{ m s}^{-2}.$

*Example 4*
A particle moving in a straight line with constant acceleration moves 26 m in the 4th second and 30 m in the 5th second of its motion. Find (a) its acceleration (b) its initial speed (c) the distance moved in the 6th second.

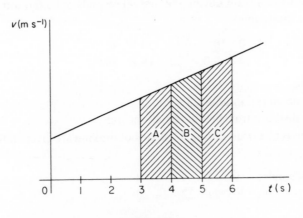

Fig. 3.6

(a) Figure 3.6 shows a sketch of the $v$–$t$ graph.
The shaded area A represents the distance travelled in the 4th second, i.e. 26 m.

$\therefore$ area of trapezium A = (value of $v$ when $t = 3\frac{1}{2}$) × 1 = 26
$\therefore$ the value of $v$ when $t = 3\frac{1}{2}$ is 26 m s$^{-1}$.
Similarly the shaded area B represents the distance moved in the 5th second, i.e. 30 m giving the value of $v$ when $t = 4\frac{1}{2}$ as 30 m s$^{-1}$.
Thus the speed has increased by 4 m s$^{-1}$ in 1 second between $t = 3\frac{1}{2}$ and $t = 4\frac{1}{2}$, i.e. the acceleration is 4 m s$^{-2}$.
(b) Using $v = u + at$ with $v = 26$, $t = 3\frac{1}{2}$, $a = 4$ gives $26 = u + 4(3\frac{1}{2})$ giving $u = 12$,
i.e. the initial speed is 12 m s$^{-1}$.
(c) The shaded area C represents the distance travelled in the 6th second. Using $v = u + at$ with $t = 5\frac{1}{2}$ gives $v = 30 + 4(1)$, i.e. $v = 34$ when $t = 5\frac{1}{2}$.
$\therefore$ the distance moved in the 6th second $= 34(1) = 34$ m.

**Exercises 3.3**
In questions 1–3 a particle moves in a straight line and after being in motion for

$t$ seconds it has acceleration $a$ m s$^{-2}$, velocity $v$ m s$^{-1}$ and is at a distance $s$ metres from a fixed point in its path.

1 Find the velocity and acceleration when $t = 6$ given that
(a) $s = 2t^3 - 3t^2 + t - 4$          (b) $s = t^2 + 6t - 5$.

2 Find the acceleration when $t = 3$ and the distance travelled in the interval between $t = 1$ and $t = 4$ given that
(a) $v = 3t^2 + 6t + 2$          (b) $v = t^3 + 4t$.

3 Find the velocity when $t = 2$ and the distance travelled in the interval between $t = 2$ and $t = 6$ given that $v = 8$ when $t = 1$ and
(a) $a = 6t + 4$          (b) $a = 4t^2 + 3t + 2$.

4 The speeds of a train during the first 40 seconds of its motion are given in the following table:

| Time (seconds) | 0 | 10 | 20 | 30 | 40 |
|---|---|---|---|---|---|
| Speed (m s$^{-1}$) | 0 | 3.0 | 8.5 | 18.2 | 38.0 |

Estimate
(a) the acceleration after 25 seconds
(b) the distance travelled in the 40 seconds.

5 The speeds of a train during 8 seconds of braking are given in the following table:

| Time (seconds) | 0 | 2 | 4 | 6 | 8 |
|---|---|---|---|---|---|
| Speed (m s$^{-1}$) | 30 | 27 | 22 | 14 | 0 |

Estimate
(a) the retardation after 5 seconds
(b) the distance travelled by the train in coming to rest.

6 A car has acceleration $a$ at time $t$. Corresponding values of $a$ and $t$ are given in the following table:

| $t$ (seconds) | 0 | 1 | 2 | 3 | 4 | 5 | 6 |
|---|---|---|---|---|---|---|---|
| $a$ (m s$^{-2}$) | 0 | 0.28 | 0.88 | 1.70 | 2.40 | 2.85 | 3.10 |

Given that the car starts from rest, estimate the speed of the car when $t = 1$, 2, 3, 4, 5, 6 seconds. Hence estimate the distances travelled in the intervals $t = 0$ to $t = 3$ seconds and $t = 3$ to $t = 6$ seconds.

7 A train starts from rest at one station and is uniformly accelerated until it reaches a speed of 25 m s$^{-1}$. It then travels with this speed until the brakes are applied and it is brought to rest with uniform retardation at the next station. The magnitude of the retardation is three times the magnitude of the acceleration. The distance between the stations is 8 km and the time taken for the journey is 6 minutes. Sketch the $v$–$t$ graph and find
(a) the distance travelled at constant speed
(b) the time for which the speed is constant
(c) the initial acceleration.

8 A train starts from rest at a station and moves with uniform acceleration

$0.5 \, \text{m s}^{-2}$ until it reaches a certain speed. It maintains this speed until the brakes are applied and the train is brought to rest at the next station with uniform retardation $0.8 \, \text{m s}^{-2}$. Sketch the $v$–$t$ graph. If the distance between the stations is 4 km and the journey takes 3 minutes, find the constant speed and the time for which the train travels at this speed.

9 A lift starts from one floor and is uniformly accelerated until a speed of $v \, \text{m s}^{-1}$ is reached. The lift is then immediately subjected to a uniform retardation which brings the lift to rest at a higher floor. Sketch the velocity–time graph and use it to find the value of $v$ given that the distance between the floors is 20 m and the time taken for the journey is 4 seconds.

10 A train is uniformly accelerated from rest for a distance of 2 km. It then travels at constant speed for 3 km and is then uniformly retarded to rest for a distance of 1 km. The journey takes 5 minutes. Find the constant speed of the second stage of the journey and the time for which the train travels at this speed.

11 A train is travelling at $108 \, \text{km h}^{-1}$ when the brakes are applied, producing a uniform retardation $3f \, \text{m s}^{-2}$. When the speed has been reduced to $54 \, \text{km h}^{-1}$, the train travels at this speed for a certain distance. It is then uniformly accelerated at $f \, \text{m s}^{-2}$ until a speed of $108 \, \text{km h}^{-1}$ is again reached. From the instant the brakes are applied to the instant when the speed again reaches $108 \, \text{km h}^{-1}$ the time taken is 6 minutes and in this time the train travels 8 km. Find the value of $f$ and the distance travelled at $54 \, \text{km h}^{-1}$.

12 A particle moving in a straight line with uniform acceleration travels 12.5 m in the 2nd second and 15.5 m in the 3rd second of its motion. Sketch the $v$–$t$ graph and hence find
(a) the acceleration of the particle
(b) its initial speed
(c) the distance it travels in the 5th second.

13 A car is travelling on a straight road when the brakes are applied producing a uniform retardation. In the 2nd and 5th seconds after the application of the brakes the car travels distances of 37 m and 31 m respectively. Sketch the $v$–$t$ graph and hence find
(a) the retardation of the car
(b) its initial speed
(c) the time it takes the car to come to rest.

## 3.4 Further motion in a straight line with variable acceleration

Some cases of non-uniformly accelerated motion have been considered. With the usual notation, $s$, $v$ and $a$ have been expressed in terms of $t$ or, alternatively, corresponding values of $s$ and $t$, $v$ and $t$ and $a$ and $t$ have been given. Now cases will be considered where the relationship between (a) $v$ and $s$ (b) $a$ and $s$ (c) $a$ and $v$ is known.

(a) Velocity–distance.

If the velocity is given in terms of the distance, the acceleration at a given distance can be obtained by using

$$\text{acceleration, } a = v\frac{dv}{ds}.$$

The gradient of the velocity–distance graph at a particular point will give the value of $\frac{dv}{ds}$ for that value of $s$. If this value of $\frac{dv}{ds}$ is multiplied by the velocity for that value of $s$, the acceleration at that particular distance is obtained.

To find the time taken to cover a particular distance, consider the graph obtained by plotting $1/v$ against $s$ (Fig. 3.7).

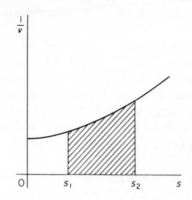

Fig. 3.7

The shaded area represents $\displaystyle\int_{s_1}^{s_2} \frac{1}{v}\,ds.$

Since $\qquad v = \dfrac{ds}{dt}, \quad \text{and} \quad \dfrac{1}{v} = \dfrac{dt}{ds}, \quad \displaystyle\int \frac{1}{v}\,ds = \int \frac{dt}{ds}\,ds = \int dt$

so that $\qquad \displaystyle\int_{s_1}^{s_2} \frac{1}{v}\,ds = \int_{t_1}^{t_2} dt = \Big[\,t\,\Big]_{t_1}^{t_2} = t_2 - t_1,$

i.e. the time taken to cover the distance $s_2 - s_1$ can be obtained by evaluating the area enclosed by the graph of $1/v$ against $s$, the $s$-axis and the ordinates $s = s_1$ and $s = s_2$.

(b) Acceleration–distance.

If the acceleration–distance graph is drawn (Fig. 3.8), the shaded area represents

$$\int_{s_1}^{s_2} a\,ds.$$

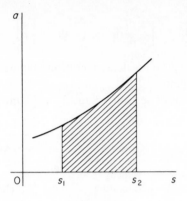

Fig. 3.8

Since $a = v\dfrac{dv}{ds}$

shaded area $= \displaystyle\int_{s_1}^{s_2} a\,ds = \int_{s_1}^{s_2} v\frac{dv}{ds}ds = \int_{v_1}^{v_2} v\,dv = \left[\tfrac{1}{2}v^2\right]_{v_1}^{v_2} = \tfrac{1}{2}(v_2^2 - v_1^2),$

i.e. the area 'under' the acceleration–distance graph gives the change in the value of $\tfrac{1}{2}v^2$ between $s = s_1$ and $s = s_2$.

(c) Acceleration–velocity.

If the graph of the reciprocal of the acceleration, $1/a$, against the velocity is drawn (Fig. 3.9) the shaded area represents

$$\int_{v_1}^{v_2} \frac{1}{a}dv.$$

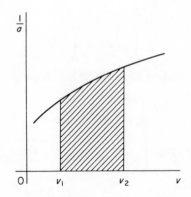

Fig. 3.9

Since $a = \dfrac{dv}{dt}$ gives $\dfrac{1}{a} = \dfrac{dt}{dv}$

shaded area $= \int_{v_1}^{v_2} \frac{1}{a} dv = \int_{v_1}^{v_2} \frac{dt}{dv} dv = \int_{t_1}^{t_2} dt = \left[ t \right]_{t_1}^{t_2} = t_2 - t_1,$

i.e. the area 'under' the graph of 1/(acceleration) against velocity gives the time taken as the velocity changes from $v_1$ to $v_2$.

The graphical methods of Section 3.3 and 3.4 are summarised in the following table.

| Graph | Gradient | Area under curve |
|-------|----------|------------------|
| $s-t$ | velocity | |
| $v-t$ | acceleration | change in distance |
| $a-t$ | | change in velocity |
| $v-s$ | $\dfrac{\text{acceleration}}{\text{velocity}}$ | |
| $(1/v)-s$ | | change in time |
| $a-s$ | | change in $\frac{1}{2}$(velocity)$^2$ |
| $(1/a)-v$ | | change in time |

Example 1

With the usual notation, $v = \sqrt{(2s + 3)}$ and $s = 0$ when $t = 0$. Find the time taken for $s$ to increase from 0 to 11 and find the acceleration when $s = 11$.

$v = (2s + 3)^{1/2} \quad \therefore \quad \dfrac{dv}{ds} = \frac{1}{2}(2s + 3)^{-1/2}(2) = \dfrac{1}{\sqrt{(2s + 3)}}.$

When $s = 11$, $v = 5$ and $\dfrac{dv}{ds} = \dfrac{1}{5}.$

$\therefore$ acceleration $= v\dfrac{dv}{ds} = 5\left(\dfrac{1}{5}\right) = 1.$

Also $\int_{s_1}^{s_2} \dfrac{1}{v} ds = \int_{t_1}^{t_2} dt$

so that, if $T$ is the time taken for $s$ to increase from 0 to 11,

$$\int_{0}^{T} dt = \int_{0}^{11} \dfrac{1}{\sqrt{(2s + 3)}} ds$$

$\therefore \qquad T = \left[ \sqrt{(2s + 3)} \right]_{0}^{11} = 5 - \sqrt{3}.$

Example 2

A particle moving with variable acceleration has speed $v$ m s$^{-1}$ after travelling a distance $s$ m. Corresponding values of $v$ and $s$ are given in the following table:

| $s$ | 0 | 5 | 10 | 15 | 20 | 25 | 30 |
|-----|-----|-----|-----|-----|-----|-----|-----|
| $v$ | 1.0 | 2.2 | 2.9 | 3.4 | 3.5 | 3.3 | 3.0 |

Estimate the time taken to travel 30 m.

Plotting $1/v$ against $s$, the area between the graph, the $s$-axis and the ordinates at $s = s_1$ and $s = s_2$ will give the required time. Corresponding values of $s$ and $1/v$ are as follows:

| $s$ | 0 | 5 | 10 | 15 | 20 | 25 | 30 |
|-----|-----|------|------|------|------|------|------|
| $1/v$ | 1.0 | 0.45 | 0.34 | 0.29 | 0.29 | 0.30 | 0.33 |

The required area could be obtained approximately by plotting values of $1/v$ against $s$ and evaluating the area under the curve by counting squares. Here the required area is evaluated by the trapezium rule to give the required time as

$$\tfrac{1}{2}(5)[(1.0 + 0.33) + 2(0.45 + 0.34 + 0.29 + 0.29 + 0.30)] \approx 11.7$$

i.e. the time taken to travel 30 m is approximately 11.7 seconds.

*Example 3*
A particle starts from rest at a point O and moves in a straight line, having velocity $v$ m s$^{-1}$ and being $s$ m from O when it has been in motion for $t$ seconds. The acceleration, $a$ m s$^{-2}$, is given by $a = 3s^2 + 2s$. Find the velocity when $s = 10$.

$$a = v\frac{dv}{ds} = 3s^2 + 2s.$$

$$\therefore \quad \int v\frac{dv}{ds}ds = \int (3s^2 + 2s)ds$$

i.e. $$\int v\, dv = \int (3s^2 + 2s)ds$$

$$\therefore \quad \left[\tfrac{1}{2}v^2\right]_0^{v_1} = \left[s^3 + s^2\right]_0^{10}, \quad \text{where } v = v_1 \quad \text{when } s = 10,$$

$$\therefore \quad \tfrac{1}{2}v_1^2 = 1000 + 100 = 1100, \text{ giving } v_1^2 = 2200 \text{ and } v_1 = 46.9,$$

i.e. when $s = 10$, the velocity of the particle is approximately 46.9 m s$^{-1}$.

*Example 4*
A particle moving in a straight line has speed $v$ m s$^{-1}$ and acceleration $a$ m s$^{-2}$ at time $t$ seconds when it is at a distance $s$ m from O, a point in its path. Corresponding values of $a$ and $v$ are given in the following table:

| $v$ | 10 | 20 | 30 | 40 | 50 |
|-----|-----|-----|-----|-----|-----|
| $a$ | 1.4 | 1.8 | 2.2 | 2.6 | 3.0 |

Estimate the time taken for the speed to increase from 10 m s$^{-1}$ to 50 m s$^{-1}$.

If the graph of $1/a$ against $v$ is plotted, the area between the curve and the $v$-axis between $v = 10$ and $v = 50$ will give the required time. Corresponding values of

$v$ and $1/a$ are as follows:

| $v$ | 10 | 20 | 30 | 40 | 50 |
|-----|-----|-----|-----|-----|-----|
| $1/a$ | 0.71 | 0.56 | 0.45 | 0.38 | 0.33 |

Using Simpson's rule with 5 ordinates, the required time is given by

$$\int_{10}^{50} \frac{1}{a}\,dv = \tfrac{1}{3}(10)[(0.71 + 0.33) + 4(0.56 + 0.38) + 2(0.45)]$$

$$= 19$$

i.e. the time taken for the speed to increase from 10 m s$^{-1}$ to 50 m s$^{-1}$ is approximately 19 seconds.

## Exercises 3.4

In each of the following questions a particle moves in a straight line through a point O and, $t$ seconds after passing through O, it has velocity $v$ m s$^{-1}$, acceleration $a$ m s$^{-2}$ and is at a distance $s$ m from O.

1  Corresponding values of $v$ and $s$ are given in the following table:

| $s$ | 0 | 10 | 20 | 30 | 40 | 50 |
|-----|-----|-----|-----|-----|-----|-----|
| $v$ | 5 | 14 | 21 | 25 | 28 | 30 |

Estimate the time taken for the velocity of the particle to increase from 5 m s$^{-1}$ to 30 m s$^{-1}$. Find also the acceleration when $s = 20$.

2  Corresponding values of $s$ and $a$ are as follows:

| $s$ | 10 | 20 | 30 | 40 | 50 |
|-----|-----|-----|-----|-----|-----|
| $a$ | 0 | 0.3 | 1.0 | 2.1 | 4.0 |

Estimate the velocity of the particle when $s = 30$, given that $v = 5$ when $s = 10$.

3  Corresponding values of $a$ and $v$ are as follows:

| $v$ | 0 | 5 | 10 | 15 | 20 | 25 | 30 |
|-----|-----|-----|-----|-----|-----|-----|-----|
| $a$ | 2.0 | 1.1 | 0.7 | 0.8 | 1.6 | 2.8 | 4.0 |

Estimate the time taken as the velocity increases from
(a) 0 to 15 m s$^{-1}$   (b) 10 to 30 m s$^{-1}$

4  Given that $v = \dfrac{1}{1 + s}$, find the time taken in travelling a distance of 20 m from O.

5  Given that $v = \sec(\pi s/12)$, find the time taken in travelling a distance of 6 m from O.

6  Given that $a = s^2 + 3$ and $v = 0$ when $s = 0$, find the velocity of the particle after it has moved a distance of 10 m.

7  Given that $a = 3s^2 + 2s + 1$ and $v = 4$ when $s = 0$, find the velocity of the particle when it is 8 m from O.

8   Given that $a = v + 2$ and $v = 6$ when $t = 0$ and $s = 0$, find (a) the time taken (b) the distance travelled, as the velocity increases from $6 \text{ m s}^{-1}$ to $12 \text{ m s}^{-1}$.

9   Given that $a = v^2 + 1$ and $v = 0$ when $t = 0$, find (a) the time taken for the particle to attain a speed of $1 \text{ m s}^{-1}$ (b) the distance travelled during this time.

## 3.5   Motion in a plane

So far only motion in a straight line has been considered. Each displacement from an origin O in that line was represented by a directed number, with a positive number representing a displacement to one side of O and a negative number representing a displacement to the opposite side of O. Similarly, velocities and accelerations were represented by directed numbers. Thus, although displacement, velocity and acceleration are vector quantities which require magnitude, direction and sense to be defined, it has been possible to represent these vector quantities by directed numbers. This is simply because the directed number gave the magnitude and sense, the direction being given by the straight line in which the motion took place. Next, motion in a plane will be considered, i.e. two-dimensional motion.

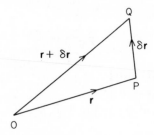

Fig. 3.10

Suppose a particle is at point P at time $t$ and at point Q at time $(t + \delta t)$. Let P and Q have position vectors $\mathbf{r}$ and $(\mathbf{r} + \delta \mathbf{r})$ respectively (Fig. 3.10). In the interval of time $\delta t$ the particle has undergone a displacement

$$\overrightarrow{PQ} = (\mathbf{r} + \delta \mathbf{r}) - \mathbf{r} = \delta \mathbf{r}.$$

The average velocity over the interval of time $\delta t$ is $\dfrac{\delta \mathbf{r}}{\delta t}$ and the velocity at P is defined as

$$\lim_{\delta t \to 0} \left( \frac{\delta \mathbf{r}}{\delta t} \right) = \frac{d\mathbf{r}}{dt}.$$

Also, if the particle has velocity $\mathbf{v}$ at P and velocity $(\mathbf{v} + \delta \mathbf{v})$ at Q, the average acceleration in the interval of time $\delta t$ is $\dfrac{\delta \mathbf{v}}{\delta t}$. The acceleration at P is defined as

$$\lim_{\delta t \to 0} \left( \frac{\delta \mathbf{v}}{\delta t} \right) = \frac{d\mathbf{v}}{dt}.$$

When $\mathbf{v}$ represents velocity, $v$ represents $|\mathbf{v}|$, the magnitude of the velocity, i.e. the speed. Similarly when $\mathbf{a}$ represents acceleration, $a$ represents $|\mathbf{a}|$, the magnitude of the acceleration.

Velocities and accelerations, being vectors, naturally obey the laws of vector addition and subtraction. For instance, a particle having two components of velocity $\mathbf{v}_1$ and $\mathbf{v}_2$, not necessarily at right angles, will have a resultant velocity $\mathbf{V} = \mathbf{v}_1 + \mathbf{v}_2$ given by the law for vector addition. Thus, if, in Fig. 3.11, $\overrightarrow{OA}$ and $\overrightarrow{AB}$ represent $\mathbf{v}_1$ and $\mathbf{v}_2$ respectively, then $\overrightarrow{OB}$ represents $\mathbf{V}$, where $\mathbf{V} = \mathbf{v}_1 + \mathbf{v}_2$.

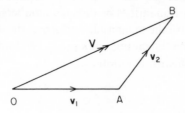

Fig. 3.11

Also, if
$$\mathbf{v}_1 = a_1\mathbf{i} + a_2\mathbf{j}$$
and
$$\mathbf{v}_2 = b_1\mathbf{i} + b_2\mathbf{j},$$
then
$$\mathbf{V} = \mathbf{v}_1 + \mathbf{v}_2 = (a_1 + b_1)\mathbf{i} + (a_2 + b_2)\mathbf{j}.$$

Figure 3.12 shows a rectangular tray ABCD which lies on a table. A fly starts at A and crawls on the tray with constant velocity $\mathbf{v}_F$ along AD. At the same time the tray is moved across the table with constant velocity $\mathbf{v}_T$ in the direction AB. Relative to the table the fly now has two components of velocity, $\mathbf{v}_F$ and $\mathbf{v}_T$, and its resultant velocity $\mathbf{V}$ is given by $\mathbf{v}_F + \mathbf{v}_T$ (Fig. 3.13).

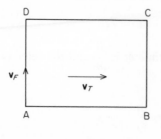

Fig. 3.12

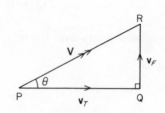

Fig. 3.13

If **i** and **j** are unit vectors in the directions AB and AD respectively,

then $\qquad\qquad\qquad \mathbf{v}_F = a\mathbf{j}$ where $a = |\mathbf{v}_F|$

and $\qquad\qquad\qquad \mathbf{v}_T = b\mathbf{i}$ where $b = |\mathbf{v}_T|$.

Then $\qquad\qquad\qquad \mathbf{V} = \mathbf{v}_F + \mathbf{v}_T = b\mathbf{i} + a\mathbf{j}$

and $\qquad\qquad\qquad |\mathbf{V}| = \sqrt{(b^2 + a^2)}$ and $\tan\theta = \dfrac{a}{b}$.

A practical example is a boat sailing on a river, where the velocity of the boat through the water corresponds to the velocity of the fly, $\mathbf{v}_F$, and the velocity of the water current corresponds to the velocity of the tray, $\mathbf{v}_T$. Another example is an aircraft flying in a wind. The aircraft has a velocity through the air and at the same time the air is moving, its velocity being the velocity of the wind. Thus the velocity of the aircraft is the resultant of its velocity through the air and the wind velocity.

*Example 1*
A river which is 40 m wide flows from west to east between parallel banks with speed $4 \text{ m s}^{-1}$. A boat sails across the river with a speed relative to the water of $6 \text{ m s}^{-1}$ due north. Find the velocity of the boat relative to the banks and the distance downstream of the point where the boat reaches the opposite bank.

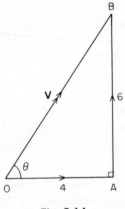

Fig. 3.14

Figure 3.14 shows the velocity diagram for the boat. The boat has two components of velocity, $4 \text{ m s}^{-1}$ in the direction OA and $6 \text{ m s}^{-1}$ in the direction AB. The resultant velocity, i.e. relative to the banks, of the boat is **V**. From the triangle OAB

$$V = |\mathbf{V}| = \sqrt{(4^2 + 6^2)} = \sqrt{52}$$

and $\qquad\qquad\qquad \tan\theta = \dfrac{6}{4}$ giving $\theta = 56°\,19'$, i.e. the velocity of the boat relative to the banks is $\sqrt{52}$ m s$^{-1}$ at an angle 56° 19′ to the banks.

Figure 3.15 shows the path of the boat from point P on one bank to point Q on the other bank.

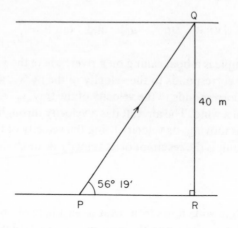

Fig. 3.15

Triangle PQR gives

$$PR = 40 \cot 56° 19′ = 40\left(\frac{4}{6}\right) = 26\tfrac{2}{3},$$

i.e. Q is $26\tfrac{2}{3}$ m downstream of P.

*Example 2*
A light aircraft has an airspeed of 100 knots and there is a north-westerly wind (i.e. blowing *from* the north west) of 40 knots. The aircraft flies horizontally through the air in a north-easterly direction. Find the velocity of the aircraft relative to the ground.

Let **w** = wind velocity and **u** = velocity of the aircraft through the air. The resultant velocity, i.e. relative to the ground, is **V**, where **V** = **u** + **w**. Figure 3.16 shows the velocity diagram for the aircraft.

If the triangle OAB is drawn to scale, OB will give the magnitude of **V** and (45° + $\phi$) will give the bearing of the path of the aircraft relative to the ground. From Fig. 3.16

$$V^2 = u^2 + w^2 = 100^2 + 40^2 = 20^2(5^2 + 2^2)$$

giving $\qquad V = 20\sqrt{29} \quad$ or $\quad 107.7$.

Also $\quad \tan \phi = \dfrac{40}{100} = 0.4$ giving $\phi = 21°48′$.

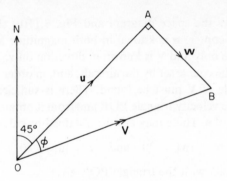

Fig. 3.16

Thus over the ground the aircraft has a velocity of 107.7 knots on a bearing of $(45° + 21°48')$, i.e. $66°48'$.

*Example 3*
A helicopter flies from A to B, where B is 100 km north east of A. The helicopter has an airspeed of $80 \text{ km h}^{-1}$ and there is a westerly wind blowing at $30 \text{ km h}^{-1}$. Assuming that the flight path of the helicopter is horizontal, find the course set by the helicopter and the time it takes to fly from A to B.

Let **u** be the velocity of the helicopter relative to the air, **w** the wind velocity and **V** the helicopter's resultant velocity, i.e. its velocity relative to the ground. **V** is in the direction AB, i.e. north east.

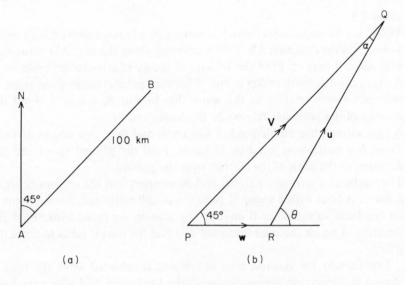

Fig. 3.17

Figure 3.17(a) shows the space diagram and Fig. 3.17(b) shows the velocity diagram for the helicopter. **w** is known in both magnitude and direction, **u** is known in magnitude only and **V** is known in direction only. The direction of **u** must be found, i.e. the course set by the aircraft, and, in order to find the time of flight, the magnitude of **V** must be found. There is sufficient information to construct to scale the velocity triangle PQR and from it obtain the magnitude of **V** and the direction of **u**. These may also be obtained by calculation as follows:

$$w = |\mathbf{w}| = 30 \quad \text{and} \quad u = |\mathbf{u}| = 80.$$

Using the sine formula with the triangle PQR gives

$$\frac{30}{\sin \alpha} = \frac{80}{\sin 45°} = \frac{V}{\sin \theta}.$$

$$\therefore \qquad \sin \alpha = \frac{30 \sin 45°}{80} = 0.2652 \quad \text{giving} \quad \alpha = 15° 23'.$$

$$\therefore \qquad \theta = 45° + \alpha = 60° 23'.$$

Thus the course set by the helicopter is on a bearing of $90° - 60° 23'$, i.e. $29° 37'$.

Also
$$V = \frac{80 \sin 60° 23'}{\sin 45°} = 98.4,$$

i.e. the speed of the helicopter over the ground is 98.4 km h$^{-1}$.

Thus the time of flight from A to B $= \dfrac{100}{98.4}$ h $\approx$ 61 minutes.

**Exercises 3.5**

1  ABCD is a rectangular tray which is moving on a horizontal table with speed 4 cm s$^{-1}$ in the direction AB. A fly is crawling along the edge AD of the tray with speed 3 cm s$^{-1}$. Find the velocity of the fly relative to the table.

2  A river runs due north at 200 m min$^{-1}$ between parallel banks 80 m apart. A swimmer swims relative to the water due east at 50 m min$^{-1}$. Find the velocity of the swimmer relative to the banks.

3  A light aircraft has an airspeed of 120 knots and flies on a course of 140°. There is a north-west wind of 35 knots. Find the ground speed and the direction of the path of the aircraft over the ground.

4  The banks of a river are parallel and 50 m apart and the current flows at 8 m s$^{-1}$. A boat with a speed of 10 m s$^{-1}$ in still water sails from a point A on one bank to the point B directly opposite on the other bank. Find the direction in which the boat is steered and find the time it takes to cross the river.

   Explain why the shortest time of crossing is achieved when the boat is steered in a direction perpendicular to the banks and find where the boat reaches the opposite bank in this case.

5  A ship sails from A to B and back. B is 60 nautical miles south west of A. Throughout the voyage there is a constant current of 12 knots flowing due north to south. If the ship has a maximum speed of 20 knots in still water, find the direction in which the ship should be steered for each leg of the voyage for the voyage to be completed in the shortest possible time and find this time.

6  A boat crosses a river which flows at $6 \text{ m s}^{-1}$ between parallel banks 30 m apart. The boat has a speed of $8 \text{ m s}^{-1}$ in still water.
(a) If the boat is steered in a direction perpendicular to the banks, find the time taken for the crossing and the point of landing on the opposite bank.
(b) If the boat sails to reach the opposite bank at a point directly opposite to its point of departure, find the direction in which the boat is steered and the time taken for the crossing.

## 3.6  Relative velocity

Consider the case of two flies crawling on the tray ABCD which is on a horizontal table. The first fly has a velocity $v_1$ relative to the tray and the second fly has a velocity $v_2$ relative to the tray. The tray now starts to move on the table with velocity $-v_2$. Each fly now has two components of velocity relative to the table.

The first fly has components $v_1$ and $-v_2$ and
the second fly has components $v_2$ and $-v_2$.
The resultant velocities, i.e. relative to the table, of the two flies are
first fly       $v_1 + (-v_2) = v_1 - v_2$,
second fly     $v_2 + (-v_2) = 0$.
Thus the second fly has zero velocity, i.e. is at rest, relative to the fixed table. The first fly has velocity $v_1 - v_2$ relative to the table. Since the second fly is at rest relative to the table the first fly has velocity $v_1 - v_2$ relative to the second fly. Similarly the velocity of the second fly relative to the first fly is $v_2 - v_1$.

If any two objects A and B have velocities $v_A$ and $v_B$ respectively,
the velocity of A relative to B $= v_A - v_B$ and
the velocity of B relative to A $= v_B - v_A$.

*Example 1*
Particle A has velocity $3i + 4j$ and particle B has velocity $6i - 2j$. Find
(a) the velocity of A relative to B
(b) the velocity of B relative to A.

Velocity of A relative to B $= v_A - v_B = (3i + 4j) - (6i - 2j) = -3i + 6j$.
Velocity of B relative to A $= v_B - v_A = (6i - 2j) - (3i + 4j) = 3i - 6j$.

*Example 2*
Particle A has velocity $2i + j$, particle B has velocity $3i + 5j$ and particle C has velocity $ai + bj$. The velocity of C relative to A is in the direction $i - j$ and the

velocity of C relative to B is in the direction $3\mathbf{i} + \mathbf{j}$. Find $a$ and $b$.

$$\mathbf{v_A} = 2\mathbf{i} + \mathbf{j}, \ \mathbf{v_B} = 3\mathbf{i} + 5\mathbf{j}, \ \mathbf{v_C} = a\mathbf{i} + b\mathbf{j}.$$

Velocity of C relative to A $= \mathbf{v_C} - \mathbf{v_A} = (a\mathbf{i} + b\mathbf{j}) - (2\mathbf{i} + \mathbf{j})$
$$= (a - 2)\mathbf{i} + (b - 1)\mathbf{j}$$

but this is in the direction $\mathbf{i} - \mathbf{j}$.

$$\therefore \qquad \frac{a - 2}{1} = \frac{b - 1}{-1},$$

i.e. $\qquad\qquad\qquad\qquad\qquad a + b = 3 \qquad\qquad\qquad\qquad\qquad\ \ldots(1)$

Velocity of C relative to B $= \mathbf{v_C} - \mathbf{v_B} = (a\mathbf{i} + b\mathbf{j}) - (3\mathbf{i} + 5\mathbf{j})$
$$= (a - 3)\mathbf{i} + (b - 5)\mathbf{j}$$

but this is in the direction $3\mathbf{i} + \mathbf{j}$.

$$\therefore \qquad \frac{a - 3}{3} = \frac{b - 5}{1},$$

i.e. $\qquad\qquad\qquad\qquad\qquad a - 3b = -12. \qquad\qquad\qquad\qquad \ldots(2)$

Solving equations (1) and (2) gives $a = -3/4, b = 15/4$.

*Example 3*
Ship A is sailing north east at a speed of 20 knots and ship B is sailing on a bearing of 300° at a speed of 12 knots. Find the velocity of B relative to A.

If, at 1400 h, B is 60 nautical miles due north of A, find the least distance between the ships. Find also the time when they are nearest to each other.

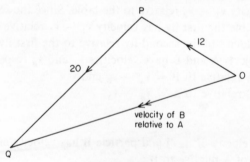

Fig. 3.18

The velocity of B relative to A is the vector sum of the velocity of B and a velocity equal in magnitude but opposite in direction to the velocity of A. These two velocities are represented by $\overrightarrow{OP}$ and $\overrightarrow{PQ}$ in the velocity diagram drawn to scale in Fig. 3.18.

The velocity of B relative to A is represented by $\overrightarrow{OQ}$, i.e. the velocity of B relative to A is approximately 26 knots on a bearing of approximately 252°.

Figure 3.19 shows the space diagram, drawn to scale, showing the path of B relative to A.

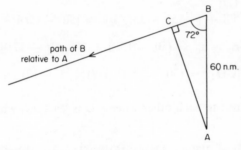

Fig. 3.19

The shortest distance between the ships is represented by AC giving the shortest distance between the ships as approximately 57 nautical miles.
The time taken to reach the position of closest approach is given by

$$\frac{\text{distance BC}}{\text{speed of B relative to A}} = \frac{19 \text{ nautical miles}}{26 \text{ knots}} \approx 0.73 \text{ h}$$

i.e. approximately 44 minutes, so that this position is reached at 1444 h.

*Example 4*
At 0900 h a cargo ship is sailing due north at a constant speed of 12 knots. A destroyer which, at this time, is 50 nautical miles due east of the cargo ship, is sailing at a constant speed of 30 knots on a bearing of 330°. Assuming that both ships maintain their courses and speeds, find the least distance between the ships in the subsequent motion. Find also the time when the ships are nearest to each other.

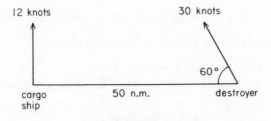

Fig. 3.20

Figure 3.20 shows the situation at 0900 h. Taking the unit of length as the nautical mile and the position of the cargo ship at 0900 h as origin and **i** and **j** as unit vectors due east and north respectively the velocity of the cargo ship is 12**j**.

Velocity of the destroyer $= -30 \cos 60°\mathbf{i} + 30 \sin 60°\mathbf{j}$
$$= -15\mathbf{i} + 15\sqrt{3}\mathbf{j}.$$

After $t$ hours the cargo ship is at point C and the destroyer is at point D with position vectors

$$\mathbf{r}_C = 12t\mathbf{j}$$
$$\mathbf{r}_D = 50\mathbf{i} + (-15\mathbf{i} + 15\sqrt{3}\mathbf{j})t = (50 - 15t)\mathbf{i} + 15\sqrt{3}t\mathbf{j}$$

$\therefore$ $\qquad \overrightarrow{CD} = \mathbf{r}_D - \mathbf{r}_C = (50 - 15t)\mathbf{i} + (15\sqrt{3}t - 12t)\mathbf{j}$

$\therefore$ $\qquad CD^2 = |\overrightarrow{CD}|^2 = (50 - 15t)^2 + (15\sqrt{3} - 12)^2t^2.$

The ships are nearest to each other when CD is least, i.e. when $\dfrac{d}{dt}(CD^2) = 0.$

$$\frac{d}{dt}(CD^2) = 2(50 - 15t)(-15) + (15\sqrt{3} - 12)^2(2t) = 0$$

giving $t = 1.78$ and $CD^2 = 1162$, i.e. $CD \approx 34$.
Thus the least distance between the ships is approximately 34 nautical miles. This occurs after 1.78 h, i.e. at 1047 h.

An alternative method of solution of this problem is as follows:

velocity of the destroyer $= -15\mathbf{i} + 15\sqrt{3}\mathbf{j}$
velocity of the cargo ship $= 12\mathbf{j}$
velocity of the destroyer relative to the cargo ship $= -15\mathbf{i} + (15\sqrt{3} - 12)\mathbf{j}$.
Figure 3.21(a) shows the path of the destroyer relative to the cargo ship.
Figure 3.21(b) shows the velocity diagram for the velocity of the destroyer relative to the cargo ship.

From Fig. 3.21(b)

$$\tan \theta = \frac{15\sqrt{3} - 12}{15} = 0.9321 \quad \text{giving} \quad \theta = 42°\,59'.$$

The least distance between the ships is $s$ (Fig. 3.21(a)), where

$$s = 50 \sin \theta = 50 \sin 42°\,59' \approx 34 \text{ nautical miles.}$$

The speed of the destroyer relative to the cargo ship

$$= \sqrt{[15^2 + (15\sqrt{3} - 12)^2]} = 20.51 \text{ knots.}$$

The time taken to reach the position when the ships are closest

$$= \frac{LM}{20.51} = \frac{50 \cos \theta}{20.51} = 1.78 \text{ h} = 1 \text{ h } 47 \text{ min,}$$

i.e. the least distance between the ships is approximately 34 nautical miles at 1047 h.

This problem could also have been solved by drawing to scale the diagrams in Figs. 3.21(a) and (b). The least distance could then have been obtained from the length MN in Fig. 3.21(a) and the speed of the destroyer relative to the cargo ship from the length PQ in Fig. 3.21(b).

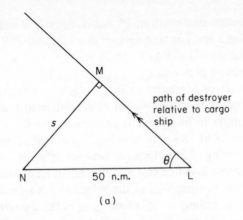

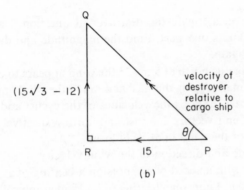

Fig. 3.21

## Exercises 3.6

1 The velocity of a particle A is **u** and the velocity of a particle B is **v**. Find the velocity of A relative to B and the velocity of B relative to A when

(a) **u** = 2**i** + **j**,      **v** = **i** − **j**

(b) **u** = 4**i** − **j**,      **v** = 2**i** + 3**j**

(c) **u** = **i** + **j**,      **v** = 5**i** − 3**j**

(d) **u** = −3**i** − 4**j**,      **v** = −2**i** + 3**j**

(e) **u** = **i** + 2**j** + 3**k**,      **v** = 2**i** − 5**j** + 6**k**

(f) **u** = −**i** + **j** − 2**k**,      **v** = 3**i** + 2**j** − **k**.

2 Particles A and B have velocities 2**i** + **j** and 4**i** − 2**j** respectively. Find the velocity of C if the velocities of C relative to A and B are respectively in the directions

(a) −**i**      and      −**i** + **j**

(b) 4**i** + 7**j**    and      **i** + 5**j**

(c) −5**i** + 2**j**   and     −12**i** + 7**j**.

3  Ship A is sailing due north at 20 knots. Ship B is sailing due east at 15 knots. Taking **i** and **j** as unit vectors due east and due north respectively, express, in terms of **i** and **j**,
   (a) the velocities of A and B
   (b) the velocity of A relative to B.
   Find the magnitude of the velocity of A relative to B and the direction in which A appears to be sailing to an observer on B.

4  Ship A is sailing at 18 knots on a bearing of 030°. Ship B is sailing at 24 knots on a bearing of 240°. Express these velocities in terms of unit vectors **i** and **j** in the directions due east and due north respectively. Hence find the magnitude and the direction of the velocity of A relative to B.

5  A destroyer is sailing at 25 knots in a north-westerly direction and a merchant ship is sailing at 15 knots to the south west. Find the direction in which the destroyer appears to be moving to an observer on the merchant ship.

6  To an observer on a frigate the destroyer in question 5 appears to be travelling at 20 knots due west. Find the magnitude and direction of the velocity of the frigate.

7  To a cyclist riding due north at 5 m s$^{-1}$ the wind appears to come from the north east. To a man jogging due east at 4 m s$^{-1}$ the wind appears to come from the south east. Express the velocities of the cyclist and the jogger in terms of **i** and **j**, unit vectors due east and north respectively. Hence find
   (a) the velocity of the wind in $a\mathbf{i} + b\mathbf{j}$ form
   (b) the magnitude and direction of the wind velocity.

8  Ship A is sailing with a speed of $v_1$ knots on a bearing of $\alpha°$ and ship B is sailing with speed $v_2$ knots on a bearing of $\beta°$. Find graphically the velocity of A relative to B in magnitude and direction given that
   (a) $v_1 = 20,$  $v_2 = 15,$  $\alpha = \phantom{0}90,$  $\beta = 160$
   (b) $v_1 = 12,$  $v_2 = 18,$  $\alpha = \phantom{0}60,$  $\beta = 210$
   (c) $v_1 = 25,$  $v_2 = 15,$  $\alpha = 112,$  $\beta = 253$
   (d) $v_1 = 26,$  $v_2 = 16,$  $\alpha = 304,$  $\beta = 206.$

9  Ship P is sailing with speed $u$ knots on a bearing of $\theta°$. To an observer on ship P the ship Q appears to be sailing on a bearing of $\phi°$ with speed $v$ knots. Find graphically, in magnitude and direction, the velocity of Q given that
   (a) $u = 20,$  $\theta = \phantom{00}0,$  $v = 30,$  $\phi = 270$
   (b) $u = 18,$  $\theta = \phantom{0}90,$  $v = 18,$  $\phi = \phantom{0}30$
   (c) $u = 15,$  $\theta = 160,$  $v = 25,$  $\phi = \phantom{0}40$
   (d) $u = 23,$  $\theta = 243,$  $v = 17,$  $\phi = 321.$

10 Ship A is sailing at a steady speed of 12 knots due east and ship B is sailing due north at a steady speed of 18 knots. Find graphically, in magnitude and direction, the velocity of A relative to B.
   If B is 40 nautical miles due east of A at 1000 h, find the least distance between the ships in the subsequent motion. Find also the time when the

distance between the ships is least.

11   A destroyer is sailing with a constant speed of 28 knots on a bearing of 120°. A frigate is sailing on a bearing of 200° at a constant speed of 22 knots. Find graphically, or otherwise, the magnitude and the direction of the velocity of the frigate relative to the destroyer.

If the destroyer is 60 nautical miles due south of the frigate at 0800 h, find the least distance between the ships subsequently. Find also the time when the distance between the ships is least.

## 3.7   Kinematics in two and three dimensions

In Section 3.5 $\mathbf{v}$ was defined as $\dfrac{d\mathbf{r}}{dt}$ and $\mathbf{a}$ as $\dfrac{d\mathbf{v}}{dt}$.

$\dfrac{d\mathbf{r}}{dt}$ is often expressed as $\dot{\mathbf{r}}$ and $\dfrac{d\mathbf{v}}{dt} = \dfrac{d}{dt}\left(\dfrac{d\mathbf{r}}{dt}\right)$ as $\ddot{\mathbf{r}}$.

Consider a particle moving so that, at time $t$, its position vector $\mathbf{r}$ is given by

$$\mathbf{r} = f(t)\mathbf{i} + \phi(t)\mathbf{j}.$$

Then   $\mathbf{v} = \dfrac{d\mathbf{r}}{dt} = \dot{\mathbf{r}} = f'(t)\mathbf{i} + \phi'(t)\mathbf{j}$

and   $\mathbf{a} = \dfrac{d\mathbf{v}}{dt} = \ddot{\mathbf{r}} = f''(t)\mathbf{i} + \phi''(t)\mathbf{j}.$

### Example 1
A particle moves in a plane so that it has position vector $\mathbf{r}$ at time $t$, where $\mathbf{r} = (t^2 + 2t)\mathbf{i} + (t^3 + t^2 + t)\mathbf{j}$. Find the velocity and acceleration of the particle when $t = 2$.

$$\mathbf{r} = (t^2 + 2t)\mathbf{i} + (t^3 + t^2 + t)\mathbf{j}$$

Differentiating              $\dot{\mathbf{r}} = (2t + 2)\mathbf{i} + (3t^2 + 2t + 1)\mathbf{j}$

Differentiating again    $\ddot{\mathbf{r}} = 2\mathbf{i} + (6t + 2)\mathbf{j}$

When $t = 2$, velocity    $\dot{\mathbf{r}} = 6\mathbf{i} + 17\mathbf{j}$

and acceleration           $\ddot{\mathbf{r}} = 2\mathbf{i} + 14\mathbf{j}.$

### Example 2
A particle moves so that, at time $t$, its acceleration is $2\mathbf{i} + t\mathbf{j}$. Given that, when $t = 0$, it has position vector $4\mathbf{i} + 7\mathbf{j}$ and velocity $2\mathbf{i} - \mathbf{j}$, find its position vector and its velocity when $t = 4$.

$$\ddot{\mathbf{r}} = 2\mathbf{i} + t\mathbf{j}.$$

Integrating              $\dot{\mathbf{r}} = 2t\mathbf{i} + \tfrac{1}{2}t^2\mathbf{j} + \mathbf{C}_1.$

When $t = 0$, $\dot{\mathbf{r}} = 2\mathbf{i} - \mathbf{j}$ giving $2\mathbf{i} - \mathbf{j} = \mathbf{C}_1.$

∴              $\dot{\mathbf{r}} = 2t\mathbf{i} + \tfrac{1}{2}t^2\mathbf{j} + 2\mathbf{i} - \mathbf{j}$
$$= (2t + 2)\mathbf{i} + (\tfrac{1}{2}t^2 - 1)\mathbf{j}.$$

Integrating $\qquad \mathbf{r} = (t^2 + 2t)\mathbf{i} + (\tfrac{1}{6}t^3 - t)\mathbf{j} + \mathbf{C}_2.$

When $t = 0$, $\mathbf{r} = 4\mathbf{i} + 7\mathbf{j}$ giving $4\mathbf{i} + 7\mathbf{j} = \mathbf{C}_2.$

$\therefore \qquad\qquad\qquad \mathbf{r} = (t^2 + t)\mathbf{i} + (\tfrac{1}{6}t^3 - t)\mathbf{j} + 4\mathbf{i} + 7\mathbf{j}$
$\qquad\qquad\qquad\qquad = (t^2 + 2t + 4)\mathbf{i} + (\tfrac{1}{6}t^3 - t + 7)\mathbf{j}.$

When $t = 4$, $\qquad\qquad \mathbf{r} = 28\mathbf{i} + \dfrac{41}{3}\mathbf{j}$

and $\qquad\qquad\qquad\qquad \dot{\mathbf{r}} = 10\mathbf{i} + 7\mathbf{j},$

i.e. at time $t = 4$, the particle has velocity $10\mathbf{i} + 7\mathbf{j}$ and position vector $28\mathbf{i} + \dfrac{41}{3}\mathbf{j}.$

*Example 3*
A particle moves so that, at time $t$, it has position vector $\mathbf{r}$, where $\mathbf{r} = \sin \omega t\mathbf{i} + \cos \omega t\mathbf{j} + t\mathbf{k}$ and $\omega$ is a constant.

Find the velocity of the particle when $t = \pi/(3\omega)$ and show that the magnitude of its acceleration is constant.

$\qquad\qquad\qquad\qquad \mathbf{r} = \sin \omega t\mathbf{i} + \cos \omega t\mathbf{j} + t\mathbf{k}.$
Differentiating $\qquad \dot{\mathbf{r}} = \omega \cos \omega t\mathbf{i} - \omega \sin \omega t\mathbf{j} + \mathbf{k}$
and $\qquad\qquad\qquad \ddot{\mathbf{r}} = -\omega^2 \sin \omega t\mathbf{i} - \omega^2 \cos \omega t\mathbf{j}.$
When $t = \pi/(3\omega) \qquad \dot{\mathbf{r}} = \omega \cos (\pi/3)\mathbf{i} - \omega \sin (\pi/3)\mathbf{j}$
$\qquad\qquad\qquad\qquad\quad = \tfrac{1}{2}\omega\mathbf{i} - \dfrac{\sqrt{3}}{2}\omega\mathbf{j}.$

$\qquad\qquad |\ddot{\mathbf{r}}|^2 = (-\omega^2 \sin \omega t)^2 + (-\omega^2 \cos \omega t)^2$
$\qquad\qquad\qquad\; = \omega^4(\sin^2 \omega t + \cos^2 \omega t) = \omega^4$

i.e. the magnitude of the acceleration is constant.

**Exercises 3.7**
1  A particle has position vector $\mathbf{r}$ at time $t$. Find the velocity and the acceleration of the particle
   (a) when $t = 2$, given that $\mathbf{r} = t^2\mathbf{i} + t^3\mathbf{j}$
   (b) when $t = 4$, given that $\mathbf{r} = (t^2 + 2)\mathbf{i} + (t - 1)\mathbf{j}$
   (c) when $t = 3$, given that $\mathbf{r} = (t^3 + 2t^2 + 1)\mathbf{i} + (t^2 - 2)\mathbf{j}$
   (d) when $t = 1$, given that $\mathbf{r} = e^t\mathbf{i} + e^{2t}\mathbf{j}$
   (e) when $t = 2$, given that $\mathbf{r} = t^3\mathbf{i} + t^2\mathbf{j} + t\mathbf{k}$

   (f) when $t = 3$, given that $\mathbf{r} = \sin \dfrac{\pi}{6}t\mathbf{i} + \cos \dfrac{\pi}{6}t\mathbf{j} + t^2\mathbf{k}.$

2  The acceleration of a particle at time $t$ is $\ddot{\mathbf{r}}$. Given that $\dot{\mathbf{r}} = \mathbf{j}$ and $\mathbf{r} = \mathbf{i}$ when $t = 0$, find the speed of the particle and its position vector
   (a) when $t = 1$, given that $\ddot{\mathbf{r}} = 2\mathbf{i} + \mathbf{j}$

(b) when $t = 2$, given that $\ddot{\mathbf{r}} = -3\mathbf{i} + t\mathbf{j}$

(c) when $t = 3$, given that $\ddot{\mathbf{r}} = -t\mathbf{i} + 4\mathbf{j}$

(d) when $t = 4$, given that $\ddot{\mathbf{r}} = e^{2t}\mathbf{i} + e^{3t}\mathbf{j}$

(e) when $t = 2$, given that $\ddot{\mathbf{r}} = \sin\frac{\pi}{4}t\mathbf{i} + 2\cos\frac{\pi}{8}t\mathbf{j}$

(f) when $t = 3$, given that $\ddot{\mathbf{r}} = 6\mathbf{i} - 12\mathbf{j} + 3\mathbf{k}$

(g) when $t = 4$, given that $\ddot{\mathbf{r}} = 9\mathbf{i} - 6t\mathbf{j} + \mathbf{k}$.

## 3.8 Projectiles

Figure 3.22 shows the path of a particle projected with speed $u$ at an angle of elevation $\alpha$ at time $t = 0$ from a point O on a horizontal plane. At time $t$ the particle is at P and is moving with speed $v$ at an angle of elevation $\theta$. The highest point of the path is A and the particle returns to the horizontal plane through O at the point B. Air resistance is assumed to be negligible.

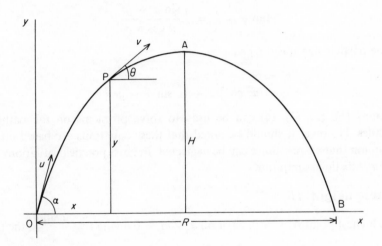

Fig. 3.22

Taking OB as the $x$-axis and the vertical through O as the $y$-axis, P has coordinates $(x, y)$. Initially the particle has components of velocity of magnitude $u \cos \alpha$ horizontally along the $x$-axis and $u \sin \alpha$ along the $y$-axis.

Since air resistance is negligible, the horizontal component of velocity of the particle remains constant,

i.e. $\dot{x} = u \cos \alpha$, where $\dot{x}$ represents $\dfrac{\mathrm{d}x}{\mathrm{d}t}$, the horizontal component of velocity

and $x = (u \cos \alpha)t$, since $x = 0$ when $t = 0$.

In a vertical direction the particle has the acceleration due to gravity which is constant. Thus the equations (1), (2), (3) and (4) of Section 3.1 apply.

$$\therefore \quad \dot{y} = u \sin \alpha - gt, \text{ where } \dot{y} \text{ represents } \frac{dy}{dt}, \text{ the vertical component of}$$

velocity

and $\quad y = (u \sin \alpha)t - \frac{1}{2}gt^2 \quad$ since $\dot{y} = u \sin \alpha$ and $y = 0$ when $t = 0$.

Thus at time $t$ the particle is at P which has coordinates

$$x = (u \cos \alpha)t \qquad \qquad \dots(1)$$

and $\qquad \qquad y = (u \sin \alpha)t - \frac{1}{2}gt^2. \qquad \qquad \dots(2)$

At P the particle has a velocity with horizontal and vertical components

$$\dot{x} = u \cos \alpha \qquad \qquad \dots(3)$$

and $\qquad \qquad \dot{y} = u \sin \alpha - gt. \qquad \qquad \dots(4)$

At P the path of the particle makes an angle $\theta$ with the horizontal, as shown in Fig. 3.22, where

$$\tan \theta = \frac{\dot{y}}{\dot{x}} = \frac{u \sin \alpha - gt}{u \cos \alpha}$$

and the particle has speed $v$ given by

$$v^2 = \dot{x}^2 + \dot{y}^2$$
$$= u^2 \cos^2 \alpha + (u \sin \alpha - gt)^2.$$

Equations (1), (2), (3), (4) can be used to solve problems on the paths of projectiles. However it should be noted that these equations are based on the assumption that air resistance can be neglected. In most practical situations this is an unrealistic assumption.

## Greatest height, H

At the highest point A, $\dot{y} = 0$, i.e. $u \sin \alpha - gt = 0$ giving $t = \dfrac{u \sin \alpha}{g}$, the time taken for the particle to move from O to A. Substituting this value of $t$ in the equation $y = (u \sin \alpha)t - \frac{1}{2}gt^2$ (2) gives

$$H = \left( u \sin \alpha \right)\left( \frac{u \sin \alpha}{g} \right) - \frac{1}{2}g\left( \frac{u \sin \alpha}{g} \right)^2$$

$$= \frac{u^2 \sin^2 \alpha}{2g}$$

Alternatively using $v^2 = u^2 + 2as$ gives $0 = u^2 \sin^2 \alpha - 2gh$, giving $H$ as before.

## Time of flight

At B, $y = 0$ and using $y = (u \sin \alpha)t - \frac{1}{2}gt^2$ (2) gives $0 = (u \sin \alpha)t - \frac{1}{2}gt^2$,

giving $t = 0$ (at O) or $t = \dfrac{2u \sin \alpha}{g}$ (at B).

$$\therefore \qquad \text{Time of flight from O to B} = \frac{2u \sin \alpha}{g},$$

which is twice the time from O to A.

## Horizontal range $R$

Using $x = (u \cos \alpha)t$ gives

$$R = \left( u \cos \alpha \right)\left( \frac{2u \sin \alpha}{g} \right) = \frac{2u^2 \sin \alpha \cos \alpha}{g}$$

$$= \frac{u^2 \sin 2\alpha}{g}.$$

Since the largest possible value of $\sin 2\alpha$ is 1, the maximum horizontal range for a given speed $u$ is $u^2/g$, when $\sin 2\alpha = 1$, i.e. $2\alpha = \pi/2$ and $\alpha = \pi/4$. Thus the maximum horizontal range is $u^2/g$ and this is achieved when the angle of projection is $\pi/4$ (or $45°$).

For a particular horizontal range $d$, where $d < u^2/g$,

$$d = \frac{u^2 \sin 2\alpha}{g} \quad \text{giving} \quad \sin 2\alpha = \frac{gd}{u^2}.$$

Let $\phi$ be the acute angle such that $\sin \phi = gd/u^2$. Since $\sin (\pi - 2\alpha) = \sin 2\alpha$, $\phi = 2\alpha$ or $\pi - 2\alpha$, i.e. $\alpha = \phi/2$ or $\pi/2 - \phi/2$, where $\sin \phi = gd/u^2$, so that, for a particular horizontal range $d$, the particle may be projected at an angle $\phi/2$ or $\pi/2 - \phi/2$ to the horizontal, i.e. at $\phi/2$ to the horizontal or at $\phi/2$ to the vertical.

## Equation of path

Eliminating $t$ from
$$x = (u \cos \alpha)t \qquad \qquad \dots(1)$$
and
$$y = (u \sin \alpha)t - \tfrac{1}{2}gt^2 \qquad \qquad \dots(2)$$

gives
$$y = \left( u \sin \alpha \right)\left( \frac{x}{u \cos \alpha} \right) - \tfrac{1}{2}g\left( \frac{x}{u \cos \alpha} \right)^2$$

i.e.
$$y = x \tan \alpha - \frac{gx^2}{2u^2 \cos^2 \alpha} \qquad \qquad \dots(5)$$

or
$$y = x \tan \alpha - \frac{gx^2}{2u^2} \sec^2 \alpha \qquad \qquad \dots(6)$$

or
$$y = x \tan \alpha - \frac{gx^2}{2u^2}(1 + \tan^2 \alpha). \qquad \qquad \dots(7)$$

Each of the equations (5), (6) and (7) is a form of the cartesian equation of the path of the particle referred to horizontal and vertical axes through the point of projection (see Fig. 3.22). These equations are of the form $y = cx - dx^2$, where $d$ is positive, i.e. $$y = x(c - dx).$$

This is the equation of a parabola of the form shown in Fig. 3.23. Thus the path of the particle is a parabola with its vertex upwards.

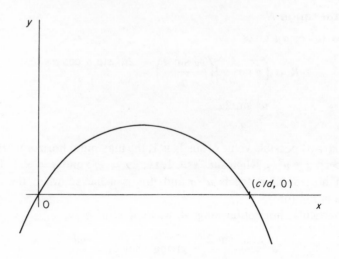

Fig. 3.23

*Example 1*
A particle is projected with speed 20 m s$^{-1}$ at an angle of elevation of 30° from a point on level ground. Find
(a) the greatest height reached
(b) the time of flight
(c) the horizontal range
(d) the direction in which the particle is moving after
(i) $\frac{1}{2}$ second (ii) $1\frac{1}{2}$ seconds.
(Take $g$ to be 10 m s$^{-2}$.)

(a) With the usual notation $u = 20$, $\alpha = 30°$.
The initial horizontal speed $= u \cos \alpha = 20 \cos 30° = 10\sqrt{3}$ m s$^{-1}$.
The initial vertical speed $\quad = u \sin \alpha = 20 \sin 30° = 10$ m s$^{-1}$.
Using $v = u + at$ for the vertical motion, the time taken to reach the greatest height is $10/10 = 1$ second.

Using $s = \left(\dfrac{u + v}{2}\right)t$ with $t = 1$ for the vertical motion gives

$$\text{greatest height reached} = \left(\frac{10 + 0}{2}\right)1 = 5 \text{ metres.}$$

(b) The time of flight is twice the time to reach the greatest height and so the time of flight is 2 seconds.

(c) The horizontal range $= (u \cos \alpha)t = (10\sqrt{3})(2) = 20\sqrt{3}$ metres.
These results could of course have been obtained by substituting $u = 20$ and $\alpha = 30°$ in the formulae for greatest height, time of flight and horizontal range.
(d) (i) When $t = \frac{1}{2}$,

$$\dot{x} = u \cos \alpha = 20(\sqrt{3}/2) = 10\sqrt{3}$$
$$\dot{y} = u \sin \alpha - gt = 20(\tfrac{1}{2}) - 10(\tfrac{1}{2}) = 5.$$

$\therefore \qquad \tan \theta = \dfrac{\dot{y}}{\dot{x}} = \dfrac{5}{10\sqrt{3}} = 0.2887 \quad$ giving $\quad \theta = 16°\ 6'.$

(ii) When $t = 1\frac{1}{2}$,

$$\dot{x} = u \cos \alpha = 10\sqrt{3} \quad \text{as before}$$
$$\dot{y} = u \sin \alpha - gt = 20(\tfrac{1}{2}) - 10(\tfrac{3}{2}) = -5.$$

$\therefore \qquad \tan \theta = \dfrac{\dot{y}}{\dot{x}} = \dfrac{-5}{10\sqrt{3}} = -0.2887 \quad$ giving $\quad \theta = -16°\ 6'.$

Thus after $\frac{1}{2}$ second the path is at an angle of $16°\ 6'$ above the horizontal and after $1\frac{1}{2}$ seconds the path is at an angle of $16°\ 6'$ below the horizontal.

*Example 2*
A particle is projected with speed $60\ \mathrm{m\,s}^{-1}$ at an angle arctan $(4/3)$ to the horizontal from a point O on a horizontal plane. Find
(i) when the particle is at a height of 99 metres above the plane
(ii) the horizontal distances from O of the particle at the times when it is at a height of 99 m.
(Take $g$ to be $10\ \mathrm{m\,s}^{-2}$.)

With the usual notation $u = 60$, $\tan \alpha = 4/3$, $y = 99$.
(i) Using $\qquad y = (u \sin \alpha)t - \frac{1}{2}gt^2$ $\qquad\qquad\qquad\qquad\qquad$ (2)

$$99 = 60(4/5)t - 5t^2$$

i.e. $\qquad 5t^2 - 48t + 99 = 0 \quad$ or $\quad (t - 3)(5t - 33) = 0$

giving $\qquad\qquad\qquad t = 3 \quad$ or $\quad 33/5\ (= 6.6),$

i.e. 3 seconds after projection and again 6.6 seconds after projection the particle is at a height of 99 metres.
(ii) Using $\qquad\qquad\qquad x = (u \cos \alpha)t$

when $t = 3$, $\qquad\qquad\quad x = 60(3/5)3 = 108$

when $t = 33/5$, $\qquad\quad x = 60(3/5)(33/5) = 237.6,$

i.e. after 3 seconds the particle is at a horizontal distance 108 m from O and after 6.6 seconds is at a horizontal distance 237.6 m from O.

*Example 3*

A particle is projected with speed 50 m s$^{-1}$. When it has travelled a horizontal distance of 20 m it is at a height of 9 m above the point of projection. Find the tangents of the two possible angles of projection. (Take $g$ to be 10 m s$^{-2}$.)

Putting $x = 20$, $y = 9$ and $u = 50$ in the equation of the path in the form

$$y = x \tan \alpha - \frac{gx^2}{2u^2}(1 + \tan^2 \alpha) \tag{7}$$

gives
$$9 = 20 \tan \alpha - \frac{10(400)}{2(2500)}(1 + \tan^2 \alpha),$$

i.e.
$$4 \tan^2 \alpha - 100 \tan \alpha + 49 = 0$$
$$(2 \tan \alpha - 1)(2 \tan \alpha - 49) = 0$$
∴
$$\tan \alpha = \tfrac{1}{2} \text{ or } 49/2,$$

i.e. the two possible angles of projection are $\tan^{-1}(\tfrac{1}{2})$ and $\tan^{-1}(49/2)$.

## Exercises 3.8

For the following questions a particle is projected with speed $u$ m s$^{-1}$ at an angle of elevation $\alpha$. Take $g$ to be 10 m s$^{-2}$.

1  Find (i) the time of flight (ii) the horizontal range (iii) the greatest height reached, when
(a) $u = 40$,  $\alpha = 30°$
(b) $u = 60$,  $\alpha = 60°$
(c) $u = 50$,  $\alpha = 45°$
(d) $u = 80$, $\tan \alpha = 3/4$
(e) $u = 60$, $\tan \alpha = 4/3$
(f) $u = 52$, $\tan \alpha = 5/12$
(g) $u = 78$, $\tan \alpha = 12/5$.

2  Find the direction in which the particle is moving $t$ seconds after projection when
(a) $u = 40$,  $\alpha = 30°, t = 1$
(b) $u = 40$,  $\alpha = 30°, t = 4$
(c) $u = 40$,  $\alpha = 30°, t = 3$
(d) $u = 60$, $\tan \alpha = 3/4, t = 2$
(e) $u = 60$, $\tan \alpha = 3/4, t = 5$.

3  Find when the particle is at a height $h$ m if
(a) $u = 40$,  $\alpha = 30°, h = 15$
(b) $u = 40$,  $\alpha = 30°, h = 20$
(c) $u = 40$,  $\alpha = 30°, h = 18\tfrac{3}{4}$
(d) $u = 60$, $\tan \alpha = 4/3, h = 76$
(e) $u = 60$, $\tan \alpha = 4/3, h = 112$
(f) $u = 60$, $\tan \alpha = 4/3, h = 115$.

4  Find the horizontal and vertical distances travelled by the particle after

*t* seconds if

(a) $u = 60, \alpha = 30°, \quad t = 2$
(b) $u = 60, \alpha = 30°, \quad t = 4$
(c) $u = 40, \alpha = \arctan \frac{3}{4}, t = 2$
(d) $u = 40, \alpha = \arctan \frac{3}{4}, t = 3$.

5   Find the angle of projection if the greatest height reached is $H$ m when

(a) $u = 40, H = 20$
(b) $u = 60, H = 60$
(c) $u = 50, H = 40$.

6   Find the possible angles of projection for horizontal range $R$ m when

(a) $u = 20, R = 30$
(b) $u = 40, R = 100$
(c) $u = 60, R = 240$.

7   The particle passes through a point whose horizontal and vertical distances from the point of projection are $x$ m and $y$ m respectively. Find the possible angles of projection when

(a) $u = 40, x = \quad 80, y = \quad 40$
(b) $u = 50, x = 120, y = \quad 96$
(c) $u = 60, x = 270, y = -22.5$.

## 3.9   Range on an inclined plane

A particle is projected from a point O on a plane inclined at an angle $\beta$ to the horizontal as shown in Fig. 3.24. The initial speed of the particle is $u$ and the particle is projected at angle of elevation $\alpha$, where $\alpha > \beta$. The path of the particle lies in a vertical plane through a line of greatest slope of the inclined plane. The particle strikes the plane at the point A. The time of flight from O to A and the distance OA, the range on the inclined plane, will now be found.

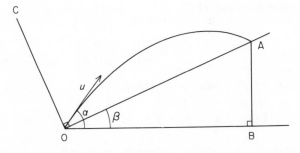

Fig. 3.24

In Section 3.8 frequent use was made of equations which related to motion in horizontal and vertical directions. Now it is convenient to consider motion in the direction OA and in the direction OC.

The acceleration due to gravity has components of magnitude $g \sin \beta$ in the direction AO and $g \cos \beta$ in the direction CO, or $-g \sin \beta$ in the direction OA and $-g \cos \beta$ in the direction OC.

Initially the velocity of the particle has components $u \cos (\alpha - \beta)$ in the direction OA and $u \sin (\alpha - \beta)$ in the direction OC.

Using the equation $s = ut + \frac{1}{2}at^2$ for motion in the direction OC, $s = 0$ at O and A giving

$$0 = [u \sin (\alpha - \beta)]t - \tfrac{1}{2}(g \cos \beta)t^2$$

$\therefore$ $\qquad\qquad t = 0$ (at O)$\quad$ or $\quad t = \dfrac{2u \sin (\alpha - \beta)}{g \cos \beta}$ (at A),

i.e. $\qquad$ the time of flight from O to A $= \dfrac{2u \sin (\alpha - \beta)}{g \cos \beta}$.

In this time the particle travels a horizontal distance OB, where

$$\text{OB} = (u \cos \alpha)t = \left(u \cos \alpha\right)\left[\frac{2u \sin (\alpha - \beta)}{g \cos \beta}\right] = \frac{2u^2 \sin (\alpha - \beta) \cos \alpha}{g \cos \beta}.$$

Also $\qquad\qquad \text{OA} = \dfrac{\text{OB}}{\cos \beta} = \dfrac{2u^2 \sin (\alpha - \beta) \cos \alpha}{g \cos^2 \beta}$,

the range on the inclined plane.

Using the identity $2 \cos A \sin B = \sin (A + B) - \sin (A - B)$ gives

$$2 \sin (\alpha - \beta) \cos \alpha = \sin (2\alpha - \beta) - \sin \beta,$$

so that $\qquad\qquad \text{OA} = \dfrac{u^2}{g \cos^2 \beta}[\sin (2\alpha - \beta) - \sin \beta].$

For different values of $\alpha$, the largest possible value of $\sin (2\alpha - \beta)$ is 1.

$\therefore$ $\qquad$ Maximum value of OA $= \dfrac{u^2}{g \cos^2 \beta}[1 - \sin \beta]$

$$= \frac{u^2(1 - \sin \beta)}{g(1 - \sin^2 \beta)} = \frac{u^2}{g(1 + \sin \beta)},$$

which is the maximum range up the plane.

$\sin (2\alpha - \beta) = 1$ gives $2\alpha - \beta = \pi/2$ and $\alpha - \beta = \dfrac{\pi}{2} - \alpha$,

i.e. for the maximum range up the plane the particle is projected in a direction which bisects the angle between the plane and the vertical.

Next the case where the particle is projected down the inclined plane will be considered.

The particle is projected from O and strikes the plane at C (Fig. 3.25). Replacing $\beta$ by $-\beta$ in the expression for OA gives

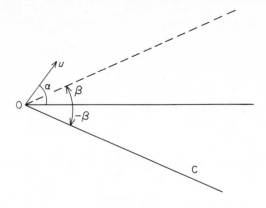

Fig. 3.25

range down inclined plane $OC = \dfrac{2u^2 \sin (\alpha + \beta) \cos \alpha}{g \cos^2 \beta}$

$= \dfrac{u^2 [\sin (2\alpha + \beta) + \sin \beta]}{g \cos^2 \beta}.$

For different values of $\alpha$, the largest possible value of $OC$, i.e. the maximum range down the inclined plane, is

$$\frac{u^2}{g(1 - \sin \beta)}.$$

*Example 1*
A particle is projected at an angle of $45°$ to the horizontal from a point on a plane inclined at an angle of $30°$ to the horizontal. The path of the particle lies in a vertical plane containing a line of greatest slope of the plane. The speed of projection is $40 \text{ m s}^{-1}$. Find the range (a) up the plane (b) down the plane. Find also
(i) the maximum ranges up and down the plane with this speed of projection
(ii) the angles at which the particle should be projected to achieve these maximum ranges.
(Take $g$ to be $10 \text{ m s}^{-2}$.)

(a) Figure 3.26 shows the situation when the particle is projected up the plane. Considering motion perpendicular to OA, using $s = ut + \frac{1}{2}at^2$,

$$0 = (40 \sin 15°)t - \tfrac{1}{2}(g \cos 30°)t^2$$

giving $t = \dfrac{2(40) \sin 15°}{10 \cos 30°} = 2.391$ (or $t = 0$).

$$OB = 40 \cos 45°(2.391) = 67.63$$

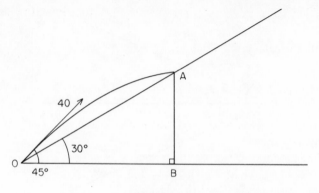

Fig. 3.26

$$OA \doteq \frac{OB}{\cos 30°} = 78.1,$$

i.e. the particle has a range up the plane of approximately 78.1 metres.
(b) Figure 3.27 show the situation when the particle is projected down the plane.

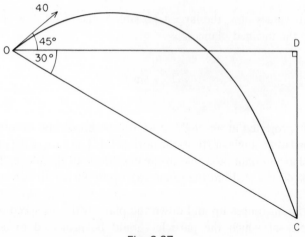

Fig. 3.27

Using $s = ut + \frac{1}{2}at^2$ for the motion perpendicular to OC,

$$0 = (40 \sin 75°)t - \frac{1}{2}(g \cos 30°)t^2$$

giving $t = \dfrac{2(40) \sin 75°}{10 \cos 30°} = 8.923$ (or $t = 0$).

$$OD = 40 \cos 45°(8.923) = 252.4$$

$$OC = \frac{OD}{\cos 30°} = 291.4,$$

i.e. the particle has a range of approximately 291.4 m down the plane.

(i) Maximum range up the plane $= \dfrac{u^2}{g(1 + \sin \beta)} = \dfrac{1600}{10(1 + \sin 30°)} = 106\frac{2}{3}.$

Maximum range down the plane $= \dfrac{u^2}{g(1 - \sin \beta)} = \dfrac{1600}{10(1 - \sin 30°)} = 320,$

i.e. the maximum ranges up and down the plane are $106\frac{2}{3}$ m and 320 m respectively.

(ii) For the maximum range up the plane $2\alpha - \beta = \pi/2$
$\therefore\ 2\alpha = \pi/2 + \pi/6 = 2\pi/3$ giving $\alpha = \pi/3$ or $60°$.
For the maximum range down the slope $2\alpha + \beta = \pi/2$
$\therefore\ 2\alpha = \pi/2 - \pi/6 = \pi/3$ giving $\alpha = \pi/6$ or $30°$,
i.e. for the maximum range up the plane the particle is projected at $60°$ to the horizontal and for the maximum range down the plane the particle is projected at $30°$ to the horizontal.


## Exercises 3.9
In these questions take $g$ to be $10\ \mathrm{m\,s^{-2}}$.

1  A particle is projected with speed $u$ m s$^{-1}$ from a point on a plane inclined at an angle $\alpha$ to the horizontal. Find the maximum possible range (i) up the plane (ii) down the plane, when
   (a) $u = 60,\qquad \alpha = 30°$
   (b) $u = 50,\ \tan \alpha = 3/4$
   (c) $u = 80,\ \tan \alpha = 5/12$
   (d) $u = 60,\ \tan \alpha = 2.4$
   (e) $u = 40,\ \tan \alpha = 7/24.$

2  Find the angles of projection for the particle to achieve the maximum possible ranges in questions 1 (a), (b), (c), (d), (e).

3  A particle is projected from a point on a plane inclined at an angle $\beta$ to the horizontal. The particle is projected with speed $u$ m s$^{-1}$ at an angle $\alpha$ to the horizontal. Find the range (i) up the plane (ii) down the plane, when
   (a) $u = 60,\ \tan \alpha = 12/5,\ \tan \beta = 3/4$
   (b) $u = 80,\ \tan \alpha = 4/3,\quad \tan \beta = 3/4$
   (c) $u = 40,\ \tan \alpha = 4/3,\quad \tan \beta = 5/12$
   (d) $u = 50,\qquad \alpha = 60°,\qquad \beta = 30°$
   (e) $u = 60,\qquad \alpha = 45°,\qquad \beta = 30°.$

4  A particle is projected at an angle $\alpha$ to the horizontal from a point on a plane inclined at an angle $\beta$ to the horizontal. The path of the particle lies in a vertical plane through a line of greatest slope of the plane. The particle strikes the plane when moving horizontally. Show that

$$\tan \alpha = 2 \tan \beta.$$

5  If the particle in question 4 strikes the plane when moving in a direction

perpendicular to the plane, show that

$$2 \tan (\alpha - \beta) \tan \beta = 1.$$

## Miscellaneous exercises 3

1  A particle moving in a straight line with constant acceleration passes through points O, A and B at times $t = 0$, 1 and 2 seconds respectively, where A and B are on the same side of O and OA $= 10$ m, OB $= 60$ m. Find the magnitudes and senses (from O to B or B to O) of both the acceleration and the velocity of the particle at time $t = 0$.  [JMB]

2  A car is travelling along a motorway at $110 \text{ km h}^{-1}$. At a point A the driver observes a sign warning of road works ahead and ordering a reduction of speed to $v \text{ km h}^{-1}$. The driver immediately applies the brakes so that the speed of the car is uniformly retarded in $t$ minutes to $v \text{ km h}^{-1}$. The car maintains this speed until a 'road clear' sign is reached. The driver then accelerates uniformly for $3t$ minutes until the car reaches a speed of $110 \text{ km h}^{-1}$ at a point B. The distance travelled at $v \text{ km h}^{-1}$ is $2\frac{1}{2}$ km, AB $= 6\frac{1}{2}$ km and the car takes 6 minutes to travel from A to B. Sketch the velocity–time graph and hence, or otherwise, show that
(i) $v(6 - 4t) = 150$ (ii) $(110 + v)t = 120$.
Hence show that $v = 50$.  [AEB]

3  A lift travels vertically a distance of 22 m from rest at the basement to rest at the top floor. Initially the lift moves with constant acceleration $x \text{ m s}^{-2}$ for a distance of 5 m; it continues with constant speed $u \text{ m s}^{-1}$ for 14 m; it is then brought to rest at the top floor by a constant retardation $y \text{ m s}^{-2}$. State, in terms of $u$ only, the times taken by the lift to cover the three stages of its journey.
    If the total time that the lift is moving is 6 seconds, calculate
(i) the value of $u$ (ii) the values of $x$ and $y$.
(Take the acceleration due to gravity to be $10 \text{ m s}^{-2}$.)  [AEB]

4  After passing a police radar check-point at $60 \text{ km h}^{-1}$, a sports car immediately began to retard uniformly until its speed was $40 \text{ km h}^{-1}$ and it continued to move at this speed until it was passed by a police car 1 km from the check-point. This police car had started from rest at the check-point at the same instant as the sports car had passed the check-point and had then moved with constant acceleration until it had passed the sports car. Assuming that the time taken by the sports car in slowing down from $60 \text{ km h}^{-1}$ to $40 \text{ km h}^{-1}$ was equal to the time that it travelled at constant speed before being passed by the police car, find, by using a velocity–time sketch, or otherwise,
(i) the time taken by the police car to reach the sports car
(ii) the speed of the police car at the instant when it passed the sports car
(iii) the time, measured from the check-point, when the speeds of the two cars were equal.  [AEB]

5  (a) A particle is uniformly accelerated from A to B, a distance of 96 m, and is then uniformly retarded from B to C, a distance of 30 m. The speeds of the particle at A and B are 6 m s$^{-1}$ and $u$ m s$^{-1}$ respectively and the particle comes to rest at C. Express, in terms of $u$ only, the times taken by the particle to move from A to B and from B to C.

   Given that the total time taken by the particle to move from A to C is 18 seconds, find
   (i) the value of $u$
   (ii) the acceleration and the retardation of the particle.
   (b) A particle P is moving due east at 6 m s$^{-1}$. A second particle Q is moving on a bearing of 210° (S 30° W) at 10 m s$^{-1}$. Find the magnitude and the direction of the velocity of P relative to Q.           [AEB]

6  Two trains, P and Q, travel by the same route from rest at station A to rest at station B. Train P has constant acceleration $f$ for the first third of the time, constant speed for the second third and constant retardation $f$ for the last third of the time. Train Q has constant acceleration $f$ for the first third of the distance, constant speed for the second third and constant retardation $f$ for the last third of the distance. Show that the times taken by the two trains are in the ratio $3\sqrt{3}$:5.           [L]

7  A particle has speed $v$ m s$^{-1}$ after being in motion for $t$ seconds. Corresponding values of $v$ and $t$ are given in the following table.

| $t$ | 1.0 | 1.5 | 2.0 | 2.5 | 3.0 |
|-----|-----|-----|-----|-----|-----|
| $v$ | 16.0 | 14.7 | 14.0 | 13.6 | 13.3 |

   Estimate graphically
   (a) the acceleration when $t = 2$
   (b) the distance moved in the interval between $t = 1$ and $t = 3$.

8  The speed of a car at one second intervals is given in the following table.

| Speed (km h$^{-1}$) | 0 | 19.5 | 34.5 | 47.0 | 56.3 | 64.2 | 71.0 | 75.7 | 80.0 |
|---------------------|---|------|------|------|------|------|------|------|------|
| Time (seconds) | 0 | 1 | 2 | 3 | 4 | 5 | 6 | 7 | 8 |

   Estimate graphically
   (a) the initial acceleration
   (b) the acceleration after 4 seconds
   (c) the distance moved in the first 4 seconds
   (d) the distance moved in the next 4 seconds.

9  Show that the time taken to travel a given distance can be found by drawing the graph of the reciprocal of the speed against the distance covered and evaluating the area under the graph.

   The speeds of a train at distances from a signal box are noted and recorded in the following table:

| Distance (mile) | 0 | 0.1 | 0.2 | 0.3 | 0.4 | 0.5 | 0.6 |
|-----------------|---|-----|-----|-----|-----|-----|-----|
| Speed (mile h$^{-1}$) | 10 | 18.8 | 24.4 | 29.4 | 33.3 | 37 | 40 |

Find, graphically or otherwise, the time taken by the train to travel 0.6 mile from the signal box. [AEB]

10 Show how the time taken by a particle to travel a given distance can be found from the graph of the reciprocal of the speed against the distance covered by the particle.

The speeds of a car at distances from a check-point are noted and recorded in the following table:

| Distance (km) | 0 | 0.2 | 0.4 | 0.6 | 0.8 | 1.0 | 1.2 |
|---|---|---|---|---|---|---|---|
| Speed ($km\,h^{-1}$) | 20 | 37.6 | 48.8 | 58.8 | 66.6 | 74.0 | 80.0 |

Find, graphically or otherwise, the time taken by the car to travel 1.2 km from the check-point. [AEB]

11 (i) A particle moves in a straight line with initial acceleration 10 $m\,s^{-2}$. The acceleration decreases uniformly with time until after ten seconds the acceleration is 5 $m\,s^{-2}$, and from then onwards it remains constant. If the initial velocity is 100 $m\,s^{-1}$, find when the velocity has doubled, and sketch the graph of velocity against time.

(ii) A point P moves along the $x$-axis with acceleration away from the origin O inversely proportional to $OP^2$. It is initially at rest with $OP = a$, and when $OP = 2a$ its velocity is $V$. Show that when $OP = ka$, where $k$ is large, the velocity of P is approximately $(2k - 1) V/(k\sqrt{2})$. [L]

12 A particle moves in a straight line with initial speed $u$ at time $t = 0$, and with acceleration inversely proportional to $(t + t_0)^3$, where $t_0$ is positive and constant. Show that the speed of the particle approaches a limiting value. If this value is $2u$, show that after time $t$ the particle will have travelled a distance

$$ut(2t + t_0)/(t + t_0). \qquad [L]$$

13 A particle of mass $m$ starts from rest at A with acceleration $(6t + 6)\,m\,s^{-2}$ and moves in the horizontal line AB towards B, where $t$ seconds is the time after motion begins. At the same instant, $(t = 0)$, a second particle of mass $2m$ starts from rest at B with an acceleration $(12t + 4)\,m\,s^{-2}$ towards A.

(i) If the speeds of the two particles are equal when they collide, calculate the distance AB.

(ii) If, however, AB = 8 m, show that the particles will collide after one second. [AEB]

14 Two particles start simultaneously from the point O and move in the same straight line. The first particle starts from rest and after $t$ seconds its acceleration is $(12t - 8)\,m\,s^{-2}$. The second particle has an initial velocity of 2 $m\,s^{-1}$ and, after $t$ seconds, its acceleration is $2v\,m\,s^{-2}$, where $v\,m\,s^{-1}$ is the velocity of this particle at that instant. Show that the first particle returns to O after 2 seconds.

After $T$ seconds, where $T > 2$, the second particle is $k$ times the distance of the first particle from O. Show that $e^{2T} - 1 = 2kT^2(T - 2)$.

Find the value of $k$ when $T = 3$.

Sketch the velocity–time graph for each particle. [AEB]

15   A particle starts from rest at time $t = 0$ and moves in a straight line with variable acceleration $f$ $m\,s^{-2}$, where

$$f = \frac{t}{5} \qquad (0 \leqslant t < 5)$$

$$f = \frac{t}{5} + \frac{10}{t^2} \quad (t \geqslant 5),$$

$t$ being measured in seconds. Show that the velocity is $2\frac{1}{2}\,m\,s^{-1}$ when $t = 5$ and $11\,m\,s^{-1}$ when $t = 10$.

Show also that the distance travelled by the particle in the first 10 seconds is $(43\frac{1}{3} - 10\ln 2)$ m. [L]

16   A particle moves along a horizontal straight line with acceleration proportional to $\cos \pi t$, where $t$ is the time. When $t = 0$ the velocity of the particle is $u$, and when $t = \frac{1}{2}$ its velocity is $2u$. Find the distance that the particle has travelled when $t = 2$, and draw the velocity–time graph for the interval $0 \leqslant t \leqslant 2$. [L]

17   Show that, with the usual notation, $v = \dfrac{dv}{ds} = \dfrac{dv}{dt}$.

A particle moves in a straight line away from a fixed point O in the line in such a way that its speed, $v\,m\,s^{-1}$, at a distance $s$ metres from O is given by

$$v = \frac{4}{2s + 3}.$$

If the particle passes through the point A, where OA $= 5$ m, calculate
(i) the time taken to cover the distance OA
(ii) the acceleration of the particle at A
(iii) the distance OB, where B is the position of the particle 1 second after leaving O. [AEB]

18   A particle leaves a point A at time $t = 0$ with speed $u$ and moves towards a point B with a retardation $\lambda v$, where $v$ is the speed of the particle at time $t$. The particle is at a distance $s$ from A at time $t$. Show that
(i) $v = u - \lambda s$ (ii) $\log_e (u - \lambda s) = \log_e u - \lambda t$.

At $t = 0$ a second particle starts from rest at B and moves towards A with acceleration $2 + 6t$. The particles collide at the mid-point of AB when $t = 1$. Find the distance AB and the speeds of the particles on impact. [AEB]

19   A particle P is projected vertically upwards from a point O on a horizontal plane and returns to O after 4 seconds.
(i) Calculate the speed of projection and the total distance travelled by P.

A second particle Q is projected from O and strikes the horizontal plane at E, where OE = 240 m, after 8 seconds.

(ii) Calculate the angle of projection of Q.

(iii) Given that P and Q are projected simultaneously, find the magnitude and the direction of the velocity of Q relative to P at the instant of projection.

(Take the acceleration due to gravity to be 10 m s$^{-2}$)       [AEB]

20  A, B and C are three buoys in the sea at the corners of an equilateral triangle of side 1 km. A man can swim at a speed of 1.5 km h$^{-1}$ in still water. Find, by calculation or drawing, how long the man takes to swim round the course ABCA when there is a current of 0.5 km h$^{-1}$ in a direction perpendicular to AB.

(Whatever your method, clearly labelled vector diagrams must be shown.)       [C]

21  Three fixed buoys A, B and C form an equilateral triangle of side 8 kilometres. The buoy B is due east of A and the buoy C is to the north of the line AB. A steady current of speed 4 kilometres per hour flows from west to east. A motor-boat which has a top speed of 12 kilometres per hour in still water does the triangular journey ABCA at top speed. Find, graphically or otherwise, the time taken on each leg of the journey giving your answers in minutes.       [C]

22  An aircraft whose speed in still air is 800 km h$^{-1}$ flies at a constant height and describes a horizontal circuit which is in the form of an equilateral triangle ABC of side 1000 km, where A is north of BC and B is due west of C. A wind of speed 100 km h$^{-1}$ is blowing from the west. The aircraft starts from A and flies round the circuit in the sense ABC. Show that the times taken for the aircraft to cover the legs AB and CA are equal. Neglecting the time taken to turn at the corners, show also that the time taken for the aircraft to fly once round the circuit is approximately 3.8 hours.       [L]

23  A speed-boat which can travel at 20 knots in still water starts from the corner X of an equilateral triangle XYZ of side 10 nautical miles and describes the complete course XYZX in the least possible time. A tide of 5 knots is running in the direction $\overrightarrow{ZX}$.

Find

(a) the speed along XY,

(b) to the nearest minute, the time taken by the speed-boat to traverse the complete course XYZX.       [L]

24  A river flows at 5 m s$^{-1}$ from west to east between parallel banks which are at a distance 300 metres apart. A man rows a boat at a speed of 3 m s$^{-1}$ in still water.

(i) State the direction in which the boat must be steered in order to cross the river from the southern bank to the northern bank in the shortest possible time. Find the time taken and the actual distance covered by the boat for this crossing.

(ii) Find the direction in which the boat must be steered in order to cross the river from the southern bank to the northern bank by the shortest possible route. Find the time taken and the actual distance covered by the boat for this crossing. [AEB]

25 Two particles A and B move with constant velocity vectors $(4\mathbf{i} + \mathbf{j} - 2\mathbf{k})$ and $(6\mathbf{j} + 3\mathbf{k})$ respectively, the unit of speed being the metre per second. At time $t = 0$, A is at the point with position vector $(-\mathbf{i} + 20\mathbf{j} + 21\mathbf{k})$ and B is at the point with position vector $(\mathbf{i} + 3\mathbf{k})$, the unit of distance being the metre.

Find the value of $t$ for which the distance between A and B is least and find also this least distance. [L]

26 A particle A starts from the point P whose position vector is $(a\mathbf{i} + 3\mathbf{j} + 2\mathbf{k})$ with constant velocity vector $(\mathbf{i} - \mathbf{j} + \mathbf{k})$. Simultaneously a second particle B leaves the point Q whose position vector is $(7\mathbf{i} - 2\mathbf{j} + 4\mathbf{k})$ with constant velocity vector $(-2\mathbf{i} + \mathbf{j} + 4\mathbf{k})$. Find the velocity vector of A relative to B and the position vector of A relative to B at $t$ units of time after the start. Find also the distance between A and B at this time. If the distance AB is least when $t = 2$, find the value of $a$. [L]

27 One particle A has velocity vector $4\mathbf{i} + 3\mathbf{j} + 2\mathbf{k}$ and another particle B has velocity vector $2\mathbf{i} - \mathbf{j} + 4\mathbf{k}$. The velocity of a third particle C relative to A is $4\mathbf{i} - 10\mathbf{j} + 4\mathbf{k}$. Find the velocity vector of C and the velocity of C relative to B.

If A is at the origin when B is at the point with position vector $8\mathbf{i} + 16\mathbf{j} - 8\mathbf{k}$, show that A and B will collide. Determine the position vector of the point of collision. [L]

28 At time $t = 0$ the position vectors of two particles P and Q are $\mathbf{i} + \mathbf{j} + 3\mathbf{k}$ and $4\mathbf{i} + 5\mathbf{j} + \mathbf{k}$ respectively. The particles have constant velocity vectors $2\mathbf{i} + \mathbf{j} + 2\mathbf{k}$ and $-4\mathbf{j} + 3\mathbf{k}$ respectively. Find the position vector of Q relative to P when $t = T$.

Show that the distance between the two particles is a minimum when $t = 14/15$ and find the minimum distance. Also find the position vector of Q relative to P at this instant. [L]

29 Two ships A and B have constant velocity vectors $20\mathbf{i} + 16\mathbf{j}$ and $4\mathbf{i} + 7\mathbf{j}$ respectively, the speeds being in $\mathrm{m\,s^{-1}}$. A third ship C moves relative to A in a direction $5\mathbf{i} - 6\mathbf{j}$ and relative to B in a direction $7\mathbf{i} + \mathbf{j}$. Calculate the velocity vector of C and show that its speed is $5\sqrt{29}\ \mathrm{m\,s^{-1}}$.

Initially $\overrightarrow{AB}$ is $100\mathbf{i}$, where the unit of distance is the metre. Assuming that the velocities remain constant, show that the shortest distance between A and B subsequently is $(900/\sqrt{337})$ m. [AEB]

30 Distances being measured in nautical miles and speeds in knots, a motor-boat sets out at 11 a.m. from a position $-6\mathbf{i} - 2\mathbf{j}$ relative to a marker buoy, and travels at a steady speed of magnitude $\sqrt{53}$ on a direct course to intercept a ship. The ship maintains a steady velocity vector $3\mathbf{i} + 4\mathbf{j}$ and at 12 noon is at a position $3\mathbf{i} - \mathbf{j}$ from the buoy. Find the velocity vector of the

motor-boat, the time of interception, and the position vector of the point of interception from the buoy. [L]

31 A cyclist observes that when his velocity is $u\mathbf{i}$ the wind appears to come from the direction $\mathbf{i} + 2\mathbf{j}$, but when his velocity is $u\mathbf{j}$ the wind appears to come from the direction $-\mathbf{i} + 2\mathbf{j}$. Prove that the true velocity of the wind is $(\frac{3}{4}\mathbf{i} - \frac{1}{2}\mathbf{j})u$.

Find the speed and the direction of motion of the cyclist when the wind velocity appears to be $u\mathbf{i}$. [AEB]

32 Two ships A and B move with constant velocities and have the following velocity and position vectors at time $t = 0$:

|        | Velocity vector | Position vector |
|--------|-----------------|-----------------|
| Ship A | $-\mathbf{i} + 3\mathbf{j}$ | $5\mathbf{i} + \mathbf{j}$ |
| Ship B | $2\mathbf{i} + 5\mathbf{j}$ | $3\mathbf{i} - 3\mathbf{j}$ |

Show that any point on the line $\mathbf{r} = 4\mathbf{i} - \mathbf{j} + \lambda(2\mathbf{i} - \mathbf{j})$, where $\lambda$ is a scalar, is equidistant from the ships A and B at time $t = 0$. Find the position vector of the point which is equidistant from both ships at time $t = 0$ and from the point with position vector $\mathbf{i} - \mathbf{j}$.

Calculate the value of $t$ when the distance between A and B is least. [AEB]

33 The banks of a river are straight and parallel and at a distance $a$ apart and the river flows with a constant uniform speed $v$. A and B are two points on opposite banks of the river with the line AB at right angles to the banks. Two swimmers set off at the same instant from A and B and swim with a steady speed $u$ relative to the water, where $u > v$ so that they approach each other along the line AB. Find the time taken for the swimmers to meet. [JMB]

34 The banks of a river 40 m wide are parallel and A and B are points on opposite banks. The distance AB is 50 m and B is downstream of A. There is a constant current of $4 \text{ m s}^{-1}$ flowing. What is the minimum speed at which a motor-boat must be able to move in still water in order to cross this river from A to B?

If a boat sails from A to B with constant velocity in $7\frac{1}{2}$ seconds, find its speed relative to the water and the direction in which it is steered.

Whilst this boat is sailing from A to B a man runs across a bridge which is at right angles to the banks of the river. To this man the boat appears to be travelling parallel to the banks of the river. Find the speed at which the man is running. [AEB]

35 A particle P moves from the origin O of rectangular cartesian axes and travels along the positive $x$-axis with speed $30 \text{ m s}^{-1}$. A second particle Q moves along the positive $y$-axis with speed $40 \text{ m s}^{-1}$ in a direction away from O. Find the magnitude and direction of the velocity of P relative to Q.

A third particle R moves directly towards O from the point with coordinates $(-3, -4)$ with constant speed $V$. Find the value of $V$ which makes the magnitude of the velocity of R relative to Q as small as possible. (A graphical method may be used.) [C]

36 A ship is sailing northwards at a constant speed of 20 km per hour. At noon, when the ship is 10 km due east of a shore station A, a helicopter takes off from A to intercept the ship. There is a wind of speed 65 km per hour blowing from the direction $\alpha$ E of S, where $\tan \alpha = 5/12$. The helicopter flies with a constant speed of 80 km per hour relative to the air in a constant direction relative to the air. Show that this direction must be 60° E of S for the helicopter to intercept the ship. Calculate the time (to the nearest tenth of a minute) at which the helicopter reaches the ship.

Calculate also the direction of flight relative to the air and the time of take-off which would enable the helicopter to intercept the ship with the shortest time of flight. [JMB]

37 A ship X, whose maximum speed is 30 km h$^{-1}$, is due south of a ship Y which is sailing at a constant speed of 36 km h$^{-1}$ on a fixed course of 150°. If X sails immediately to intercept Y as soon as possible, find the required course.

If initially the ships are 50 km apart, find to the nearest minute how long it takes X to reach Y. [L]

38 A ship A is travelling north-east at 30 kilometres per hour and a ship B is travelling north-west at 40 kilometres per hour. At noon B is 20 kilometres due east of A. Given that both ships maintain constant course and speed, calculate the least distance between them and the time (to the nearest minute) at which they are nearest to one another.

Assuming that 30 kilometres per hour is the maximum speed of A, show that A could have intercepted B by changing course at noon, and determine the direction which A should have taken in order to intercept B as early as possible. [JMB]

39 At 0800 hours a ship A is steaming due north at a constant speed of 9 knots and a ship B, which is 10 nautical miles due west of A, is steaming at a constant speed of 15 knots. If the ships maintain their speeds and courses they will collide. Find the direction in which B is steaming.

If, at 0800 hours, ship B changes direction to steam on a course 030°, find graphically, or otherwise,
(a) the distance between the ships when they are closest to each other
(b) the time when they are closest to each other. [L]

40 A ship A is steaming due south at 20 knots and another ship B is steaming due east at 15 knots. At noon they are each 50 nautical miles from the point of intersection of their courses and are both steaming towards that point. Show that if the ships maintain their respective courses and speeds the least distance between the ships will be 10 nautical miles, and find the time at which they will be in that position.
(1 knot = 1 nautical mile per hour.) [L]

**41** An aircraft carrier A is at a point O at 0700 hours and is steaming south at 30 knots. An enemy battleship B is 8 sea-miles to the east and appears to an observer on the carrier to be moving north-west at $30\sqrt{2}$ knots. Determine the true velocity of B and the bearing of A from B when they are closest together.

If A is liable to be hit by a shot fired from B when the two ships are not more than $5\sqrt{2}$ sea-miles apart, prove that A is within range from 0702 hours until 0714 hours. [AEB]

**42** The ship A is moving along a course of 150° (S 30°E) with speed 10 knots. The ship B is moving along a course of 260° (S 80°W) with speed 14 knots. At noon the ship A is 5 nautical miles due west of the ship B. By a graphical method, or otherwise, find
(i) the magnitude and direction of the velocity of A relative to B
(ii) the time when the two ships are closest together.

If visibility is limited to 3 nautical miles, find the period of time for which the ships remain in visual contact. [AEB]

**43** A river flows at $4 \, \mathrm{m\,s}^{-1}$ between parallel banks, which are 50 m apart. A boat starts from a point A on one bank and moves with constant speed along the straight line AB where B is on the other bank directly opposite A. If the boat takes 10 seconds to cross the river, find, graphically or otherwise,
(i) the speed of the boat relative to the water
(ii) the angle between AB and the direction in which the boat is steered.

A floating ball is moving at the speed of the current parallel to the banks and halfway between them. If the ball and the points A and B all lie in the same line at the instant the boat leaves A, find by further drawing, or otherwise,
(iii) the magnitude and the direction of the velocity of the ball relative to the boat
(iv) the shortest distance between the boat and the ball. [AEB]

**44** A boat can move in still water at $5 \, \mathrm{m\,s}^{-1}$. A point A is situated on the southern bank of a river which is flowing steadily at $2 \, \mathrm{m\,s}^{-1}$ from east to west. A buoy, moored in the river at the point B, is 1000 m from A on a bearing of 030°. Using a graphical method, or otherwise, find the least time which would be required for the boat to move from A to B and the direction in which the boat should be headed to achieve this.

The boat returns to the southern bank from B by the straight line path which requires the *least* time to complete the crossing and lands at the point C. Find the least time and the distance AC.

A car is moving eastwards at $10 \, \mathrm{m\,s}^{-1}$ along a road running parallel to the southern bank during the return journey of the boat from B to C. What is the magnitude and the direction of the velocity of the boat relative to the car? [AEB]

**45** (a) The helmsman of a ferry boat moving due south at 6 knots observes that the wind appears to blow from due west. He doubles the speed of his boat and observes that the wind now appears to blow from the south-west. Find the true speed and direction of the wind.

(b) At a particular instant when a small ship is moving due south at 12 knots a dinghy is 2 nautical miles south-west from the ship. The dinghy has an outboard motor and is moving south-eastwards at 3 knots. Assuming that the ship and the dinghy continue to move with these velocities, and assuming that the water is still and there is no wind, find the minimum distance apart.

(*Graphical solutions will be accepted.*) [AEB]

**46** The wind is blowing steadily at $w$ km h$^{-1}$ from $\alpha°$ west of north. To a cyclist travelling due east at 10 km h$^{-1}$ the wind appears to blow from due north. Sketch a velocity triangle to show this. Name what each side in your triangle represents and mark the direction of each vector with a clear arrow.

The cyclist increases his speed to 15 km h$^{-1}$ and the wind appears to blow from 30° east of north. Sketch a second velocity triangle to show this and mark it clearly.

Find, graphically or otherwise, the values of $w$ and $\alpha$.

Find the apparent speed and the apparent direction of the wind to a motorist who is travelling due north at 60 km h$^{-1}$. [AEB]

**47** The equation of the path of a particle P is

$$\mathbf{r} = \mathbf{i}t + \mathbf{k}t^2,$$

where $t$ is the time. Show that the acceleration of P is constant.

The velocity of another particle Q relative to P is $(\mathbf{i} - \mathbf{j})$ and when $t = 0$, $\overrightarrow{PQ} = \mathbf{j}$. Find the equation of the path of Q and the time at which Q is nearest to P. [L]

**48** The velocity vectors of the particles $P_1$, $P_2$ are

$$u_1\mathbf{i} + v_1\mathbf{j}, \quad u_2\mathbf{i} + v_2\mathbf{j}$$

respectively. Their relative velocity has the same magnitude as that of the velocity of $P_1$. If the velocity of one particle is reversed, the magnitude of the relative velocity is doubled. Find the ratio of the speeds of $P_1$ and $P_2$ and the sine of the angle between their directions. [L]

**49** The acceleration $\mathbf{a}$ of a particle A moving in the $xy$-plane is given by $\mathbf{a} = 2\mathbf{i}$, where distances are measured in metres and time in seconds. Given that the particle starts with a velocity $\mathbf{i} + \mathbf{j}$ from the origin at time $t = 0$, find the velocity vector and position vector of the particle at time $t$. Show that the particle moves along the curve $x = y^2 + y$.

A particle B moves with constant velocity in the direction $\mathbf{j}$ and is at the point $6\mathbf{i} - 4\mathbf{j}$ at $t = 1$. If the particles A and B collide, find the speed of particle B and the time of collision. [AEB]

**50** Two particles A and B move so that, at time $t(\geqslant 0)$, their respective position vectors are $(a \cos \omega t)\mathbf{j} + (a \sin \omega t)\mathbf{k}$ and $(at^2 - \frac{1}{4}a)\mathbf{i} + (2at)\mathbf{j}$, where $a$ and $\omega$ are positive constants. Find the cartesian equations of each path. Show that the paths intersect and that the least value of $\omega$ for which the particles collide is $4\pi$.

Given that $\omega = 4\pi$, find the set of values of $t$ for which the velocities of A and B are perpendicular and find also the cosine of the angle between the velocities when $t = \frac{1}{3}$. [L]

**51** The position vectors of two particles A and B at time $t$ are

$$\text{A:} \quad \mathbf{r} = (a \cos \omega t)\mathbf{i} + (a \sin \omega t)\mathbf{j},$$
$$\text{B:} \quad \mathbf{r} = a(1 - \omega^2 t^2)\mathbf{i} - (a\,\omega^2 t^2)\mathbf{j},$$

where $a$ and $\omega$ are constants. Describe the nature of their paths and give their equations in cartesian form. Find the position vectors of the points in which the paths intersect.

Find the times at which B crosses the path of A. Find also the values of $t$ when the particles are moving in directions which are (a) parallel (b) perpendicular, to each other. [L]

**52** Sketch the curves

$$\mathbf{r} = (2a \cos \theta)\mathbf{i} + (2a \sin \theta)\mathbf{j},$$
$$\mathbf{r} = 3ap^2\mathbf{i} + 3ap\mathbf{j}.$$

Find the position vectors of the points in which they meet.

A particle describes the first curve in such a way that $\theta = \omega t$, where $\omega$ is constant and $t$ is the time. Another particle describes the second curve in such a way that $p = kt$, where $k$ is constant. Show that the acceleration of each particle is constant in magnitude and find the times at which these accelerations are at right angles. [L]

**53** Two particles A and B have position vectors $(2 \sin \omega t)\mathbf{i} + (2 \cos \omega t)\mathbf{j}$, where $\omega > 0$, and $2t\mathbf{i} + t^2\mathbf{j}$ respectively at time $t$. Find the cartesian equations of the paths followed by A and B.
Find
(a) the magnitude of the velocity of A relative to B when $t = 0$
(b) the magnitude of the acceleration of A relative to B when $t = \pi/(2\omega)$.
Find also the values of $t$ for which
(c) the accelerations of A and B are parallel and in the same sense
(d) the accelerations of A and B are parallel and in opposite senses. [L]

**54** At time $t = 0$ a particle A is at the point $3\mathbf{i} - 2\mathbf{j} - 4\mathbf{k}$, where $\mathbf{i}, \mathbf{j}$ and $\mathbf{k}$ are unit vectors of magnitude one metre in the directions of the cartesian axes $Ox, Oy, Oz$ respectively. The particle A moves with constant acceleration $2\mathbf{j} - 2\mathbf{k}$ and has a velocity $4\mathbf{i} - 3\mathbf{j} + 4\mathbf{k}$ at time $t = 0$. A second particle B moves with constant velocity $-\mathbf{i} + 4\mathbf{j} + 3\mathbf{k}$ and at time $t = 1$ is at the point $7\mathbf{i} + 2\mathbf{j}$.
Find (i) the position of each particle at time $t$

(ii) the position of B when the particles are moving at right angles to each other.

Show that the particle A just reaches the plane $z = 0$.      [AEB]

55 A particle is launched at time $t = 0$ so that it follows the path

$$\mathbf{r} = (-15 + 5t)\mathbf{i} + (70 + 30t - 5t^2)\mathbf{k}.$$

(Distances are measured in metres and time in seconds.)

Find

(i) the position vector of the point of projection and the velocity vector of the particle at time $t = 0$.

(ii) the speed of the particle at time $t$.

The particle meets the line through the point with position vector $5\mathbf{i}$ and gradient $\mathbf{i} + 19\mathbf{k}$ at the point A. Show that this occurs when $t = 5$. Find the position vector of the point A and its distance from the point of projection.

Find also the time at which the particle is moving at right angles to its initial direction of motion.      [AEB]

56 A particle is projected from ground level with speed $V$ at an angle of elevation $\theta$, and it moves freely under gravity. Find expressions for the greatest height reached and the distance between the point of projection and the point where the particle returns to ground level.

The particle is at a height $h$ or more above ground level for exactly half the total time of flight. Show that $h$ is three-quarters of the greatest height reached.

If $\theta = 45°$ find the magnitude of the velocity (in terms of $V$) and its inclination to the horizontal when the height of the particle is one-half of the greatest height.      [C]

57 A particle is projected from a point A on horizontal ground with speed $V$ at an angle of elevation $\alpha$. The particle attains a maximum vertical height of $h$ above A and it hits the ground again at B, where $AB = R$. Show that

(i) $R = \dfrac{V^2 \sin 2\alpha}{g}$

(ii) $\tan \alpha = 4h/R$

(iii) $V^2 = g(R^2 + 16h^2)/(8h)$.

If $R = 2h\sqrt{2}$, calculate the angle which the path of the particle makes with AB at time $\sqrt{(h/g)}$ after projection.      [AEB]

58 If a particle is projected with speed $u$ at an angle of elevation $\alpha$, show that the horizontal range is $u^2 \sin 2\alpha/g$ and the maximum height attained is $u^2 \sin^2 \alpha/(2g)$.

A golf ball is struck so that it leaves a point A on the ground with speed $49 \text{ m s}^{-1}$ at an angle of elevation $\alpha$. If it lands on the green which is the same level as A, the nearest and furthest points of which are 196 m and 245 m respectively from A, find the set of possible values of $\alpha$. Find also the maximum height the ball can reach and still land on the green.

There is a tree at a horizontal distance 24.5 m from A and to reach the green the ball must pass over this tree. Find the maximum height of the tree if this ball can reach any point on the green.

(Assume the point A, the green and the base of the tree to be in the same horizontal plane.) [AEB]

(Take $g$ to be $9.8 \text{ m s}^{-2}$.)

**59** Two particles, P and Q, are projected simultaneously under gravity from a point O with equal speeds $V$, at angles $\alpha$ and $90° - \alpha$ above the horizontal respectively, where $\alpha < 45°$. The particles move in the same vertical plane which contains the horizontal and upward vertical axes $Ox$ and $Oy$ respectively. Find the coordinates of the positions of the two particles at time $t$ after projection. Show that the line joining these positions makes an angle of $135°$ with $Ox$.

State the coordinates of P when it is at the highest point of its path. Show that, at this instant, the distance between P and Q is

$$\frac{V^2 \sin \alpha}{g}(\cos \alpha - \sin \alpha)\sqrt{2}.$$

Show further that this distance is greatest when $\alpha = 22\frac{1}{2}°$. [AEB]

**60** A particle is projected with speed $u$ at an angle of elevation $\alpha$, where $\tan \alpha = 3$. Find, in terms of $u$ and $g$, the height of the particle when its speed is $\frac{1}{2}u$. Find also the directions in which the particle is moving at this height.

If the velocity of the particle makes an angle of $45°$ with the horizontal at times $t_1$ and $t_2$ (where $t_2 > t_1$) after projection, show that $t_2 = 2t_1$. [L]

**61** A particle is to be projected under gravity from O, the centre of a horizontal circle of radius 70 m. Calculate the minimum velocity of projection of the particle if its trajectory is to pass through the circumference of the circle.

Another particle is projected from O with a velocity of $108 \text{ km h}^{-1}$ at an angle of $40°$ to the horizontal. This particle just clears the top of a vertical post on the circumference of the circle. Find, to the nearest tenth of a metre, the height of the post.

(Take $g$ as $9.81 \text{ m s}^{-2}$.) [L]

**62** A missile is launched under gravity from a point O and passes through the points A($6h$, $2h$) and B($12h$, $3h$) referred to horizontal and upward vertical axes OX and OY respectively. If the velocity of projection is $V$ at an angle $\theta$ to the horizontal, show that

(i) $\tan \theta = 5/12$

(ii) $2V = 13\sqrt{(gh)}$.

Find the direction of the missile after it has travelled from O for a time $2\sqrt{(h/g)}$. [AEB]

**63** Two particles, P and Q are simultaneously projected from a point O with the same speed but at different angles of elevation, and they both pass through a point C which is at a horizontal distance $2h$ from O and at a height $h$ above the level of O. The particle P is projected at an angle $\tan^{-1} 2$ above the horizontal. Find the speed of projection in terms of $g$ and $h$, and show that Q is projected at an angle $\tan^{-1}(\frac{4}{3})$ to the horizontal.

Show that the time interval between the arrivals of the two particles at C is

$$(3 - \sqrt{5})\sqrt{\left(\frac{2h}{3g}\right)}.$$

[JMB]

64 A particle is projected from a point A at an angle of elevation $\alpha$. The particle passes through a point B when travelling upwards in a direction which is at an angle $\beta$ to the horizontal. If AB makes an angle $\theta$ with the horizontal, show that

$$2 \tan \theta = \tan \alpha + \tan \beta.$$

A particle is projected from a point A. The particle passes through a point C, whose horizontal and vertical distances from A are respectively 80 m and 20 m, when travelling downwards at an angle $\tan^{-1}(5/6)$ to the horizontal. Find the speed and angle of projection.

Find also the speed with which another particle must be projected from A, if C is the highest point of its trajectory. [L]
(Take $g$ as $10 \, \mathrm{m\,s^{-2}}$.)

65 A particle is projected at an angle of elevation $\alpha$ from a point O on a horizontal plane. The particle passes through a point P when travelling upwards at an angle $\theta$ to the horizontal. If the line PO is at an angle $\beta$ to the horizontal, show that $\tan \alpha$, $\tan \beta$ and $\tan \theta$ are in arithmetic progression.

The particle passes through points $P_1$ and $P_2$ for which $\beta = \beta_1$ and $\beta = \beta_2$ respectively. When passing through $P_1$ the particle is travelling upwards at an angle $\phi$ to the horizontal and when passing through $P_2$ is travelling downwards at an angle $\phi$ to the horizontal. If $\tan \phi = \frac{3}{4}$ and $2 \tan \beta_2 = \tan \beta_1$, show that $\tan \alpha = 2\frac{1}{4}$. [AEB]

66 If a projectile has an initial velocity $V$ at an angle $\theta$ to the horizontal, show that the equation of its trajectory can be expressed in the form

$$y = x \tan \theta - \frac{gx^2 \sec^2 \theta}{2V^2}.$$

Two particles are projected in the same vertical plane from the point O with the same speed $\sqrt{(ag)}$ in directions making acute angles $\alpha$ and $\beta$ with the horizontal. If their paths cross at the point P, show that the horizontal distance from O to P is $2a/(\tan \alpha + \tan \beta)$.

Show that P is at a higher level than O if $\tan \alpha \tan \beta > 1$. [L]

67 A particle is projected from a point O with speed $V$ and angle of elevation $\theta$, and it moves freely under gravity. Show that, referred to horizontal and vertical axes through O, the equation of the trajectory is

$$y = x \tan \theta - \frac{gx^2}{2V^2} \sec^2 \theta.$$

The particle passes through the points $(a, 2h)$ and $(2a, 3h)$. Show that $5h = 2a \tan \theta$, and that the particle also passes through the point $(5a, 0)$.

[C]

**68** A particle is projected from a point O with speed $V$ at an angle of elevation $\alpha$. Prove that the equation of its path referred to horizontal and vertical axes $Ox$, $Oy$ is

$$y = nx - \frac{gx^2}{2V^2}(1 + n^2),$$

where $n = \tan \alpha$.

There is a vertical wall perpendicular to $Ox$ at a distance $b$ from O and the particle hits it at a height $h$ above the level of O. Show that, for a fixed value of $V$, $h$ is maximum when $\alpha$ is chosen so that $ngb = V^2$, and that in this case the particle is travelling in a direction making an angle $\alpha$ with the wall when it hits it. Find the speed of the particle at the moment of impact in terms of $V$, $b$ and $g$.  [JMB]

**69** A particle is projected from a point O on horizontal ground with speed $V$ at an angle of elevation $\theta$, and passes through a point P which is a horizontal distance $x$ from O and at a height $y$ above the ground. Prove that

$$y = x \tan \theta - \frac{gx^2(1 + \tan^2 \theta)}{2V^2}.$$

A ball is thrown with a speed of $10 \text{ m s}^{-1}$ from the top of a tower which is $15 \text{ m}$ high. The ball strikes the ground at the same level as the base of the tower and at a horizontal distance of $20 \text{ m}$ from the point where the ball was thrown.

Calculate

(i) the angle of elevation at which the ball was thrown

(ii) the time of flight of the ball

(iii) the direction in which the ball is moving at the instant when it strikes the ground.

(Take the acceleration due to gravity to be $10 \text{ m s}^{-2}$.)  [AEB]

**70** A particle is projected from a point O with a velocity $u$ at an angle $\alpha$ above the horizontal. Show that the equation of its path can be written in the form

$$y = -\tfrac{1}{2}\frac{g \sec^2 \alpha}{u^2}\left(x - \frac{u^2}{2g} \sin 2\alpha\right)^2 + \frac{u^2}{2g} \sin^2 \alpha,$$

where O is the origin and the $y$-axis is vertically upwards. Hence, or otherwise, determine, in terms of $u$, $\alpha$ and $g$, the coordinates $(b, h)$ of the highest point of the path, and show that $\tan \alpha = 2h/b$. In the case when $\alpha \neq \tfrac{1}{4}\pi$ find a second angle of projection such that a particle projected with the same initial speed from O attains its maximum height when $x = b$, and express this maximum height in terms of $h$ and $b$.  [JMB]

**71** A particle is projected with speed $u$ at an angle of elevation $\alpha$. After travelling a horizontal distance $x$ the particle is at a height $y$. Express $x$ and $y$ in terms of $t$, the time from the moment of projection, and deduce that

$$\frac{2u^2 \cos^2 \alpha}{g} = \frac{x^2}{x \tan \alpha - y}.$$

A small ball is struck from a point A on level ground and passes at a height of 2 metres over a point on the ground at a distance of 12 metres from A. The ball is caught at a height of 1 metre by a boy who is 24 metres from A. Find the speed and direction of motion of the ball as it leaves A. Find also the speed of the ball when it is caught by the boy.
(Take $g$ to be $10 \text{ m s}^{-2}$.)                                                                [L]

72  A particle is projected upwards with a speed of $20 \text{ m s}^{-1}$ at an angle $\theta$ to the horizontal from a point which is situated at a height $h$ metres above a horizontal plane. The particle takes $t$ seconds to reach the plane. Taking the acceleration due to gravity to be $10 \text{ m s}^{-2}$, show that $t$ is the positive root of the equation $5t^2 - 20t \sin \theta - h = 0$. The particle is moving horizontally after 1 second, and the total time of flight is 3 seconds. Calculate the value of $\theta$ in degrees and show that $h = 15$. Hence calculate the total horizontal distance travelled by the particle and find the angle at which the particle strikes the plane. Find when the particle is moving at $45°$ to the horizontal.                                                                [AEB]

73  A projectile is fired with an initial velocity of magnitude $V$ inclined at an angle $\alpha$ above the horizontal. Find the equation of the trajectory referred to horizontal and vertical axes through the point of projection.

A projectile is fired horizontally from a point O, which is at the top of a cliff, so as to hit a fixed target in the water, and it is observed that the time of flight is $T$. It is found that, with the same initial speed, the target can also be hit by firing at an angle $\alpha$ above the horizontal. Show that the distance of the target from the point at sea-level vertically below O is $\frac{1}{2}gT^2 \tan \alpha$.
                                                                [C]

74  A particle is projected with velocity $V$ and elevation $\alpha$ from a point O. Show that the equation of the path of the particle, referred to horizontal and vertical axes $Ox$ and $Oy$ respectively in the plane of the path, is

$$y = x \tan \alpha - gx^2(1 + \tan^2 \alpha)/(2V^2).$$

A particle P is projected with velocity $70 \text{ m s}^{-1}$ from the top of a vertical tower, of height 40 m, standing on a horizontal plane. The particle strikes the plane at a distance 200 m from the foot of the tower. Find the two possible angles of projection.
If these angles are $\alpha_1$, $\alpha_2$ where $\alpha_1 > \alpha_2$, calculate
(a) the greatest height correct to the nearest metre above the top of the tower when P is projected at inclination $\alpha_1$
(b) the time of flight, correct to the nearest tenth of a second, when P is projected at inclination $\alpha_2$. (Take $g$ as $9.8 \text{ m s}^{-2}$.)                                [L]

**75** (i) Particles are projected from a fixed point O with speed $u$ at different angles of elevation. All the particles move in the same vertical plane $Oxy$, where $Ox$ is horizontal and $Oy$ is vertically upwards.

Find the coordinates of the highest point reached by a particle which is projected at an angle of elevation $\alpha$. Prove that the highest points of the paths of all the particles lie on the curve with equation

$$gx^2 + 4gy^2 - 2u^2 y = 0.$$

(ii) Particles are projected simultaneously from a fixed point in different directions but in the same vertical plane and with the same speed $V$. Prove that at time $t$ after projection the positions of all the particles will lie on the circumference of a circle of radius $Vt$. [AEB]

**76** A tennis ball, struck at ground level, leaves the ground with velocity $\sqrt{(2ga)}$ at an angle $\theta$ to the horizontal and strikes a vertical wall at a horizontal distance $b(\leqslant 2a)$ from the point of projection. Prove that the height $h$ of the point of impact up the wall is given by

$$h = b \tan \theta - \frac{b^2}{4a}(1 + \tan^2 \theta).$$

Show that the greatest height which can be reached up the wall by varying $\theta$ is $a - \dfrac{b^2}{4a}$. Show also that in this case when the particle hits the wall it is moving in a direction making an angle $\theta$ with the wall. [AEB]

**77** A particle is projected at an angle $\alpha$ to the horizontal from a point A on a plane inclined at an angle $\arctan(1/2)$ to the horizontal. The particle moves in a vertical plane through a line of greatest slope and strikes the plane again at a point B higher up the plane. Find $\tan \alpha$ if

(a) the particle is moving horizontally at B

(b) the particle strikes the plane normally at B. [L]

**78** An inclined plane makes an angle $\beta$ with the horizontal. A particle is projected up the slope from a point on the plane with a velocity $V$ at an angle $\alpha$ to the plane, the direction of projection being in the vertical plane containing a line of greatest slope. Show that the time of flight is given by $2V \sin \alpha/(g \cos \beta)$ and find the range of the particle along the plane.

If the particle is travelling horizontally when it strikes the plane, show that

$$\tan \alpha = \frac{\sin \beta \cos \beta}{2 - \cos^2 \beta}.$$ [AEB]

**79** A particle is projected under gravity from a point A on an inclined plane which makes an angle $\alpha$ with the horizontal. The velocity of projection is $V$ at an angle $\tan^{-1}(\frac{1}{2})$ to an upward line of greatest slope of the plane. The motion takes place in the vertical plane through a line of greatest slope and the particle hits the plane at the point B which is further up the plane than

A. Find the time taken for the particle to travel between the points A and B in terms of $V$, $g$ and $\alpha$.

Find also the possible values of tan $\alpha$ if the particle is moving at an angle $\alpha$ to the horizontal when it strikes the plane at B.          [AEB]

80  A particle is projected with speed $u$ from a point on a plane which is inclined at an angle $\beta$ to the horizontal. The particle is projected at an angle $\alpha$ to the horizontal in a vertical plane through a line of greatest slope of the plane. Show that the range up the plane is

$$\frac{2u^2 \sin (\alpha - \beta) \cos \alpha}{g \cos^2 \beta}.$$

Deduce that the range up the plane is a maximum when $\alpha = \frac{1}{4}\pi + \frac{1}{2}\beta$.

If the *maximum* range down the plane is twice the maximum range up the plane, find the angle $\beta$.          [L]

# 4 Statics of particles

## 4.1 Force

In Chapter 3 the motion of a point in two dimensions was discussed without any consideration being given to the means by which this motion was produced. The next two chapters introduce 'Newtonian mechanics' i.e. a study of the physical world as governed by Newton's Laws of Motion, formulated by Isaac Newton in the seventeenth century.

For simplification the concept of a 'particle' is used. A particle by definition is a body whose size is negligible, so that it may be regarded as located at a point.

The mass of a particle is the quantity of matter that it contains. Mass is a scalar quantity. The unit of mass is the kilogram, the abbreviation for which is kg. The kilogram is defined to be the mass of the standard cylinder of platinum kept at Sèvres.

Newton's first law states that

Every particle continues in a state of rest or of motion with constant speed in a straight line unless it is acted upon by a force.

This law, in effect, defines a force as 'that which changes, or tends to change, the state of rest or the uniform motion of a particle'.

Newton's second law introduces the concept of the momentum of a particle which is defined as the product of the mass, $m$, of the particle and its velocity, $\mathbf{v}$. Velocity is a vector quantity and so momentum, being the product of a scalar quantity and a vector quantity, is a vector quantity. A statement of Newton's second law is

The rate of change of the momentum of a particle is proportional to the force acting on the particle and is in the same direction as this force.

In effect, this law states

$$\mathbf{F} = \frac{\mathrm{d}}{\mathrm{d}t}(m\mathbf{v}),$$

with a suitable choice of units.

Only particles of constant mass will be considered. An example of variable mass is a rocket. When the mass is constant

$$\mathbf{F} = m\frac{\mathrm{d}\mathbf{v}}{\mathrm{d}t}$$

or
$$\mathbf{F} = m\mathbf{a}$$

where **a** is the acceleration of the particle. If **a** has magnitude 1 metre/second$^2$ and the particle has mass 1 kilogram, then the magnitude of **F** is 1 newton, the unit of force. A newton is that force which gives to a particle of mass 1 kg an acceleration of 1 $m\,s^{-2}$. The abbreviation for newton is N. The abbreviation $m\,s^{-2}$ has been used for metre per second per second.

Clearly since displacement is, by definition, a vector quantity, then velocity, as the derivative of displacement, is also a vector and hence acceleration is a vector. Thus force is a vector quantity. When representing a force by a letter bold type will be used, e.g. **F**, as with other vector quantities. The italic letter $F$ will then represent the magnitude of the vector **F**, i.e. $|\mathbf{F}|$.

Unlike velocity, which is specified by a knowledge of its magnitude and direction and so corresponds to a 'free vector', a force is only completely specified when the line along which it acts is known, i.e. a force corresponds to a 'localised vector' (known sometimes as a 'sliding' vector since a force may be regarded as acting at any point in its line of action).

## 4.2 Combination of two forces

Given two forces **P** and **Q**, each acting on a particle O, the single force **R** equivalent to these two forces can be found in two ways (Fig. 4.1). This force **R** which is equivalent to the two forces **P** and **Q** is known as the resultant of **P** and **Q**.

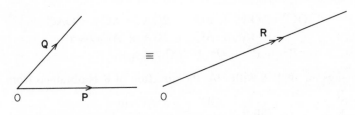

Fig. 4.1

(a) The first method is to draw the parallelogram OACB (Fig. 4.2) in which $\overrightarrow{OA}$ completely represents **P**, i.e. $\overrightarrow{OA}$ represents **P** in magnitude, direction and line of action, and $\overrightarrow{OB}$ similarly completely represents **Q**.

Then $\overrightarrow{OC}$ completely represents **R**, the resultant of **P** and **Q**.

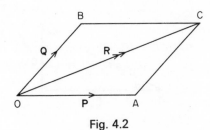

Fig. 4.2

(b) The second method is to draw the triangle OAC (Fig. 4.3) in which $\overrightarrow{OA}$ represents **P** in magnitude and direction only and similarly $\overrightarrow{AC}$ represents **Q** in magnitude and direction only. Then $\overrightarrow{OC}$ represents **R**, the resultant of **P** and **Q**, in magnitude and direction only.

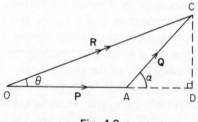

Fig. 4.3

Since the lines of action of **P** and **Q** meet at O, then **R** must act through O and so the magnitude and direction of **R** and also its line of action are known.

Thus in this second method the magnitude and direction of the resultant force are determined in one stage and then the line of action of the resultant force is considered at the next stage. This approach is frequently used in more complex situations.

With $R = |\mathbf{R}|$, $P = |\mathbf{P}|$ and $Q = |\mathbf{Q}|$, then, using the cosine rule on the triangle OAC (Fig. 4.3),

$$OC^2 = OA^2 + AC^2 - 2OA \times AC \cos OAC$$
$$= OA^2 + AC^2 + 2OA \times AC \cos \alpha$$

giving $$R^2 = P^2 + Q^2 + 2PQ \cos \alpha.$$

If OC makes an angle $\theta$ with OA, the direction of **R** is obtained from

$$\tan \theta = \frac{CD}{OD} = \frac{AC \sin \alpha}{OA + AC \cos \alpha}$$

$$= \frac{Q \sin \alpha}{P + Q \cos \alpha}.$$

An important special case is when $\alpha = 90°$ (Fig. 4.4).

Then $R^2 = P^2 + Q^2$ and $\tan \theta = \dfrac{Q}{P}$.

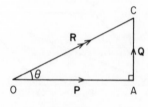

Fig. 4.4

*Example 1*

Forces of magnitude 4 N and 3 N act along the sides AB and AD respectively of a square ABCD. Find the magnitude and direction of their resultant.

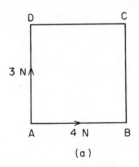

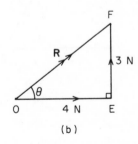

(a)                              (b)

Fig. 4.5

Figure 4.5(a) shows the square ABCD and the two forces along AB and AD. Figure 4.5(b) shows the force diagram with $\overrightarrow{OE}$ and $\overrightarrow{EF}$ representing the forces of 4 N and 3 N respectively and $\overrightarrow{OF}$ representing the resultant **R**.

Then   $R^2 = 4^2 + 3^2$, giving $R = 5$ N

and   $\tan \theta = \dfrac{3}{4}$, giving $\theta = 36° 52'$,

i.e. the resultant is of magnitude 5 N and acts in a direction at an angle of 36°52′ to AB through A, the point of intersection of the two forces.

   Alternatively **i** and **j** could be taken to be unit vectors of magnitude 1 N in the directions $\overrightarrow{AB}$ and $\overrightarrow{AD}$ respectively.

Then the force along AB = 4**i**

and   the force along AD = 3**j**

giving                    **R** = 4**i** + 3**j**

and $R^2 = |R|^2 = 4^2 + 3^2$, i.e. $R = 5$ N

and $\tan \theta = \dfrac{3}{4}$, i.e. $\theta = 36° 52'$.

*Example 2*

Forces of magnitude 5 N and 6 N act along the sides AB and BC of an equilateral triangle ABC. Find the magnitude and direction of their resultant.

Figure 4.6(a) shows the triangle ABC and the two forces along AB and BC. Figure 4.6(b) shows the force diagram with $\overrightarrow{OD}$ and $\overrightarrow{DE}$ representing the forces of 5 N and 6 N respectively and $\overrightarrow{OE}$ representing the resultant **R**.

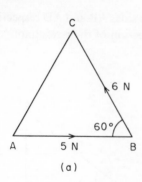

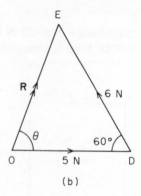

(a)                                        (b)

Fig. 4.6

Then from Fig. 4.6(b)  $R^2 = 5^2 + 6^2 - 2(5)(6) \cos 60°$
$$= 61 - 60(\tfrac{1}{2}) = 31, \text{ giving } R = \sqrt{31},$$

and  $\dfrac{R}{\sin 60°} = \dfrac{6}{\sin \theta}$, i.e. $\sin \theta = \dfrac{6 \sin 60°}{\sqrt{31}}$, giving $\theta = 68° 57'$,

i.e. the resultant has magnitude $\sqrt{31}$ N and acts in a direction at an angle of 68° 57′ to AB through B, the point of intersection of the two forces.

### Exercises 4.2
1  Forces **P** and **Q** act in the plane of a square ABCD. Find the magnitude, direction and point of application of the resultant of **P** and **Q** when
  (a) $P = 6$ N and **P** acts along AB, $Q = 8$ N and **Q** acts along AD
  (b) $P = 12$ N and **P** acts along AB, $Q = 16$ N and **Q** acts along BC
  (c) $P = 5$ N and **P** acts along CD, $Q = 12$ N and **Q** acts along BC
  (d) $P = 24$ N and **P** acts along AD, $Q = 7$ N and **Q** acts along DC
  (e) $P = 24$ N and **P** acts along CB, $Q = 10$ N and **Q** acts along CD
  (f) $P = 10$ N and **P** acts along AB, $Q = 12$ N and **Q** acts along AC
  (g) $P = 8$ N and **P** acts along AD, $Q = 15$ N and **Q** acts along BD
  (h) $P = 20$ N and **P** acts along AC, $Q = 16$ N and **Q** acts along DB.
2  ABCDEF is a regular hexagon. Find the magnitude, direction and line of action of the resultant of forces of
  (a) 8 N along AB and 10 N along AF
  (b) 6 N along AD and 8 N along BF
  (c) 10 N along AC and 10 N along AE
  (d) 12 N along BD and 6 N along ED
  (e) 20 N along BE and 10 N along CF
  (f) 30 N along CB and 20 N along CD
  (g) 40 N along AE and 30 N along ED.

## 4.3 Resolution of forces

In the last section, the resultant of two forces was found, i.e. two forces were reduced to one equivalent force. It is also possible, and often useful, to replace one force by two equivalent forces.

Consider a force **P** acting at O and at an angle $\alpha$ to Ox (Fig. 4.7). Force **P** is represented by $\overrightarrow{OC}$ and $\overrightarrow{OC} = \overrightarrow{OA} + \overrightarrow{AC}$.

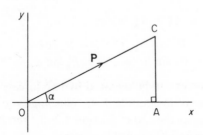

Fig. 4.7

$$OA = OC \cos \alpha \quad \text{and} \quad AC = OC \cos (90° - \alpha) = OC \sin \alpha$$

$$\therefore \qquad \overrightarrow{OC} = OC \cos \alpha \mathbf{i} + OC \sin \alpha \mathbf{j}$$

where **i** and **j** are unit vectors along OA and Oy respectively.
Since OC represents $P$, the magnitude of **P**,

$$\mathbf{P} = P \cos \alpha \mathbf{i} + P \sin \alpha \mathbf{j}$$

i.e. a force of magnitude $P$ acting at O at an angle $\alpha$ to Ox can be replaced by two forces, $P \cos \alpha \mathbf{i}$ along Ox and $P \sin \alpha \mathbf{j}$ at right angles to Ox. The force **P** is then said to have been resolved into two components, one of magnitude $P \cos \alpha$ along Ox and the other of magnitude $P \sin \alpha$ at right angles to Ox.

To find the component of a force in a particular direction, the magnitude of the force is multiplied by the cosine of the angle between the direction and the line of action of the force.

This method of resolving a force into two components at right angles to each other is much used in mechanics. For instance, the method can be used to find the resultant of several forces acting on a particle.

Consider a system of coplanar forces acting on a particle at the point O. Let the system consist of $n$ forces, $\mathbf{P}_1, \mathbf{P}_2, \ldots, \mathbf{P}_n$ with magnitudes $P_1, P_2, \ldots, P_n$ and acting at angles $\alpha_1, \alpha_2, \ldots, \alpha_n$ to the line Ox.

Then
$$\mathbf{P}_1 = P_1 \cos \alpha_1 \mathbf{i} + P_1 \sin \alpha_1 \mathbf{j}, \qquad \text{where } \mathbf{i}, \mathbf{j} \text{ are unit vectors along Ox and Oy respectively,}$$

$$\mathbf{P}_2 = P_2 \cos \alpha_2 \mathbf{i} + P_2 \sin \alpha_2 \mathbf{j}$$

$$\dotfill$$

$$\dotfill$$

$$\mathbf{P}_n = P_n \cos \alpha_n \mathbf{i} + P_n \sin \alpha_n \mathbf{j}$$

---

$$\mathbf{P}_1 + \mathbf{P}_2 + \ldots + \mathbf{P}_n = (P_1 \cos \alpha_1 + \ldots + P_n \cos \alpha_n)\mathbf{i}$$
$$+ (P_1 \sin \alpha_1 + \ldots + P_n \sin \alpha_n)\mathbf{j}$$
$$= X\mathbf{i} + Y\mathbf{j}, \text{ where}$$

$$X = P_1 \cos \alpha_1 + P_2 \cos \alpha_2 + \ldots + P_n \cos \alpha_n$$
and $$\quad Y = P_1 \sin \alpha_1 + P_2 \sin \alpha_2 + \ldots + P_n \sin \alpha_n.$$

Thus the system of $n$ forces can be replaced by two forces, one of magnitude $X$ along $Ox$ and the other of magnitude $Y$ at right angles to $Ox$.

Using the special case at the end of Section 4.2, the resultant of $\mathbf{X}$ and $\mathbf{Y}$ can be found and hence the resultant of the system:

$$R^2 = X^2 + Y^2$$
and $$\qquad \tan \theta = \frac{Y}{X},$$

where $\theta$ is the angle made by the resultant with $Ox$.

It is important to note that, if both $X = 0$ and $Y = 0$, then there is no resultant force acting on the particle.

*Example 1*

A particle at the vertex A of a square ABCD is acted upon by forces of magnitudes 3 N, 4 N and 5 N along AB, AC and AD respectively. Find the magnitude of the resultant of this system of forces and also the angle the line of action of this resultant makes with AB.

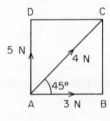

Fig. 4.8

The forces are shown in Fig. 4.8.

Let $X =$ the sum of the components of the three forces along AB,

$\quad Y =$ the sum of the components of the three forces along AD,

and let $\mathbf{i}$ and $\mathbf{j}$ be unit vectors along AB and AD respectively.

Then the force of 3 N along **AB** $= 3\mathbf{i}$
and the force of 4 N along **AC** $= 4 \cos 45°\mathbf{i} + 4 \sin 45°\mathbf{j}$
and the force of 5 N along **AD** $= 5\mathbf{j}$.
Adding we get the resultant force $= (3 + 4 \cos 45°)\mathbf{i} + (4 \cos 45° + 5)\mathbf{j}$.

$\therefore \qquad X = 3 + 4 \cos 45° \approx 5.83 \quad$ and $\quad Y = 4 \cos 45° + 5 \approx 7.83$

and $\qquad R^2 = X^2 + Y^2 = 5.83^2 + 7.83^2$, giving $R \approx 9.76$

and $\qquad \tan \theta = \dfrac{Y}{X} = \dfrac{7.83}{5.83}, \qquad\qquad$ giving $\theta \approx 53°\, 20'$,

i.e. the resultant of the system has magnitude approximately 9.76 N and its line of action is at an angle 53° 20′ to **AB**.

*Example 2*
Forces of magnitude 2 N, 3 N, 4 N, 5 N, 6 N act respectively along the lines AB, AC, AD, AE, AF, where ABCDEF is a regular hexagon. Find the magnitude of the resultant of the system and the angle its line of action makes with AD.

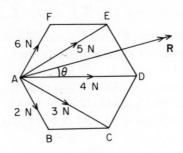

Fig. 4.9

The forces acting are shown in Fig. 4.9.
Let the resultant of the system be **R** with magnitude $R$ and components $X$ along AD and $Y$ at right angles to AD and towards EF.

Then
$$X = 2 \cos 60° + 3 \cos 30° + 4 + 5 \cos 30° + 6 \cos 60° \approx 14.93,$$
and $Y = -2 \cos 30° - 3 \cos 60° + 0 + 5 \cos 60° + 6 \cos 30° \approx 4.46.$

$\therefore \qquad R = X^2 + Y^2 = 14.93^2 + 4.46^2$, giving $R \approx 15.58$

and $\qquad \tan \theta = \dfrac{Y}{X} = \dfrac{4.46}{14.93} = 0.2987, \qquad$ giving $\theta \approx 16°\, 38'.$

Thus the resultant of the system acts at A, the point of intersection of the lines of action of the five forces and has magnitude 15.58 N approximately and acts at an angle of 16° 38′ to AD between EF and AD.

## Exercises 4.3

The triangle ABC is equilateral and M is the mid-point of BC. The following systems of forces act on a particle at A. In each case find the magnitude of the resultant of the system and the angle its line of action makes with AB.

**1**   5 N along AB,  6 N along AM,  8 N along AC.
**2**   10 N along AB,  8 N along AM, 12 N along AC.
**3**   12 N along BA, 10 N along AM, 16 N along AC.
**4**   20 N along AB, 30 N along MA, 20 N along AC.
**5**   40 N along BA, 30 N along MA, 20 N along CA.

The mid-points of the sides BC and CD of a square ABCD are L and M respectively. For each of the following systems of forces find the magnitude of the resultant and the angle its line of action makes with AB.

**6**   2 N along AB,  4 N along AC,  6 N along AD.
**7**   10 N along AB, 20 N along AL, 30 N along AD.
**8**   5 N along BA, 10 N along AL, 20 N along AM.
**9**   6 N along AB,  8 N along AL, 10 N along AC, 12 N along AM, 14 N along AD.
**10**  20 N along BA, 12 N along AL, 16 N along CA, 10 N along DA.
**11**  30 N along BA, 20 N along LA, 10 N along AC, 20 N along MA.

ABCDEF is a regular hexagon. For each of the following systems of forces find the magnitude of the resultant and the angle its line of action makes with AD.

**12**  10 N along AB, 20 N along AD, 30 N along AF.
**13**  6 N along AB, 10 N along AC, 12 N along AD, 16 N along AF.
**14**  10 N along AC, 20 N along AD, 12 N along AE, 20 N along AF.
**15**  20 N along BA, 10 N along AC, 30 N along AD, 12 N along EA, 16 N along FA.

## 4.4   Forces in equilibrium

In Section 4.2 the force **R** was the resultant of two forces **P** and **Q** acting on a particle at a point O. If the two forces **P** and **Q** are added to a third force $-$**R** i.e. a force equal in magnitude but opposite in direction to **R**, a system of three forces **P**, **Q** and $-$**R** is obtained. Since **P** + **Q** = **R**, the resultant of the three forces **P** + **Q** + ($-$**R**) = **0**, i.e. there is no resultant force.

The system of three forces **P**, **Q** and $-$**R** is then said to be in equilibrium. The particle at O is described as being in equilibrium under the action of the three forces **P**, **Q** and $-$**R**. The forces **P** and **Q** have a resultant **R**. When the force $-$**R** is added to **P** and **Q**, the system is reduced to equilibrium. The force $-$**R** is then said to be the equilibrant of the system consisting of forces **P** and **Q**.

### Triangle of forces

Consider a system of three forces $\mathbf{F}_1$, $\mathbf{F}_2$, $\mathbf{F}_3$ acting on a particle and in equilibrium. Then $\mathbf{F}_3 = -(\mathbf{F}_1 + \mathbf{F}_2)$.

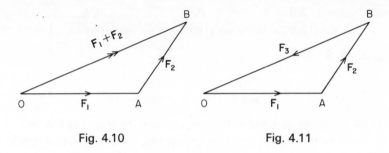

Fig. 4.10            Fig. 4.11

Let $\mathbf{F}_1$ and $\mathbf{F}_2$ be represented by $\overrightarrow{OA}$ and $\overrightarrow{AB}$. Then $(\mathbf{F}_1 + \mathbf{F}_2)$ is represented by $\overrightarrow{OB}$ (Fig. 4.10). Then $\mathbf{F}_3$ is represented by $\overrightarrow{BO}$.

The sides $\overrightarrow{OA}$, $\overrightarrow{AB}$, $\overrightarrow{BO}$ of the triangle OAB (Fig. 4.11) represent the forces $\mathbf{F}_1$, $\mathbf{F}_2$, $\mathbf{F}_3$ respectively in magnitude and direction. Notice that the arrows all go the same way round the triangle (anti-clockwise in this case).

This can be expressed as follows:

If three forces acting on a particle, i.e. at a point, are in equilibrium, they can be represented in magnitude and direction by the sides of a triangle taken in order (i.e. going the same way round the triangle).

The converse, which is also true, may be stated as follows:

If three forces acting at a point can be represented in magnitude and direction by the three sides of a triangle taken in order, the forces will be in equilibrium.

This result is known as the *Triangle of Forces*. Another theorem follows directly from this.

Consider three forces $\mathbf{F}_1$, $\mathbf{F}_2$, $\mathbf{F}_3$ acting at the point O and in equilibrium with $\alpha$, $\beta$, $\gamma$ the angles between pairs of forces as shown in Fig. 4.12(a). The triangle ABC in Fig. 4.12(b) is the triangle of forces for the three forces.

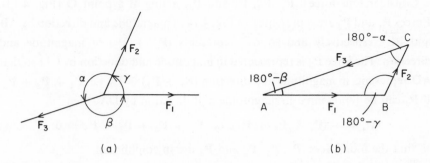

(a)            (b)

Fig. 4.12

Then $\angle ACB = 180° - \alpha$, $\angle BAC = 180° - \beta$, $\angle ABC = 180° - \gamma$. Applying the sine rule to the triangle ABC

$$\frac{AB}{\sin(180° - \alpha)} = \frac{BC}{\sin(180° - \beta)} = \frac{CA}{\sin(180° - \gamma)}$$

and since $AB:BC:CA = F_1:F_2:F_3$,

$$\frac{F_1}{\sin \alpha} = \frac{F_2}{\sin \beta} = \frac{F_3}{\sin \gamma}.$$

This result is known as Lami's theorem and may be stated as follows:
If three forces acting at a point are in equilibrium, the magnitude of each force is proportional to the sine of the angle between the other two forces.

The triangle of forces and Lami's theorem are useful in solving problems concerned with particles in equilibrium under the action of three forces.

### Polygon of forces

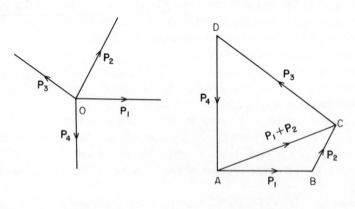

Fig. 4.13                    Fig. 4.14

Consider four forces, $P_1$, $P_2$, $P_3$ and $P_4$, acting at a point O (Fig. 4.13). Forces $P_1$ and $P_2$ are represented in Fig. 4.14 in magnitude and direction by $\overrightarrow{AB}$ and $\overrightarrow{BC}$ respectively and so $\overrightarrow{AC}$ represents $(P_1 + P_2)$ in magnitude and direction. The force $P_3$ is represented in magnitude and direction by $\overrightarrow{CD}$ so that $\overrightarrow{AD}$ represents in magnitude and direction $(P_1 + P_2) + P_3$, i.e. $P_1 + P_2 + P_3$. If $P_4$ can be represented in magnitude and direction by $\overrightarrow{DA}$, then

$$P_4 = -(P_1 + P_2 + P_3), \text{ i.e. } P_1 + P_2 + P_3 + P_4 = 0$$

so that the four forces, $P_1$, $P_2$, $P_3$ and $P_4$ are in equilibrium.

This can be extended to any number of forces acting at a point giving the theorem known as the *Polygon of Forces*:
If any number of forces acting at a point can be represented in magnitude and direction by the sides of a polygon taken in order, then the forces will be in equilibrium.

The converse is also true:

If any number of forces acting at a point are in equilibrium, then they can be represented in magnitude and direction by the sides of a polygon taken in order.

*Example 1*

In the rectangle ABCD the sides AB and BC have lengths 40 cm and 30 cm respectively. The following three forces act at A:

4 N along AB, 3 N along AD and 5 N along CA.

Show that these three forces are in equilibrium.

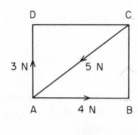

Fig. 4.15

Figure 4.15 shows the forces acting at A.

From Pythagoras' theorem the length of AC is 50 cm.

Clearly the sides $\overrightarrow{AB}$, $\overrightarrow{BC}$ and $\overrightarrow{CA}$ of the triangle ABC represent the three forces in direction.

If AB represents the force of 4 N also in magnitude, then 10 cm represent 1 N so that BC represents a force of magnitude 3 N and CA represents a force of magnitude 5 N, i.e. the sides $\overrightarrow{AB}$, $\overrightarrow{BC}$ and $\overrightarrow{CA}$ of the triangle ABC represent the three forces also in magnitude. Then by the triangle of forces, the three forces are in equilibrium.

*Example 2*

The sides AB and BC of a triangle ABC are of length 120 cm and 50 cm respectively, the angle ABC is a right angle and BD is an altitude of the triangle. The following three forces act at B:

20 N along DB, P N along BA and Q N along BC.

Given that the forces are in equilibrium, find the values of P and Q.

Figure 4.16(a) shows the forces acting at B.

Since the three forces are in equilibrium and act at a point, a triangle can be drawn in which the sides will represent the three forces in magnitude and direction.

Figure 4.16(b) shows the triangle of forces in which $\overrightarrow{EF}$ is drawn to represent the force of 20 N in magnitude and direction. The sides FG and GE are drawn

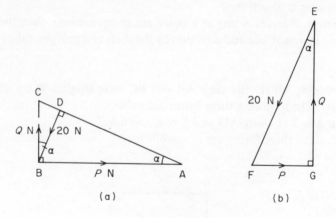

Fig. 4.16

parallel to the forces **P** and **Q** respectively. The lengths of FG and GE will give the magnitudes of these forces. If the triangle EFG is drawn to scale the values of $P$ and $Q$ can be obtained graphically.

Alternatively these values can be obtained by calculation:
In the right-angled triangle EFG

$$FG = EF \sin \alpha \quad \text{and} \quad EG = EF \cos \alpha$$

∴ $$P = 20(5/13) = 100/13 \quad \text{or} \quad 7\tfrac{9}{13}$$

and $$Q = 20(12/13) = 240/13 \quad \text{or} \quad 18\tfrac{6}{13}.$$

Alternatively Lami's theorem could be used.

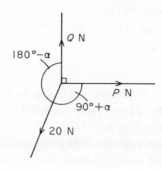

Fig. 4.17

Figure 4.17 shows the forces acting at B and the angles between these forces.

Then $$\frac{P}{\sin (180° - \alpha)} = \frac{Q}{\sin (90° + \alpha)} = \frac{20}{\sin 90°},$$

i.e.
$$\frac{P}{\sin \alpha} = \frac{Q}{\cos \alpha} = \frac{20}{1}.$$

giving $P = 20 \sin \alpha = 100/13$ and $Q = 20 \cos \alpha = 240/13$ as before.

Yet another method would be to resolve the force of 20 N along $\overrightarrow{DB}$ into two components, one of $(20 \cos \alpha)$ N in the direction $\overrightarrow{CB}$ and another of $(20 \sin \alpha)$ N in the direction $\overrightarrow{AB}$. Since the forces are in equilibrium there is no resultant force in the direction $\overrightarrow{AB}$, i.e. $20 \sin \alpha - P = 0$, and there is no resultant force in the direction $\overrightarrow{CB}$, i.e. $20 \cos \alpha - Q = 0$. As before we get $P = 100/13$ and $Q = 240/13$.

*Example 3*
The following forces act at the vertex A of a regular hexagon ABCDEF: 20 N along AB, 30 N along AC, 40 N along AE and a force **P**. The four forces are in equilibrium. Estimate graphically the magnitude and direction of the force **P**.

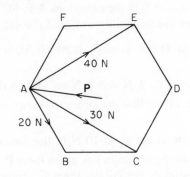

**Fig. 4.18**

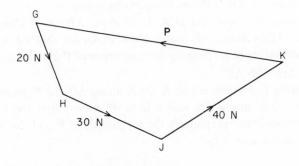

**Fig. 4.19**

Figure 4.18 shows the hexagon ABCDEF and the forces acting at A.
Figure 4.19 shows the polygon of forces GHJK drawn to scale.

In the force polygon GHJK the sides $\overrightarrow{GH}$, $\overrightarrow{HJ}$ and $\overrightarrow{JK}$ represent in magnitude and direction the forces of 20 N, 30 N and 40 N respectively. Since the four forces are in equilibrium, then $\overrightarrow{KG}$ must represent the fourth force **P** in magnitude and direction.

From the scale drawing $P \approx 72$ N is obtained and the line of action of **P** makes an angle of approximately 10° with $\overrightarrow{DA}$.

### Exercises 4.4

1  Forces of magnitude 10 N, $P$ N and $Q$ N act at the vertex A of an equilateral triangle ABC in the directions $\overrightarrow{AB}$, $\overrightarrow{BC}$ and $\overrightarrow{CA}$ respectively. Given that the forces are in equilibrium, find the values of $P$ and $Q$.

2  The triangle ABC is isosceles with $\angle ABC = 120°$. Forces of magnitude 20 N, 20 N and $20\sqrt{3}$ N act at A in the directions $\overrightarrow{AB}$, $\overrightarrow{BC}$ and $\overrightarrow{CA}$ respectively. Show that the forces are in equilibrium.

3  The triangle ABC is isosceles with $\angle ABC = 120°$. Forces of magnitude 12 N. $P$ N and $Q$ N act at B in the directions $\overrightarrow{AB}$, $\overrightarrow{BC}$ and $\overrightarrow{CA}$ respectively. Given that the forces are in equilibrium, find the values of $P$ and $Q$.

In questions 4 and 5, ABCD is a rectangle in which AB is of length 4 m and BC is of length 3 m.

4  Forces of magnitude 15 N, $X$ N and $Y$ N act at A in the directions $\overrightarrow{CA}$, $\overrightarrow{AB}$ and $\overrightarrow{AD}$ respectively. Given that the forces are in equilibrium, find the values of $X$ and $Y$.

5  The following four forces act at A: 10 N in the direction $\overrightarrow{AB}$, 20 N in the direction $\overrightarrow{AD}$, 30 N in the direction $\overrightarrow{CA}$ and a force **P**. Given that the forces are in equilibrium, estimate graphically the magnitude of **P** and the angle its line of action makes with $\overrightarrow{AB}$.

In questions 6 and 7, ABCDEF is a regular hexagon.

6  The following five forces act at A: 12 N along AB, 20 N along AC, 25 N along AD, 18 N along AF and a force **P**, Given that the forces are in equilibrium, estimate graphically the magnitude of **P** and the angle its line of action makes with $\overrightarrow{AD}$.

7  The following five forces act at A: 20 N along AC, 12 N along DA, 15 N along AE, 30 N along FA and a force **P**. Given that the forces are in equilibrium, estimate graphically the magnitude of **P** and the angle its line of action makes with $\overrightarrow{AD}$.

## 4.5  Equilibrium of a particle

In mechanics there are many different types of forces.

*Weight*  When a stone is released from rest, it falls vertically to the ground with an acceleration. The force producing this acceleration is the gravitational pull of the earth on the stone. This gravitational pull of the earth on a particle is known as the *weight* of the particle. Since the earth is not a perfectly uniform sphere, the gravitational pull on a particle varies slightly at different points on the earth's surface, i.e. the weight of a particle will vary slightly from place to place. The acceleration, $g$, due to gravity has a magnitude of approximately $9.8 \text{ m s}^{-2}$ (often taken, for ease of calculation, as $10 \text{ m s}^{-2}$).

Using Newton's second law, $\mathbf{F} = m\mathbf{a}$, where $m$ is the mass of the particle, the weight of a particle of mass 1 kg is $1 \times 9.8$ newtons approximately. Thus the weight of a particle of mass 1 kg is approximately 9.8 or 10 newtons, depending on which approximate value is taken for the acceleration due to gravity. Similarly a particle of mass $m$ kg will have a weight of magnitude $mg$ N.

*Normal reaction*  Consider a particle at rest on a horizontal table. (Fig. 4.20). Let the weight of this particle be a vertical force $\mathbf{W}$, of magnitude $W$.

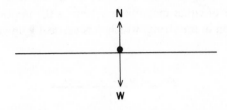

Fig. 4.20

The particle is at rest although it is known that it is being acted upon by at least one force, its own weight. This force must therefore be balanced by some other force so that there is no resultant force acting on the particle. This other force, $\mathbf{N}$, of magnitude $N$, acts vertically upwards and $N = W$. This force $\mathbf{N}$ is the force exerted by the table on the particle and is known as the *normal reaction* of the table on the particle.

Whenever a particle is in contact with a plane, the particle will exert a force on the plane at right angles to the plane and the plane will exert on the particle a force, known as the normal reaction, of equal magnitude but opposite in direction.

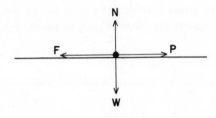

Fig. 4.21

*Friction*  Consider a particle at rest on a rough horizontal plane with a small horizontal force **P** acting on the particle (Fig. 4.21).

It has been shown that in a vertical direction, the weight of the particle is balanced by the normal reaction, i.e. $N = W$. Since the particle is at rest the horizontal force **P** must be balanced by a force equal in magnitude but opposite in direction to **P**. This is the frictional force **F**, where $F = P$, which is produced by the roughness of the contact between the particle and the plane, thus preventing movement. A 'smooth' plane is one on which the frictional force is so small that it may be considered to be negligible. The force of friction will be considered later in Section 4.7.

*Tension*  If a particle is suspended from a beam by a string, the particle can hang at rest with the string vertical. From Newton's first law this means that there is no resultant force acting on the particle. However it is known that there is one force **W**, the weight of the particle, acting on it and so this must be balanced by a force of equal magnitude acting in the opposite direction. This force is **T**, the tension in the string, which acts vertically upwards and for which $T = W$ (Fig. 4.22).

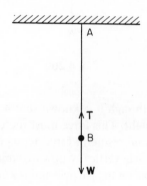

Fig. 4.22

Consider a horizontal string AB, subjected at end A to a horizontal force of magnitude $P$ to the right and at end B to a horizontal force of magnitude $P$ to the left (Fig. 4.23). The string is subjected to two forces of equal magnitude and opposite in direction and is therefore subject to no net force and is at rest.

Fig. 4.23

Consider the end A of the string. This is at rest and is subjected to a force of magnitude $P$ to the right and $\mathbf{T}$, the tension in the string, to the left, with $T = P$. Similarly, $T = P$ at the end B. The string is then said to be in tension.

Also if a 'light' rod AB is subjected to equal and opposite forces acting inwards at the ends A and B, there will be equal and opposite forces in the rod acting outwards at the two ends. The rod is then said to be in compression and the force in the rod is called a thrust.

A 'light' rod or string is one whose weight is negligible. A 'light' string, when in tension between two points, lies along the straight line joining those points.

Next will be considered some problems in which some of these forces appear.

*Example 1*
A particle of weight $W$ is at rest suspended from a beam by two light strings AB and AC. Given that AB and AC make angles of 30° and 60° respectively with the horizontal, find the tension in each string.

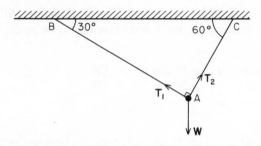

Fig. 4.24

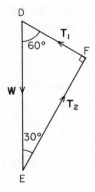

Fig. 4.25

The forces acting on the particle are as shown in Fig. 4.24, the tensions in the strings AB and AC having magnitudes $T_1$ and $T_2$ respectively.

Using Lami's theorem

$$\frac{W}{\sin 90°} = \frac{T_1}{\sin 150°} = \frac{T_2}{\sin 120°}$$

which gives $T_1 = W \sin 30° = \frac{1}{2}W$ and $T_2 = W \sin 60° = \frac{1}{2}W\sqrt{3}$.

Alternatively the triangle of forces could be used.

Figure 4.25 shows the triangle of forces DEF in which the sides $\overrightarrow{DE}$, $\overrightarrow{FD}$, $\overrightarrow{EF}$ represent the forces $\mathbf{W}$, $\mathbf{T_1}$, $\mathbf{T_2}$ respectively. The triangle is right angled and solving it gives

$$T_1 = W \sin 30° = \tfrac{1}{2}W \quad \text{and} \quad T_2 = W \sin 60° = \tfrac{1}{2}W\sqrt{3}.$$

(In numerical examples the triangle of forces can be drawn to scale and the magnitudes of the forces obtained from this scale drawing.)

Another method of solving this problem is to use the resolution of forces. Since two of the forces, the tensions in the strings AB and AC, are at right angles the weight $\mathbf{W}$ is resolved into components in the directions $\overrightarrow{AB}$ and $\overrightarrow{AC}$. Since the particle is in equilibrium, there is no resultant force in either direction and so,

resolving along AB, $\quad T_1 = W \cos 60° = \frac{1}{2}W$
resolving along AC, $\quad T_2 = W \cos 30° = \frac{1}{2}W\sqrt{3}$.

*Example 2*

A particle of weight $W$ is kept at rest on a smooth plane inclined at an angle $\alpha$ to the horizontal by a horizontal force of magnitude $P$. Find, in terms of $W$ and $\alpha$, (i) $P$ (ii) $N$, the magnitude of the normal reaction of the plane on the particle.

Since the plane is smooth there is no frictional force and so the particle is subject only to the three forces shown in Fig. 4.26. Again this problem can be solved by three methods as in the previous example.

(a) By Lami's theorem.

$$\frac{P}{\sin (180° - \alpha)} = \frac{N}{\sin 90°} = \frac{W}{\sin (90° + \alpha)}.$$

$$\therefore \quad N = \frac{W}{\cos \alpha} \quad \text{or} \quad W \sec \alpha$$

$$P = \frac{W \sin \alpha}{\cos \alpha} = W \tan \alpha.$$

(b) By the triangle of forces.

Fig. 4.27 shows ABC, the triangle of forces for the particle.

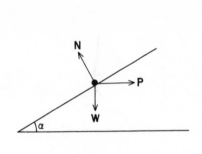

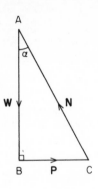

Fig. 4.26                              Fig. 4.27

From the right-angled triangle ABC,

$$P = W \tan \alpha$$

and   $\cos \alpha = \dfrac{W}{N}$   so that   $N = W/\cos \alpha = W \sec \alpha$.

(c)  By resolution of forces.
The particle is at rest so there is no resultant force acting on it.
Resolving horizontally   $P - N \sin \alpha = 0$   giving   $P = N \sin \alpha$        ...(1)
Resolving vertically        $N \cos \alpha - W = 0$   giving   $N \cos \alpha = W$    ...(2)
Solving equations (1) and (2) again gives $N = W \sec \alpha$ and $P = W \tan \alpha$.

*Example 3*
A ring R of mass $m$ is threaded on a smooth wire which is bent into the form of
a circle of radius $a$ and centre C. The wire is fixed in a vertical plane and the ring
is kept at rest by a light string, one end of which is attached to the ring and the
other end to A, the highest point of the wire. The string makes an angle $\alpha$ with
the vertical. Find
(a)  the tension in the string
(b)  the magnitude and direction of the reaction of the wire on the ring.

To construct Fig. 4.28, draw the diagram showing the wire, the ring and the
string. Then mark on the diagram the weight of the ring, and next the
tension **T** in the string. Since the string is light this will be along the line RA. The
wire is smooth and so can only exert a force **N** on the ring normal to the wire.
This normal reaction **N** must clearly be in the sense shown in Fig. 4.28 for it to
be possible for the three forces to be in equilibrium.
    As in the previous two examples there are three forces acting on the ring and
so the problem could be solved by Lami's theorem or by the triangle of forces.
However this problem will be solved by (a) the resolution of forces (b) the
triangle of forces.

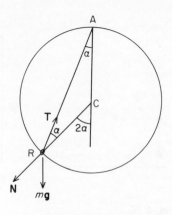

Fig. 4.28

(a) From elementary geometry $\angle ARC = \alpha$ and $\angle RCD = 2\alpha$.

Resolving horizontally $\quad N \sin 2\alpha - T \sin \alpha = 0$         ...(1)

Resolving vertically $\quad\quad T \cos \alpha - N \cos 2\alpha - mg = 0$     ...(2)

From (1) $\quad N = \dfrac{T \sin \alpha}{\sin 2\alpha}$ and putting this in (2) gives

$$T \cos \alpha - \frac{T \sin \alpha \cos 2\alpha}{\sin 2\alpha} = mg.$$

$\therefore \quad\quad\quad\quad T(\sin 2\alpha \cos \alpha - \cos 2\alpha \sin \alpha) = mg \sin 2\alpha.$

$\therefore \quad\quad\quad\quad\quad\quad\quad\quad\quad T \sin \alpha = mg \sin 2\alpha$

giving $\quad\quad T = \dfrac{mg \sin 2\alpha}{\sin \alpha} = \dfrac{mg\, 2 \sin \alpha \cos \alpha}{\sin \alpha} = 2mg \cos \alpha$

and $\quad\quad\quad N = \dfrac{T \sin \alpha}{\sin 2\alpha} = \dfrac{2mg \cos \alpha \sin \alpha}{2 \sin \alpha \cos \alpha} = mg.$

These results could also have been obtained by resolving the forces, not horizontally and vertically, but along RC and perpendicular to RC.

Resolving along RC $\quad\quad\quad\quad N - T \cos \alpha + mg \cos 2\alpha = 0$    ...(3)

Resolving perpendicular to RC $\quad T \sin \alpha - mg \sin 2\alpha = 0$    ...(4)

Equation (4) gives

$$T = \frac{mg \sin 2\alpha}{\sin \alpha} = 2mg \cos \alpha \text{ at once.}$$

Using this in (3) gives

$$N = 2mg \cos^2 \alpha - mg \cos 2\alpha$$
$$= 2mg \cos^2 \alpha - mg(2 \cos^2 \alpha - 1)$$
$$= mg.$$

The forces could also be resolved along and perpendicular to AR.
Resolving perpendicular to AR $\quad mg \sin \alpha - N \sin \alpha = 0$
giving $\quad N = mg$ at once.
Resolving along AR $\quad mg \cos \alpha - T + N \cos \alpha = 0$
again giving $\quad T = 2mg \cos \alpha.$
(b) The sides AC, CR and RA of the triangle ACR are in the directions of the three forces acting on the ring. Thus if AC also represents the weight $mg$ in magnitude, then the triangle ACR is the triangle of forces for the three forces acting on the ring (Fig. 4.29). Since the triangle is isosceles it can be seen at once that

$$N = mg \quad \text{and} \quad T = 2mg \cos \alpha.$$

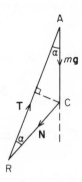

Fig. 4.29

These examples indicate that various methods for solving a problem are often available and that consideration should always be given to finding the most expeditious approach.

When solving a problem by the resolution of forces it will be possible to achieve a solution by resolving in a variety of directions. Often resolving horizontally and vertically will be appropriate but if this leads to complicated equations, then other directions should be considered. For instance, for a particle on an inclined plane resolving along and at right angles to a line of greatest slope of the plane will often be appropriate. With a particle suspended by two strings resolving at right angles to one string will give the tension in the other one at once.

*Example 4*
Figure 4.30 shows a framework ABCD consisting of four light smoothly-jointed rods. The framework is attached to a vertical wall at A and D so that AB and CD are horizontal and in the same vertical plane. The rods AB, BC and CA are of equal length. A load of 100 N is attached to the framework at B. Find the magnitude of the force in each of the four rods.

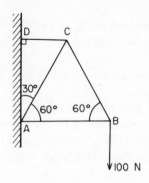

Fig. 4.30

There are three forces acting at B as shown in Fig. 4.31.
These forces are in equilibrium and so,

resolving vertically    $T_1 \cos 30° = 100$   giving   $T_1 = \dfrac{200}{\sqrt{3}}$ N   and

resolving horizontally   $T_2 = T_1 \cos 60° = \dfrac{200}{\sqrt{3}}(\tfrac{1}{2}) = \dfrac{100}{\sqrt{3}}$ N.

There are three forces acting at C as shown in Fig. 4.32.

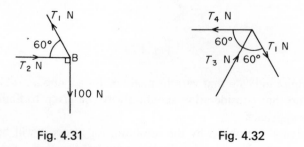

Fig. 4.31                          Fig. 4.32

These forces are in equilibrium and so,

resolving vertically    $T_3 \cos 30° = T_1 \cos 30°$, i.e. $T_3 = T_1 = \dfrac{200}{\sqrt{3}}$ N, and

resolving horizontally   $T_4 = T_1 \cos 60° + T_3 \cos 60° = \dfrac{200}{\sqrt{3}}$ N.

Thus the forces in the rods are:

AB: $\dfrac{100}{\sqrt{3}}$ N,   BC: $\dfrac{200}{\sqrt{3}}$ N,   CA: $\dfrac{200}{\sqrt{3}}$ N,   CD: $\dfrac{200}{\sqrt{3}}$ N.

The rods BC and CD are in tension; the rods AB and AC are in compression.
(See p. 137.)

## Exercises 4.5

Figure 4.33, which applies to questions 1–3, shows a particle C suspended by strings CA and CB from a horizontal beam.

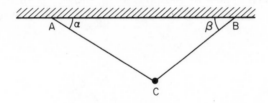

Fig. 4.33

1   Given that $\alpha = \beta = 45°$ and that the particle has weight $W$, sketch the triangle of forces and hence calculate the tensions in the strings.

2   Given that $\alpha = 40°$, $\beta = 50°$ and the particle has mass 2 kg, draw the triangle of forces to scale and hence obtain, to the nearest newton, the tensions in the strings. (Take $g$ to be 10 m s$^{-2}$.)

3   Given that $\tan \alpha = \frac{3}{4}$, $\tan \beta = \frac{4}{3}$ and the mass of the particle is $m$, find the tensions in the strings by resolving the forces (a) horizontally and vertically (b) along and perpendicular to AC.

In questions 4–7 a particle of mass 3 kg is kept at rest on a smooth plane inclined at an angle $\alpha$ to the horizontal by a force **P** whose line of action makes an angle $\beta$ with a line of greatest slope of the plane (Fig. 4.34). (Take $g$ to be 10 m s$^{-2}$.)

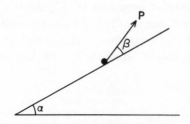

Fig. 4.34

4   When $\alpha = 60°$ and $\beta = 0$,
(a) sketch the triangle of forces and hence calculate the magnitude of **P** and of the normal reaction of the plane on the particle
(b) resolve along and perpendicular to the plane and hence calculate the magnitude of **P** and of the normal reaction
(c) estimate the magnitude of **P** and of the normal reaction from a scale drawing of the triangle of forces.

5   Find the magnitude of **P** and of the normal reaction when $\alpha = \beta = 30°$.

**6** By means of a scale drawing of the triangle of forces, find the magnitudes of **P** and the normal reaction when $\alpha = 40°$ and $\beta = 20°$.

**7** Show that the magnitude of **P** is least when $\beta = 0$.

Figure 4.35, which applies to questions 8–10, shows a particle A of weight $W$ suspended from a fixed point B by a light string AB and kept at rest by a force of magnitude $P$.

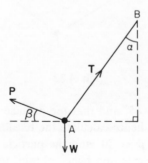

Fig. 4.35

**8** Make a scale drawing of the triangle of forces and from it find $P$ and the tension in the string when $\alpha = 30°$ and $\beta = 60°$.

**9** Calculate $P$ and the tension in the string when
(a) $\alpha = 30°, \beta = 30°$ (b) $\alpha = 30°, \beta = 0°$.

**10** Find the least possible value of $P$ and the corresponding angle $\beta$, given that $\alpha = 30°$.

Figure 4.36, which applies to questions 11–13, shows a ring R threaded on a smooth wire which is bent into the form of a circle of radius $a$ and which is fixed in a vertical plane. The ring is attached to A, the highest point of the wire, by a light string AR which makes an angle $\alpha$ with the vertical. (Take $g$ to be $10 \text{ m s}^{-2}$.)

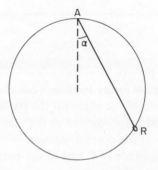

Fig. 4.36

Find the tension in the string and the magnitude and direction of the normal reaction of the wire on the ring

**11** when $\alpha = 30°$ and the ring has weight 10 N
**12** when $\alpha = 60°$ and the mass of the ring is 0.5 kg
**13** when $\alpha = \pi/4$ and the mass of the ring is 0.4 kg.
**14** Find the magnitude and sense of the force in each of the rods in the light frameworks shown in Fig. 4.37.

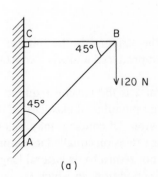

(a)

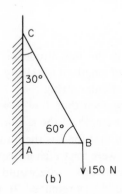

(b)

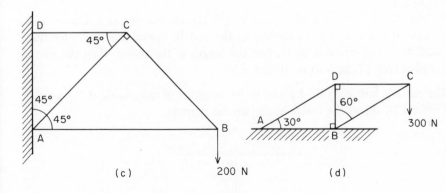

(c)          200 N          (d)

Fig. 4.37

## 4.6 Tension in an elastic string

Consider a scale pan suspended from a fixed point by a light string. If the string is elastic, it will stretch as weights are placed in the scale pan. With an ordinary piece of string, any extension produced by the weights in the scale pan will probably be very small by comparison with the unstretched length of the string. When the extension of a string is negligible by comparison with the string's unstretched length, the string is said to be inextensible. Inextensible strings figure in many problems in mechanics and the word 'inextensible' is taken to mean that, for the forces acting, the length of the string may be taken to be constant.

Elastic strings will now be considered where the extension of the string is not negligible. As long as the string returns to its unstretched length when the force stretching it is removed, experiment shows that the magnitude of the tension in an elastic string is proportional to the extension it produces. This experimental result is known as Hooke's law. It can be expressed in equation form as

$$T = \frac{\lambda}{l}x,$$

where $T$ is the magnitude of the tension in the string, $l$ is the natural or unstretched length of the string, $x$ is the extension produced and $\lambda$ is a constant for the particular string.

It can be seen that $\lambda = T$ when $x = l$ so that $\lambda$ is the tension required to double the length of the string. $\lambda$ is known as the modulus of elasticity of the string. The unit for $\lambda$ is the unit of force, the newton. Of course it may well be that if one were to stretch a string or wire until its extension equalled its natural length, the string or wire would break or would not return to its natural length after the load was removed. In the latter case the conditions in which Hooke's law applies would not exist.

*Example 1*

A light elastic string AB of natural length 120 cm has the end A fixed and a particle of mass 0.2 kg is attached to the end B. Given that the modulus of elasticity of the string is 40 N, find the length of the string when the particle hangs at rest. (Take $g$ to be 10 m s$^{-2}$.)

The particle has mass 0.2 kg and so its weight is of magnitude 0.2 $g$, i.e. 2 N. Figure 4.38 shows the forces acting on the particle.

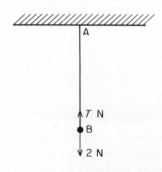

Fig. 4.38

Let the tension in the string have magnitude $T$ N.
Since the particle is at rest $T = 2$. By Hooke's law

$$T = \frac{\lambda}{l}x$$

$$\therefore \qquad 2 = \frac{40}{1.2}x \quad \text{since} \quad l = 120 \text{ cm} = 1.2 \text{ m.}$$

$$\therefore \qquad x = \left(\frac{2.4}{40}\right) \text{m} = 0.06 \text{ m} = 6 \text{ cm}$$

so that $\quad AB = l + x = (120 + 6) \text{ cm} = 126 \text{ cm.}$

*Example 2*
A light elastic string of natural length 45 cm has one end A fixed and a particle of weight 16 N attached to the other end B. A horizontal force of magnitude $P$ N is applied to the particle so that it is at rest with the string taut. The particle is then at a vertical distance of 40 cm from A and at a horizontal distance of 30 cm from A. Find the modulus of elasticity of the string.

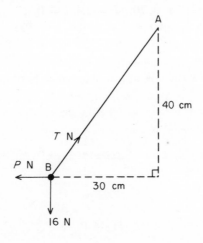

Fig. 4.39

Figure 4.39 shows the position of the string in the equilibrium position and the forces acting on the particle.
  In the right-angled triangle ABC, $AB = 50$ cm and $\cos \alpha = 4/5$.
Resolving vertically $\quad T \cos \alpha = 16$,
i.e. $\quad T(4/5) = 16 \quad$ giving $\quad T = 20$ N.
Using Hooke's law with $T = 20$ N, $l = 0.45$ m, $x = (50 - 45)$ cm $= 0.05$ m

$$T = \frac{\lambda}{l}x, \quad \text{i.e.} \quad 20 = \frac{\lambda}{0.45}(0.05) \quad \text{giving} \quad \lambda = 180 \text{ N.}$$

**Exercises 4.6**

In questions 1–8 a body of weight $W$ N is suspended from a fixed point by a light elastic string, of natural length $l$ m and modulus of elasticity $\lambda$ N. The extension produced is $x$ m.

1   If $W = 20$,   $l = 1.5$,   $\lambda = 100$, find $x$.
2   If $W = 40$,   $l = 2$,    $\lambda = 200$, find $x$.
3   If $W = 30$,   $x = 0.05$, $l = 1.2$, find $\lambda$.
4   If $W = 15$,   $x = 0.02$, $l = 0.8$, find $\lambda$.
5   If $W = 20$,   $x = 0.03$, $\lambda = 200$, find $l$.
6   If $W = 40$,   $x = 0.04$, $\lambda = 100$, find $l$.
7   If  $l = 1.6$, $x = 0.1$,   $\lambda = 150$, find $W$.
8   If  $l = 2$,   $x = 0.12$, $\lambda = 120$, find $W$.

In questions 9, 10 a particle of mass $m$ kg is attached to one end A of a light elastic string AB, of natural length $l$ m and modulus of elasticity $\lambda$ N, the other end B being fixed. The particle is kept at rest by a horizontal force of magnitude $P$ N with the string at an angle $\alpha$ to the vertical. The particle is then at a distance $d$ m from the vertical through B. Taking $g$ to be $10 \text{ m s}^{-2}$.

9   find $\lambda$ when $m = 2$, $\alpha = 30°$, $d = 0.4$, $l = 0.7$,
10  find $l$ when $m = 3$, $\tan \alpha = \frac{3}{4}$, $d = 0.6$, $\lambda = 100$.

## 4.7   Friction

Consider a particle of weight $\mathbf{W}$ at rest on a rough horizontal plane (Fig. 4.40).

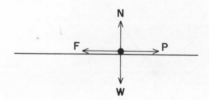

Fig. 4.40

The normal reaction $\mathbf{N}$ of the plane on the particle has magnitude $N$ which is equal to $W$. A horizontal force $\mathbf{P}$, slowly increasing in magnitude from zero, is applied to the particle. Initially the particle does not move because a frictional force $\mathbf{F}$ opposes the movement of the particle. However as $\mathbf{P}$ further increases in magnitude, experience shows that $F$, the magnitude of $\mathbf{F}$, reaches a maximum value. When the magnitude of $\mathbf{P}$ exceeds this maximum value of $F$, the particle will start to move in the direction of the force $\mathbf{P}$. The maximum value of the frictional force $\mathbf{F}$ is called the force of *limiting friction*. It is found that, in each case of the contact between the particle and the plane, the ratio

$$\frac{\text{limiting frictional force}}{\text{normal reaction}}$$

is a constant. This constant is known as the coefficient of friction between the particle and the plane and is denoted by the Greek letter $\mu$.

The properties of friction relating to particles may be summarised in the following 'laws':

1. The frictional force always acts in the direction directly to oppose the motion of the particle.
2. For a given value of the normal reaction, the frictional force which can be called into play has a maximum value known as the force of limiting friction. The particle is then said to be in *limiting equilibrium*.
3. $\dfrac{\text{force of limiting friction}}{\text{normal reaction}} = \mu$, the coefficient of friction.
4. After limiting friction has been reached and the particle starts to move, the frictional force will still act directly to oppose the motion of the particle.

Consider a particle of weight $W$ on a plane inclined at an angle $\alpha$ to the horizontal. Since the particle will tend to slide down a line of greatest slope of the plane, the frictional force **F** will act up that line of greatest slope. The particle will be subjected to three forces, its own weight **W**, the normal reaction **N** and the frictional force **F** (Fig. 4.41).

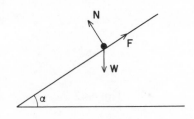

Fig. 4.41

When the particle is at rest,

resolving along the plane $\qquad\qquad F = W \sin \alpha \quad$ and

resolving perpendicular to the plane $\quad N = W \cos \alpha$.

Also from property (3) above $\quad F \leqslant \mu N \quad$ and so $\quad \tan \alpha \leqslant \mu$.

If $\alpha$ is gradually increased, $N$ will decrease and $F$ will increase until the position of limiting equilibrium is reached. Then, if $\alpha$ is increased further, $W \sin \alpha$ will exceed the force of limiting friction and the particle will slide down a line of greatest slope of the plane. The frictional force (still at its maximum value) will continue to act up that line of greatest slope.

When in limiting equilibrium

$$F = W \sin \alpha \quad \text{and} \quad N = W \cos \alpha$$

and $\qquad \mu = \dfrac{\text{force of limiting friction}}{\text{normal reaction}} = \dfrac{W \sin \alpha}{W \cos \alpha} = \tan \alpha.$

Thus the particle will be in limiting equilibrium when $\tan \alpha = \mu$, i.e. when the plane is inclined at an angle $\arctan \mu$ to the horizontal. If $\tan \alpha > \mu$, the particle will slide down the plane.

**Important note**  Only in limiting equilibrium and when the particle is in motion is $\dfrac{F}{N}$ equal to $\mu$, i.e. $F = \mu N$. Readers should avoid putting $F = \mu N$ unless it is absolutely clear that equilibrium is limiting or the particle is in motion.

### Angle of friction

Consider a particle in contact with a rough plane. The plane exerts a force on the particle and this force has two components, $N$ the normal reaction and $F$ the frictional force, the magnitude of the latter ranging from 0 to $\mu N$. These components for three possible situations are shown in Fig. 4.42.

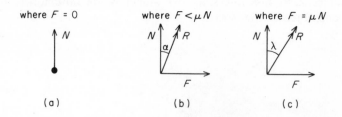

Fig. 4.42

The two components $N$ and $F$ can be combined to give a resultant reaction $R$.
In case (a), $R = N$ and $R$ is normal to the plane.
In case (c), $R$ is at an angle $\lambda$ to the normal reaction.
In case (b), $R$ is at an angle $\alpha$ to the normal reaction, where $0 < \alpha < \lambda$.
The resultant reaction will be at an angle from 0 to $\lambda$ to the normal reaction, being at an angle $\lambda$ to the normal reaction when friction is limiting. The angle $\lambda$ is known as the *angle of friction*.

In case (c), $\quad R^2 = N^2 + F^2 = N^2 + (\mu N)^2 \quad$ giving $\quad R = N\sqrt{(1 + \mu^2)}$

and $\qquad \tan \lambda = \dfrac{F}{N} = \dfrac{\mu N}{N} = \mu,$

i.e. the tangent of the angle of friction is equal to the coefficient of friction.

When a particle is in limiting equilibrium in contact with a rough plane, the resultant reaction of the plane on the particle will be at an angle $\lambda$, the angle of friction, to the normal to the plane.

*Example 1*

A particle of weight $W$ is on a rough plane inclined at an angle $\tan^{-1}(5/12)$ to the horizontal. The coefficient of friction between the particle and the plane is $\frac{3}{4}$. A force of magnitude $P$ acts on the particle along a line of greatest slope of the plane. Find $P$ if the particle is on the point of moving (a) down the plane (b) up the plane.

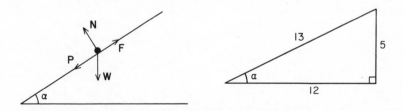

Fig. 4.43

(a) Figure 4.43 shows the forces acting on the particle in case (a).
Since the particle is on the point of moving down the plane, friction will be limiting, i.e. $F = \mu N$, and will act up the plane.

Resolving perpendicular to the plane $\quad N = W \cos \alpha$, i.e. $N = W(12/13)$.

Resolving along the plane $\quad P + W \sin \alpha = F = \mu N$
i.e. $\qquad\qquad\qquad\qquad P + W(5/13) = \frac{3}{4}N = \frac{3}{4}(12W/13)$

giving $P = 4W/13$.

(b) Figure 4.44 shows the forces acting on the particle in case (b).

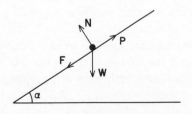

Fig. 4.44

Since the particle is on the point of moving up the plane, friction will be limiting, i.e. $F = \mu N$, and will act down the plane.
Resolving perpendicular to the plane $N = W \cos \alpha = 12W/13$
Resolving along the plane $P = F + W \sin \alpha = \mu N + W(5/13) = \frac{3}{4}N + 5W/13$
giving $P = 14W/13$.

*Example 2*

A particle of weight $W$ is on a rough plane, the coefficient of friction between the particle and the plane being $\mu$. Find the magnitude and direction of the least force required to move the particle

(a) on the plane if the plane is horizontal (b) up the plane if the plane is at an angle $\alpha$ to the horizontal.

(a) Let the force have magnitude $P$ and act at an angle $\theta$ to the vertical. Figure 4.45 shows the forces acting on the particle.

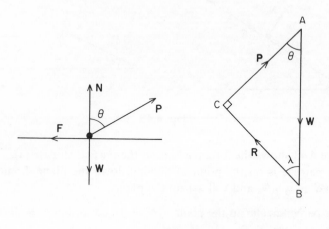

Fig. 4.45                    Fig. 4.46

Combining the normal reaction $\mathbf{N}$ and the frictional force $\mathbf{F}$ to give a resultant reaction $\mathbf{R}$, there are three forces acting on the particle. With limiting equilibrium, this resultant reaction will be at an angle $\lambda$ to the normal. Figure 4.46 shows the triangle of forces.

$\mathbf{W}$ is fixed in magnitude and direction, $\mathbf{R}$ is fixed in direction. $P$ is required to be as small as possible, i.e. in the triangle of forces the side AC must be as short as possible. AC will have its least length when it is perpendicular to BC. Then the force $\mathbf{P}$ is at right angles to the resultant reaction. In the triangle of forces ABC, $\angle$ACB $= 90°$ and $P = W \sin \lambda$ and $P$ is at $(90° - \lambda)$ to the vertical, i.e. at angle $\lambda$ to the horizontal

$$P = W \sin \lambda = \frac{\mu W}{\sqrt{(1 + \mu^2)}}.$$

(b) This case also can be dealt with by combining the normal reaction and the frictional force into a resultant reaction and deducing the least force, as in (a) above, from the triangle of forces. This is left as an exercise for the reader.

Here the alternative method of resolution of forces will be used.

Let the least force $\mathbf{P}$ act at an angle $\theta$ to the plane. Figure 4.47 shows the forces acting on the particle.

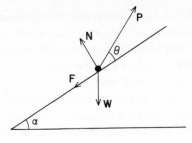

Fig. 4.47

Resolving along the plane $\qquad F + W \sin \alpha = P \cos \theta.$

Resolving perpendicular to the plane $\quad N + P \sin \theta = W \cos \alpha.$

Since equilibrium is limiting $\qquad\qquad\qquad F = \mu N.$

Solving these equations gives $P(\cos \theta + \mu \sin \theta) = W(\sin \alpha + \mu \cos \alpha)$

Putting $\mu = \tan \lambda,$ $\quad P(\cos \theta + \tan \lambda \sin \theta) = W(\sin \alpha + \tan \lambda \cos \alpha)$

$\therefore \qquad P(\cos \theta \cos \lambda + \sin \theta \sin \lambda) = W(\sin \alpha \cos \lambda + \cos \alpha \sin \lambda)$

$\therefore \qquad\qquad P \cos (\theta - \lambda) = W \sin (\alpha + \lambda)$

giving $\qquad\qquad\qquad P = \dfrac{W \sin (\alpha + \lambda)}{\cos (\theta - \lambda)}.$

Since $W, \alpha$ and $\lambda$ are constant, $P$ is least when $\cos (\theta - \lambda)$ has its greatest value, i.e. 1 when $\theta - \lambda = 0$ and $\theta = \lambda.$

Thus $P$ is least when $\mathbf{P}$ acts at an angle $\lambda$ or $\arctan \mu$ to the plane and the least value of $P$ for equilibrium is $W \sin (\alpha + \lambda),$ where $\tan \lambda = \mu.$ To move the particle, the magnitude of the force applied must exceed this value.

### Exercises 4.7
Where necessary take $g$ to be $10 \text{ m s}^{-2}.$
1 A particle of mass 0.6 kg is on a rough horizontal plane. The coefficient of friction between the particle and the plane is $\frac{1}{2}.$ Find the least force required to pull the particle along the plane if this force is (a) horizontal (b) at an angle of $30°$ to the plane (c) at an angle $\arctan \frac{1}{2}$ to the horizontal.
2 A particle of mass 2 kg is placed on a rough plane inclined at an angle of $30°$ to the horizontal. The coefficient of friction between the particle and the plane is $\frac{1}{4}.$ Find the least force required to prevent the particle slipping down the plane if this force is (a) horizontal (b) along a line of greatest slope of the plane (c) upwards at an angle of $30°$ to a line of greatest slope.
3 A particle of mass 0.8 kg is at rest on a rough horizontal table when acted upon by a force $\mathbf{P}$ of magnitude 4 N. Find the least possible value of the coefficient of friction between the particle and the table if (a) $\mathbf{P}$ is horizontal (b) $\mathbf{P}$ acts upwards at an angle of $30°$ to the horizontal.

4 A particle of weight 15 N is placed on a rough plane inclined at an angle $\tan^{-1}(4/3)$ to the horizontal. The coefficient of friction between the particle and the plane is $\frac{1}{2}$. The particle is kept at rest by a force of magnitude $P$. Find the range of possible values of $P$ if the force (a) acts up a line of greatest slope of the plane (b) is horizontal.

5 A particle of weight 20 N is kept at rest on a rough plane inclined at angle $\tan^{-1} 2.4$ to the horizontal by a force of 15 N. Find the least possible value of the coefficient of friction between the particle and the plane when the force acts up a line of greatest slope of the plane.

**Miscellaneous exercises 4**

1 Points A, B, C, D, E have the following coordinates with reference to coordinate axes $Ox$, $Oy$:

$$A(4, 3), \quad B(5, 12), \quad C(-3, 4), \quad D(-8, -6), \quad E(12, -5).$$

Find the magnitude of the resultant of each of the following sets of forces acting at O. Find also the angle the line of action of the resultant makes with $Ox$.
(a) 12 N along $Ox$, 5 N along $Oy$
(b) 10 N along OA, 26 N along OB
(c) 20 N along OC, 40 N along OD
(d) 20 N along OA, 40 N along OB, 30 N along OD
(e) 30 N along OB, 20 N along CO, 40 N along EO
(f) 39 N along OB, 30 N along OC, 40 N along OD, 26 N along EO.

2 Find in magnitude and direction the resultant of the following forces which act at a point and whose lines of action lie in a horizontal plane:
(a) 20 N due east, 30 N due north, 40 N north west
(b) 30 N due south, 20 N north east, 24 N south east
(c) 20 N on a bearing 060°, 30 N on a bearing 150°, 40 N on a bearing 270°
(d) 24 N on a bearing 120°, 18 N on a bearing 240°, 12 N on a bearing 300°.

3 In the rectangle ABCD, AB = 4 m and BC = 3 m. Three forces act at A, 40 N along CA, $P$ N along AB and $Q$ N along AD. Given that the forces are in equilibrium, find $P$ and $Q$.

4 Three forces acting at A, a vertex of an equilateral triangle ABC, are in equilibrium. One force has magnitude 20 N and acts along AB, another has magnitude 28 N and acts along AC. Find the magnitude and direction of the third force.

5 Points A, B, C, D lie on the circumference of a circle, centre O, and $\angle AOB = 120°$, $\angle BOC = 120°$, $\angle COD = 90°$, $\angle DOA = 30°$.
(a) Given that a force of 20 N along OB and forces of $P$ N and $Q$ N along OC and OD respectively are in equilibrium, find $P$ and $Q$.
(b) The following four forces act at O: 30 N along OB, 40 N along OC, $P$ N along OD and $Q$ N along OA. Given that the four forces are in equilibrium, find $P$ and $Q$.

**6** The framework ABCD (Fig. 4.48) consists of five equal light rods. The framework is at rest suspended by a rope attached at A and carries a load of 100 N at C. Find
(a) the force in each rod  (b) the tension in the rope.

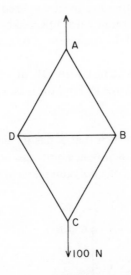

Fig. 4.48

**7** A small ring of weight $W$ is threaded on a light inextensible string of length $4a$. The ends of the string are attached to two points A and B which are in a horizontal line and at a distance $2a$ apart. The ring hangs at rest. Find, in terms of $W$, the magnitude of the tension in the string.

**8** A boat S is attached by ropes SA and SB to two points A and B on the shore. The distances AB, SA and SB are 25 m, 24 m and 7 m respectively. The wind and current exert a force of 600 N on the boat in a direction perpendicular to and away from AB. Find the tension in each rope.

**9** A particle A of weight $W$ is placed on the smooth inner surface of a fixed hemispherical bowl. The rim of the bowl is horizontal and its centre is at the point O. The particle is held at rest, with OA at an angle of 60° to the vertical, by a force **P**. Find the magnitude of **P** if it acts (a) horizontally  (b) tangentially to the surface of the bowl.

**10** A particle A of weight $W$ is placed on the surface of a fixed smooth sphere of centre O. The particle is kept at rest, with OA at an angle $\tan^{-1}\frac{3}{4}$ to the upward vertical, by a horizontal force of magnitude $P$. Find the value of $P$.
 Find also the direction in which the force should act if it is to have its least magnitude.

**11** A small ring R of weight $W$ is threaded on to a rough fixed straight rod. The coefficient of friction between the ring and the rod is $\frac{2}{3}$. Find the magnitude and direction of the least force necessary to move the ring along the rod when the rod is

(a) horizontal

(b) at an angle $\tan^{-1}$ (5/12) to the horizontal.

12  A small ring R of weight $W$ is threaded on a rough wire which is bent into the form of a circle, centre O. The wire is fixed in a vertical plane and the ring is in limiting equilibrium when OR makes an angle $\arctan \frac{3}{4}$ with the downward vertical. Find the coefficient of friction between the ring and the wire.

The ring is now held at rest with OR at an angle $\arctan$ (4/3) to the downward vertical by a force **P**. Find the least value of $P$, the magnitude of **P**, if

(a) **P** acts horizontally

(b) **P** acts in a direction such that $P$ has its least possible value.

13  If any number of forces, acting at point, can be represented in magnitude and direction by the sides of a polygon taken in order, prove that the forces are in equilibrium.

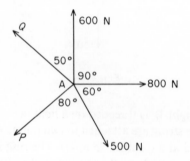

Fig. 4.49

(i) The five coplanar forces acting at the point A in Fig. 4.49 are in equilibrium. Find graphically, or otherwise, the values of $P$ and $Q$ to the nearest 5 N.

(ii) If $P = 555$ N and $Q = 815$ N, find graphically, or otherwise, the magnitude and the direction of the resultant of the five forces.     [AEB]

14  State Lami's theorem.

A light inextensible string ABCD of length 43 cm has its ends A and D attached to fixed points at the same level 25 cm apart. Weights $W$ and $kW$ are attached to the points B and C respectively, where AB = 16 cm and BC = 12 cm. The string hangs at rest with C 12 cm below AD. Calculate the value of $k$ and the tensions in the portions AB, BC, CD of the string.
[L]

15  A light inextensible string ABCD supports a weight $W$ N at B and a weight 60 N at C and the ends A, D are attached to rings of weights 10 N and 5 N

respectively which can slide along a fixed rough horizontal wire. The system is in equilibrium with AB inclined at 60° to the horizontal, BC at 30° to the horizontal and CD at 60° to the horizontal, C being below B. Find the value of $W$ and the tensions in the different portions of the string.

If the coefficient of friction $\mu$ is the same at each ring, show that the least value of $\mu$ is $(3\sqrt{3})/10$. [L]

16 A particle of mass $3m$ is tied to the end C and a particle of mass $m$ is tied at the mid-point B of a light unstretched elastic string ABC. The end A of the string is fixed and a horizontal force of magnitude $4mg$ is applied to the particle at C so that the system hangs in equilibrium as shown in Fig. 4.50.

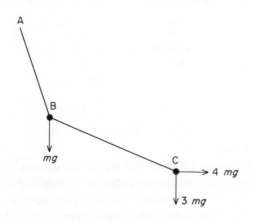

Fig. 4.50

Calculate
(i) the tensions in BC and AB
(ii) the inclinations of BC and AB to the vertical.

Given that the modulus of elasticity of the string is $6mg$, show that, for this position of equilibrium, AB:BC $= (6 + 4\sqrt{2})$:11. [AEB]

17 A particle of weight $W$ is attached to a point C of an unstretched elastic string AB, where AC $= 4a/3$, CB $= 4a/7$. The ends A and B are then attached to the extremities of a horizontal diameter of a fixed hemispherical bowl of radius $a$ and the particle rests on the smooth inner surface, the angle BAC being 30°. Show that the modulus of elasticity of the string is $W$ and determine the reaction of the bowl on the particle. [L]

18 A light elastic string of natural length $a$ and modulus of elasticity $2mg$ is stretched between two fixed points A and B; A is distant $2a$ from, and vertically above, B. A particle of mass $3m$ is fastened to the string at a point C. If, in the equilibrium position, AC $=$ CB, find the natural lengths of the portions of the strings AC and CB. What extra mass must be fastened to C so that CB becomes its natural length? [AEB]

**19** In Fig. 4.51, A and D are two fixed points, on the same horizontal level. Particles of weight $3W$ and $kW$ are attached at B and C respectively, and the system is in equilibrium in a vertical plane. The string BC is inclined at $60°$ to the vertical, and the strings AB and DC are each inclined at $30°$ to the vertical. Find the tension in each string and show that $k = 6$. If CD is an elastic string of natural length $2a$ and modulus of elasticity $6\sqrt{3}W$, calculate the length of CD in terms of $a$.                [AEB]

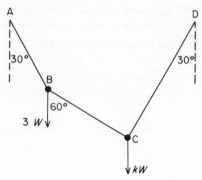

Fig. 4.51

**20** State Hooke's law.

A light elastic string AB, of modulus $KW$ and natural length $l$, has its end A tied to a fixed point and carries a particle of weight $W$ at its other end B. The particle is held in equilibrium with the string taut and making an angle $\tan^{-1}\frac{3}{4}$ with the downward vertical by the application of a force of $P$, on the particle. Find the value of $P$ and the extension in the string if the direction of $P$ is

(i) horizontal

(ii) such that the value of $P$ is as small as possible.                [AEB]

**21** A light elastic string, of unstretched length $a$ and modulus of elasticity $W$, is fixed at one end to a point on the ceiling of a room. To the other end of the string is attached a particle of weight $W$. A horizontal force $P$ is applied to the particle and in equilibrium it is found that the string is stretched to three times its natural length. Calculate (a) the angle the string makes with the horizontal (b) the value of $P$ in terms of $W$.

If, instead, $P$ is not applied horizontally find the *least* value of $P$ which in equilibrium will make the string have the same inclination to the horizontal as before. Deduce that the stretched length of the string is $\frac{3}{2}a$ in this case and find the inclination of $P$ to the vertical.                [L]

**22** A light elastic string ACB, of natural length $2l$, is made by joining an elastic string AC, of natural length $\frac{2}{3}l$ and modulus $\lambda_1$, and another elastic string CB, of natural length $1\frac{1}{3}l$ and modulus $\lambda_2$. The ends A and B are attached at two fixed points which are at the same level and are at a distance $2l$ apart. When a weight $W$ is attached at the point $C$, the string hangs at rest with the portions AC and CB at angles of $60°$ and $30°$ respectively to the

**158** Advanced Mathematics 2

horizontal. Show that $\lambda_1 = W\sqrt{3}$ and $\lambda_2 = 2W/(3\sqrt{3} - 4)$.

The weight $W$ is removed and a force applied at C in the vertical plane containing AB. Find the horizontal and vertical components of this force in terms of $W$ when the string is at rest with ABC forming an equilateral triangle. [AEB]

23 State Hooke's law.

A light elastic string ACB, of natural length $2l$, is made by joining an elastic string AC, of natural length $l$ and modulus of elasticity $\lambda W$, and an elastic string CB, of natural length $l$ and modulus of elasticity $k\lambda W$. Find the length of the string ACB when A is attached to a fixed point and a weight $W$ hangs at rest from the free end B.

Find also the extensions produced in the portions AC and CB of the string when the ends A and B are attached to fixed points at a distance $3l$ apart.

A smooth ring, of mass $mW$, is threaded on to the string and the ends A and B are joined so that the string forms an elastic band. The band is placed over two small smooth pegs at the same level and at a distance $1\frac{1}{3}l$ apart. When the ring hangs at rest the elastic band is in the form of an equilateral triangle. Find the value of $m$ in terms of $k$ and $\lambda$. [AEB]

# 5 Dynamics of a particle

## 5.1 Force and acceleration

In Chapter 3 the motion of a particle was dealt with without considering how that motion was produced. In Chapter 4 the forces acting on a particle were considered. Now the forces acting on a particle and the motion they produce will be considered.

Consider a particle of mass $m$ acted upon by a system of forces which reduces to a single resultant force $\mathbf{F}$ giving the particle an acceleration $\mathbf{a}$. Then, by Newton's second law, $\mathbf{F} = m\mathbf{a}$.

If the force $\mathbf{F}$ is constant, i.e. unchanged in both magnitude and direction, then the particle moves in a straight line with constant acceleration. Then, with the notation already adopted, $F = |\mathbf{F}|$ and $a = |\mathbf{a}|$,

and
$$F = ma,$$

i.e. $F$, the magnitude of the resultant force, can be equated to the product of the mass, $m$, of the particle and $a$, the magnitude of its acceleration.

A simple example of a particle acted upon by a single constant force is a particle falling freely under gravity. Assuming that air resistance is negligible, the particle is acted upon by a single constant force, its own weight. This produces the acceleration due to gravity of magnitude $g$. Then using $F = ma$, the weight of the particle is $mg$. If the mass $m$ is in kg and $g$, the acceleration due to gravity, is in $\mathrm{m\,s^{-2}}$, the weight will be in newtons. In particular, if the mass of the particle is 1 kg and 10 $\mathrm{m\,s^{-2}}$ is the approximate magnitude of the acceleration due to gravity, the weight of the particle is 10 newtons, i.e. a particle of mass 1 kg has a weight of approximately 10 N. Thus the unit of force, the newton, is approximately equal to the weight of a particle of mass 0.1 kg, i.e. the weight of a particle of mass 100 g. Some problems on the forces acting on a particle and the motion they produce will now be considered.

*Example 1*

A particle of mass $m$ is released from rest at a point on a smooth plane inclined at an angle $\alpha$ to the horizontal. Find the acceleration of the particle. Find also the time taken and the speed attained by the particle as it travels a distance $d$ down the plane.

The particle is acted upon by only two forces, its own weight $mg$ and the reaction, of magnitude $N$, of the plane. This reaction is perpendicular to the plane since the plane is smooth.

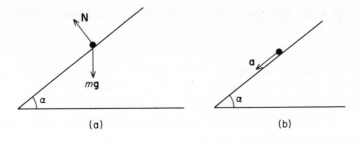

Fig. 5.1 (a) Forces (b) Acceleration

Figure 5.1(a) shows the forces acting on the particle and Fig. 5.1(b) shows the acceleration of the particle.

If $a$ denotes the acceleration down the plane, resolving down the plane and using $F = ma$ gives

$$mg \sin \alpha = ma \quad \text{giving} \quad a = g \sin \alpha.$$

Note that, since the particle slides down the plane, there is no acceleration perpendicular to the plane so that,

resolving perpendicular to the plane $\quad N - mg \cos \alpha = 0 \quad$ or $\quad N = mg \cos \alpha.$

Since the acceleration down the plane is $g \sin \alpha$, the acceleration is constant. Consequently the equations for uniformly accelerated motion developed in Chapter 3 can be used, viz.

$$v = u + at, \quad v^2 = u^2 + 2as, \quad s = ut + \tfrac{1}{2}at^2, \quad s = \tfrac{1}{2}(u + v)t.$$

Here $u = 0$, $s = d$, $a = g \sin \alpha$ and $t$ and $v$ have to be found.
Using $s = ut + \tfrac{1}{2}at^2, \quad d = \tfrac{1}{2}(g \sin \alpha)t^2,$

giving
$$t = \left[\frac{2d}{g \sin \alpha}\right]^{1/2},$$

which is the time taken by the particle to travel a distance $d$.
Using $v = u + at$ gives

$$v = g \sin \alpha \left[\frac{2d}{g \sin \alpha}\right]^{1/2} = \sqrt{(2gd \sin \alpha)}.$$

Alternatively, $v^2 = u^2 + 2as$ gives

$$v^2 = 2(g \sin \alpha)d \quad \text{giving } v \text{ as before}.$$

*Example 2*
A particle slides down a rough plane inclined at an angle of 60° to the horizontal. The coefficient of friction between the plane and the particle is $\tfrac{1}{2}$. The time taken for the speed of the particle to increase from 2 m s⁻¹ to 30 m s⁻¹ is $T$

seconds and in this time the particle travels a distance $d$ metres. Find the values of $T$ and $d$.

(Take $g$ to be $10 \text{ m s}^{-2}$.)

Let the mass of the particle be $m$ kg and let the normal reaction and the frictional force on the particle have magnitudes $N$ and $F$ newtons respectively. Let $a \text{ m s}^{-2}$ be the acceleration of the particle down the plane.

Figure 5.2(a) shows the forces acting on the particle and Fig. 5.2(b) shows the acceleration of the particle.

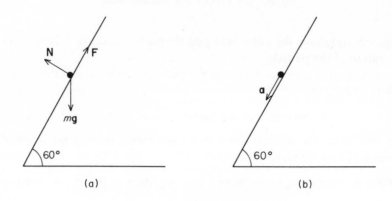

(a)                                    (b)

Fig. 5.2 (a) Forces (b) Acceleration

Resolving along the plane gives

$$mg \sin 60° - F = ma. \qquad \ldots(1)$$

Resolving perpendicular to the plane gives

$$N = mg \cos 60°, \qquad \ldots(2)$$

since there is no acceleration perpendicular to the plane.

Also, since the particle is moving, $F = \mu N = \frac{1}{2}N$. $\qquad \ldots(3)$

Solving equations (1), (2) and (3) gives

$$a = \tfrac{1}{4}(2\sqrt{3} - 1)g \approx 6.16 \text{ m s}^{-2}.$$

Thus the acceleration is constant and so the equations for uniformly accelerated motion can be used.

$u = 2$, $v = 30$, $a \approx 6.16$ and $T$ and $d$ have to be found.

Using $v = u + at$, yields $30 = 2 + 6.16T$ giving $T \approx 4.6$.

Using $v^2 = u^2 + 2as$, yields $900 = 4 + 2(6.16)d$ giving $d \approx 73$.

*Example 3*

A car of mass 1 tonne travelling on a straight level road accelerates from a speed

of 36 km h$^{-1}$ to a speed of 108 km h$^{-1}$ in 12 seconds. Given that the engine of the car exerts a constant pull of magnitude $P$ newtons and that there is a constant resistance to motion of 200 newtons, find the value of $P$.
(Take $g$ to be 10 m s$^{-2}$.)

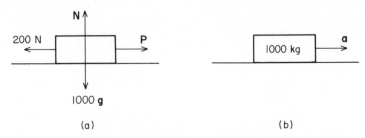

(a)                                         (b)

Fig. 5.3 (a) Forces (b) Acceleration

Figure 5.3(a) shows diagrammatically the forces in newtons acting on the car and Fig. 5.3(b) shows its acceleration.,

In this example the assumption is made that, for the linear motion, the car may be treated as a particle.
Using $F = ma$ in the direction of motion gives

$$P - 200 = 1000a \quad (1 \text{ tonne} = 1000 \text{ kg}).$$

Since $P$ is constant, the acceleration is constant and so again the equations for uniformly accelerated motion may be used.

$$36 \text{ km h}^{-1} = \frac{36 \times 1000}{60 \times 60} \text{ m s}^{-1} = \frac{36}{3.6} \text{ m s}^{-1} = 10 \text{ m s}^{-1} \quad \text{and}$$
$$108 \text{ km h}^{-1} = 3 \times 36 \text{ km h}^{-1} = 30 \text{ m s}^{-1}.$$

Using $v = u + at$ with $u = 10$, $v = 30$ and $t = 12$ gives

$$30 = 10 + 12a \quad \text{giving} \quad a = 5/3 \text{ m s}^{-2}.$$

From above $\qquad P - 200 = 1000a$

and so $\qquad\qquad P = 1000(5/3) + 200 = 1866\frac{2}{3}.$

*Example 4*
A parcel of mass 12 kg is placed on the floor of a lift. Find the reaction of the floor of the lift on the parcel when the lift is
(a) ascending with constant speed
(b) moving with acceleration 2 m s$^{-2}$ upwards
(c) moving with acceleration 2 m s$^{-2}$ downwards.
(Take $g$ to be 10 m s$^{-2}$.)

It can be assumed that, for linear motion, the parcel may be treated as a particle. There are only two forces acting on the parcel, its own weight, $12g$, i.e. 120 N and the normal reaction of magnitude $N$ newtons.

Figure 5.4(a) shows the forces acting on the parcel and Fig. 5.4(b) shows its acceleration.

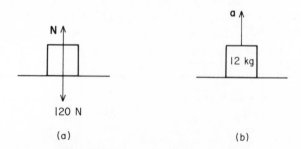

Fig. 5.4 (a) Forces (b) Acceleration

As long as the parcel is in contact with the floor of the lift, the parcel will have the speed and the acceleration of the lift.

In case (a) the speed of the lift is constant, so is the direction of motion and so the lift and the parcel have the same constant velocity, i.e. the acceleration $a = 0$.

In case (b) there is an upward acceleration of $2\ \mathrm{m\,s^{-2}}$, i.e. $a = +2$, taking the upward direction as positive.

In case (c) there is a downward acceleration of $2\ \mathrm{m\,s^{-2}}$, i.e. $a = -2$.

Then from Fig. 5.4, $F = ma$ gives

$$N - 120 = 12a,$$

i.e.               $N = 120 + 12a.$

Then in case (a)   $N = 120,$

and in case (b)   $N = 120 + 12(+2) = 144$

and in case (c)   $N = 120 + 12(-2) = 96.$

Thus when the lift is moving with constant speed the reaction of the floor of the lift on the parcel is 120 N, i.e. the reaction is equal in magnitude to the weight of the parcel. This result could have been obtained immediately from Newton's first law. Since the parcel is moving in a straight line with constant speed, there is no resultant force acting on it. Consequently the reaction must be equal in magnitude and opposite in direction to the weight of the parcel.

When the lift has an upward acceleration of $2\ \mathrm{m\,s^{-2}}$, the reaction is 144 N.

When the lift has a downward acceleration of $2\ \mathrm{m\,s^{-2}}$, the reaction is 96 N.

*Example 5*

From a point A a particle is projected with speed $24\ \mathrm{m\,s^{-1}}$ up a line of greatest slope of a rough plane inclined at an angle $\tan^{-1}\frac{3}{4}$ to the horizontal. The coefficient of friction between the particle and the plane is $\frac{1}{4}$. The particle comes to instantaneous rest at the point B. Find the distance AB and the time taken to reach B. Find also the speed of the particle when it returns to the point A. (Take $g$ to be $10 \text{ m s}^{-2}$.)

The particle is acted upon by three forces, its own weight $mg$, the normal reaction of magnitude $N$ and the frictional force of magnitude $F$.

Figure 5.5(a) shows the forces acting on the particle and Fig. 5.5(b) shows its acceleration when the particle is moving *up* the plane.

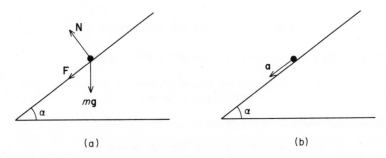

(a)                                              (b)

Fig. 5.5 (a) Forces (b) Acceleration

Since the particle is moving up the plane, the frictional force will act down the plane.
Resolving along a line of greatest slope of the plane gives

$$F + mg \sin \alpha = ma,$$

i.e.
$$F + m(10)(\tfrac{3}{5}) = ma. \qquad (1)$$

Resolving perpendicular to the plane

$$N = mg \cos \alpha = m(10)(\tfrac{4}{5}) = 8m. \qquad \ldots (2)$$

Since the particle is in motion

$$F = \mu N = \tfrac{1}{4}N = \tfrac{1}{4}(8m) = 2m. \qquad \ldots (3)$$

Solving equations (1), (2) and (3) gives $a = \tfrac{4}{5}g = 8$.
Thus the acceleration $a$ is constant and the equations for uniformly accelerated motion can be used.
Taking distances, velocities and accelerations up the plane as positive, $u = +24$, $v = 0$, $a = -8$ and $s$ has to be found.
Using $v^2 = u^2 + 2as$, yields $0 = 24^2 + 2(-8)s$ giving $s = 36$,
i.e. the particle moves 36 m up the plane in coming to rest so that AB = 36 m.

After coming to instantaneous rest at B the particle starts to slide back down the plane. When the particle is moving down the plane the frictional force will act up the plane.
Figure 5.6(a) shows the forces acting on the particle and Fig. 5.6(b) shows its acceleration when the particle is moving *down* the plane.

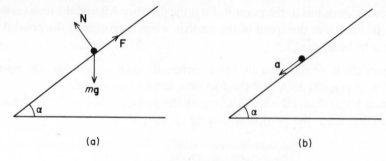

Fig. 5.6 (a) Forces (b) Acceleration

Resolving along a line of greatest slope of the plane gives

$$mg \sin \alpha - F = ma,$$

i.e.
$$m(10)(\tfrac{3}{5}) - F = ma. \qquad \qquad \ldots(4)$$

Resolving perpendicular to the plane

$$N = mg \cos \alpha = 8m \quad \text{as before.} \qquad \qquad \ldots(5)$$

Since the particle is in motion

$$F = \mu N = 2m \quad \text{as before.} \qquad \qquad \ldots(6)$$

Solving equations (4), (5) and (6) gives $a = 4$.
Again the acceleration is constant and the equations for uniformly accelerated motion may be used.
Now taking distances, velocities and accelerations down the plane to be positive, $s = \text{BA} = 36$, $u = 0$, $a = +4$ and $v$ has to be found.
Using $v^2 = u^2 + 2as$, yields $v^2 = 2(4)(36)$ giving $v = 12\sqrt{2}$,
i.e. on returning to the point A the particle has speed $12\sqrt{2} \text{ m s}^{-1}$.

**Exercises 5.1**
Where necessary take $g$ to be $10 \text{ m s}^{-2}$.
1   A particle slides down a smooth plane inclined at 30° to the horizontal. It passes point A with speed $4 \text{ m s}^{-1}$ and point B with speed $12 \text{ m s}^{-1}$. Find the distance AB and the time the particle takes to travel from A to B.
2   A particle is projected with speed $4 \text{ m s}^{-1}$ from a point X down a line of greatest slope of a smooth plane which makes an angle $\tan^{-1}(5/12)$ with the horizontal. When the particle passes through a point Y it has speed $24 \text{ m s}^{-1}$. Find the distance XY and find also the times taken to cover the distances XM and MY, where M is the mid-point of XY.
3   From a point A a particle is projected with speed $20 \text{ m s}^{-1}$ up a line of greatest slope of a smooth plane which makes an angle $\tan^{-1}(4/3)$ with the horizontal. The particle comes to instantaneous rest at the point B. Find the distance AB and the time the particle takes to go from A to B. Find also the speed of the particle as it passes through C, the mid-point of AB.

**4**  A particle is released from rest at a point A on a rough plane inclined at an angle $\tan^{-1}(4/3)$ to the horizontal. The coefficient of friction between the particle and the plane is $\frac{1}{3}$. Find the speed of the particle when it passes through the point B, where AB = 8 m, and find also the time taken in going from A to B.

**5**  At a given instant a particle is moving with speed $6 \text{ m s}^{-1}$ down a line of greatest slope of a rough inclined plane. The plane makes an angle of $60°$ with the horizontal and the coefficient of friction between the particle and the plane is $\frac{1}{2}$. Find how long it will take the particle to reach a speed of $20 \text{ m s}^{-1}$ and how far it will move in this time.

**6**  From a point A a particle is projected with speed $30 \text{ m s}^{-1}$ up a line of greatest slope of a rough plane inclined at an angle $\tan^{-1}(5/12)$ to the horizontal. The coefficient of friction between the particle and the plane is $\frac{1}{3}$. The particle comes to instantaneous rest at the point B. Find the distance AB and the time taken to reach B. Find also the speed of the particle when it returns to the point A.

**7**  A particle is released from rest at a point on a smooth plane inclined at an angle $\alpha$ to the horizontal. Given that the particle travels 40 m in the first 4 seconds, find $\alpha$.

**8**  A particle is sliding down a smooth inclined plane. It passes through point A with speed $4 \text{ m s}^{-1}$ and through point B with speed $10 \text{ m s}^{-1}$. Given that AB = 7 m, find the angle the plane makes with the horizontal.

**9**  A particle is released from rest from a point on a rough plane inclined at an angle $\tan^{-1}(4/3)$ to the horizontal. Given that it travels 10 m in the first two seconds, find the coefficient of friction between the plane and the particle.

**10**  A particle is released from rest on a rough inclined plane and takes 26 seconds to reach a speed of $20 \text{ m s}^{-1}$. Given that the coefficient of friction between the particle and the plane is $\frac{1}{3}$, find the angle the plane makes with the horizontal.

In questions 11 and 12 the engine of a car of mass $M$ kg produces a constant tractive force of magnitude $P$ N. The car is on a level road and there is a resistance to motion of magnitude $R$ N.

**11**  Given that $M = 1000, P = 600$ and $R = 200$, find the time taken to reach a speed of $30 \text{ m s}^{-1}$ from rest and the distance travelled in this time.

**12**  Given that $M = 1200, P = 800$ and $R = 180$, find the time taken and the distance travelled as the speed of the car increases from $10 \text{ m s}^{-1}$ to $20 \text{ m s}^{-1}$.

**13**  The engine of a car of mass 900 kg exerts a constant tractive force of 750 N. The car moves along a straight road inclined at an angle $\sin^{-1}(1/20)$ to the horizontal and there is a non-gravitational resistance to motion of magnitude 200 N. Find the time taken to reach from rest a speed of $25 \text{ m s}^{-1}$ and the distance travelled in reaching from rest a speed of $20 \text{ m s}^{-1}$ (a) up the road (b) down the road.

14 A train of mass 200 tonnes travels on a straight track inclined at an angle $\sin^{-1}$ (1/100) to the horizontal. The engine exerts a tractive force of magnitude 25 000 N and there is a non-gravitational resistance to motion of magnitude 2000 N. Find the acceleration of the train when travelling (a) up the track (b) down the track. Find also the distance travelled up the track in reaching from rest a speed of 108 km h$^{-1}$ and the time taken to reach from rest a speed of 96 km h$^{-1}$ when travelling down the track.

15 A crate of mass 20 kg stands on the floor of a lift. Find the reaction of the floor on the crate when the lift is ascending with (a) acceleration 3 m s$^{-2}$ (b) retardation 4 m s$^{-2}$.

16 An object of mass 5 kg is suspended from the ceiling of a lift by a spring balance. Find the reading of the spring balance when the lift is

(a) ascending with constant speed 4 m s$^{-1}$

(b) moving with acceleration 2 m s$^{-2}$ upwards

(c) moving with acceleration 1.5 m s$^{-2}$ downwards.

## 5.2 The motion of connected bodies

In Section 5.1 forces acting on a single body and the motion produced were dealt with. The motion of connected bodies will now be considered. There are many examples of this including a car towing a caravan or trailer, a locomotive pulling a set of coaches and objects connected by strings. Newton's third law states, 'To every action there is an equal and opposite reaction'. Thus the force exerted by the car on the caravan is equal in magnitude but opposite in direction to the force exerted by the caravan on the car. These forces are transmitted by the tow-bar connecting the caravan to the car. The situation is illustrated in Fig. 5.7. The tension in the tow-bar provides a force of magnitude $T$ in the direction of motion on the caravan and a force of magnitude $T$ in the opposite direction on the car.

A problem on the motion of two connected bodies can be solved by forming the equation $F = ma$

(a) for each body separately, or

(b) for one body and for the whole system.

It is essential that any diagrams should show clearly the forces acting on each part of the system.

*Example 1*

A car of mass 1000 kg tows a caravan of mass 500 kg along a straight level road. The engine of the car produces a constant tractive force of magnitude 1425 N. There is a constant resistance of magnitude 0.2 N per kg on both the car and the caravan. Find the time taken by the car and caravan in reaching a speed of 81 km h$^{-1}$ from rest. Find also the tension in the tow-bar.

(Take $g$ to be 10 m s$^{-2}$.)

The resistance on the car $= 0.2 \times 1000$ N $= 200$ N and the resistance on the caravan $= 0.2 \times 500$ N $= 100$ N.

Let the tension in the tow-bar be of magnitude $T$ N.

Figure 5.7 shows diagrammatically all the forces, in newtons, acting on the whole system consisting of the car and the caravan.

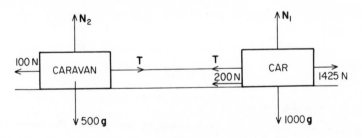

Fig. 5.7

The whole system, consisting of the car and caravan, has a total mass of 1500 kg and is subjected to a pull of magnitude 1425 N and a resistance of magnitude 300 N. Figure 5.8(a) shows diagrammatically these forces, in newtons and Fig. 5.8(b) shows the acceleration.

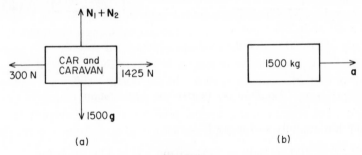

(a)           (b)

Fig. 5.8

Using $F = ma$ in the direction of motion for the whole system and assuming that the whole system can be treated as a particle,

$$1425 - 300 = 1500a \quad \text{giving} \quad a = 0.75.$$

The acceleration is constant and the final velocity is 81 km h$^{-1}$, i.e. 22.5 m s$^{-1}$. Using $v = u + at$, yields $22.5 = (0.75)t$ giving $t = 30$, i.e. a speed of 81 km h$^{-1}$ is reached from rest in 30 seconds.

Next the motion of the car only will be considered. The forces, in newtons, acting on the car are shown diagrammatically in Fig. 5.9(a) and Fig. 5.9(b) shows the acceleration.

Again resolving horizontally and using $F = ma$

$$1425 - 200 - T = 1000a = 750 \quad \text{giving} \quad T = 475.$$

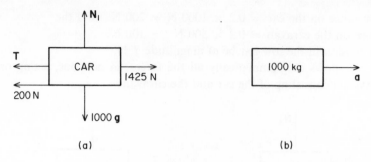

(a)                                                    (b)

Fig. 5.9

i.e. the tension in the tow-bar is 475 N.

This tension could also have been found by considering the motion of the caravan alone. Figure 5.10(a) shows the forces acting on the caravan and Fig. 5.10(b) shows the acceleration.

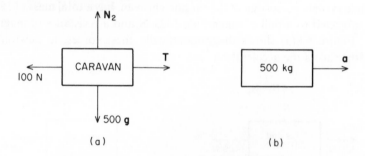

(a)                                                    (b)

Fig. 5.10 (a) Forces (b) Acceleration

Resolving horizontally and using $F = ma$

$$T - 100 = 500a = 375 \quad \text{giving} \quad T = 475 \quad \text{as before.}$$

*Example 2*
A light inextensible string passes over a fixed smooth light pulley and particles of mass $m_1$ and $m_2$ are attached to the ends A and B. Find the tension in the string and the acceleration of each particle when the system is moving freely.

Since the string is light the tension in it will be the same throughout its length. Since it is inextensible the particles will have the same speed and the same magnitude of acceleration. Since the pulley is light and smooth the tension in the string will be the same on both sides of the pulley.

Let the tension in the string have magnitude $T$ and let particle A have downward acceleration and particle B upward acceleration of magnitude $a$.
Figure 5.11 shows the system. Figure 5.12(a) shows the forces acting on the particle A. Figure 5.12(b) shows the forces acting on the particle B.

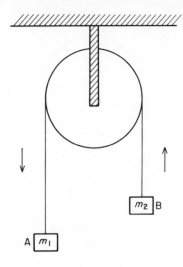

Fig. 5.11

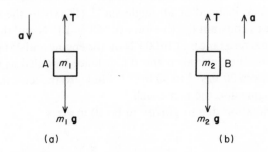

(a)                    (b)

Fig. 5.12

Using $F = ma$ for A          $m_1g - T = m_1a$          ...(1)

Using $F = ma$ for B          $T - m_2g = m_2a$          ...(2)

Adding equations (1) and (2)   $(m_1 - m_2)g = (m_1 + m_2)a$

giving

$$a = \frac{(m_1 - m_2)g}{(m_1 + m_2)}$$

and          $$T = m_2g + m_2a = \frac{2m_1m_2g}{m_1 + m_2}.$$

The forces acting on the light pulley might also be considered.

Let the beam to which the pulley is fixed exert a force vertically upwards of magnitude $P$ on the pulley. The forces acting on the pulley are shown in Fig. 5.13.

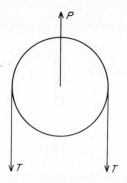

Fig. 5.13

Since the pulley is fixed $P = 2T = \dfrac{4m_1 m_2 g}{m_1 + m_2}$.

*Example 3*

A railway engine of mass 100 tonnes pulls a set of coaches of total mass 400 tonnes up a track inclined at an angle $\sin^{-1}(1/200)$ to the horizontal. The engine produces a constant tractive force of $160 \times 10^3$ N and there are constant non-gravitational resistances of $10\,000$ N on the engine and $25\,000$ N on the rest of the train. Find the time taken and the distance travelled as the speed of the train increases from $20$ m s$^{-1}$ to $40$ m s$^{-1}$. Find also the tension in the coupling between the engine and the first coach.

(Take the acceleration due to gravity to be $10$ m s$^{-2}$.)

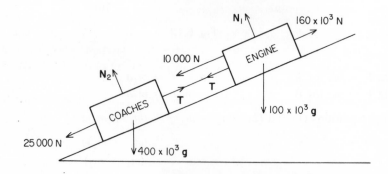

Fig. 5.14

Figure 5.14 shows diagrammatically the forces acting on the whole system consisting of the engine and coaches.

Dealing with the whole system consisting of the engine and coaches, it is a train of total mass $500 \times 10^3$ kg subjected to a pull of $160 \times 10^3$ N and to a total resistance of $35\,000$ N. Figure 5.15(a) shows diagrammatically the forces acting on the whole system and Fig. 5.15(b) shows the acceleration.

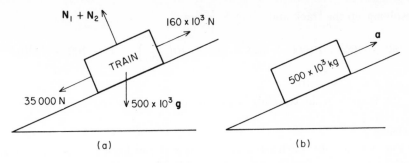

Fig. 5.15 (a) Forces (b) Acceleration

Resolving up the track and using $F = ma$

$$160 \times 10^3 - 35\,000 - 500 \times 10^3 \times 10\left(\frac{1}{200}\right) = 500 \times 10^3 a$$

∴ $$125 \times 10^3 - 25 \times 10^3 = 500 \times 10^3 a$$

giving $a = 0.2$,
i.e. the train moves with constant acceleration $0.2\ \mathrm{m\,s^{-2}}$.
Using $v = u + at$ with $u = 20$, $v = 40$ and $a = 0.2$,

$$40 = 20 + 0.2t \quad \text{giving} \quad t = 100,$$

i.e. the time taken as the speed increases from $20\ \mathrm{m\,s^{-1}}$ to $40\ \mathrm{m\,s^{-1}}$ is 100 seconds or 1 minute 40 seconds.
Using $v^2 = u^2 + 2as$ with $u = 20$, $v = 40$ and $a = 0.2$,

$$1600 = 400 + 2(0.2)s \quad \text{giving} \quad s = 3000$$

i.e. the distance travelled during this increase in speed is 3000 m or 3 km.

To find the tension in the coupling between the engine and the first coach consider the motion of either the engine alone or the set of coaches alone. The engine alone is chosen here.
Figure 5.16(a) shows diagrammatically the forces acting on the engine and Fig. 5.16(b) shows the acceleration.

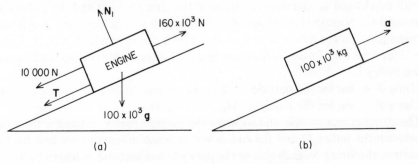

Fig. 5.16 (a) Forces (b) Acceleration

Resolving up the track and using $F = ma$

$$160 \times 10^3 - 10\,000 - 100 \times 10^3 \times 10\left(\frac{1}{200}\right) - T = 100 \times 10^3(0.2)$$

giving $T = 125 \times 10^3$,
i.e. the tension in the coupling is $125\,000$ N, i.e. $125$ kN.

*Example 4*
In Fig. 5.17 a light inextensible string has one end fixed to a beam at A and carries a particle of mass $m$ at the other end B. The string passes under a smooth pulley C of mass $M$ and over a fixed smooth pulley D. The portions of the string not in contact with the pulleys are vertical. When the system is released, the particle at the end B moves with upward acceleration $a$. Find $a$ and the tension in the string.

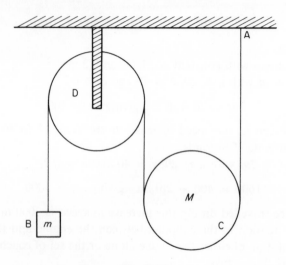

Fig. 5.17

The particle at B moves with acceleration $a$ upwards. Let the pulley C move with downward acceleration $a_1$. Because the string is light and the pulleys are smooth the tension throughout the whole length will be the same. Let it have magnitude $T$. The forces acting on the system are shown in Fig. 5.18. Figure 5.18(a) shows the forces on the particle at B. Figure 5.18(b) shows the forces on the pulley C.

Using $F = ma$ for the particle     $T - mg = ma$        ...(1)
Using $F = ma$ for the pulley     $Mg - 2T = Ma_1$      ...(2)
The string is inextensible and so when the particle rises a distance $x$ the string between the pulley D and the end A will increase in length by $x$ and the two parts of the string on each side of the pulley C will increase in length by $\frac{1}{2}x$, i.e. the particle moves twice as far as the pulley C and so

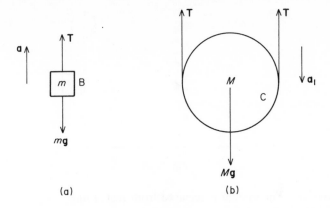

(a)                    (b)

Fig. 5.18

$$a = \frac{d^2x}{dt^2} \quad \text{and} \quad a_1 = \frac{d^2(\frac{1}{2}x)}{dt^2} = \frac{1}{2}\frac{d^2x}{dt^2}, \text{ i.e. } a = 2a_1. \qquad \dots(3)$$

Solving equations (1), (2) and (3) gives

$$a = \frac{2(M - 2m)g}{M + 4m} \quad \text{and} \quad T = \frac{3Mmg}{M + 4m}.$$

## Exercises 5.2
Where required take $g$ to be $10 \text{ m s}^{-2}$.

In questions 1–3 a light inextensible string passes over a fixed smooth pulley and has particles of masses $m$ and $M$ attached to its ends. The system is released. Find the tension in the string and the acceleration of each particle when
1   $M = 4$ kg and $m = 2$ kg.
2   $M = 3$ kg and $m = 1$ kg.
3   $M = 1.5$ kg and $m = 0.8$ kg.

In questions 4–6 a car of mass $M$ kg is towing a trailer of mass $m$ kg along a straight level road. The engine produces a tractive force of magnitude $P$ N and there is a constant resistance to the motion of the car of magnitude $R_1$ N and a constant resistance to the motion of the trailer of magnitude $R_2$ N. Calculate the tension in the tow-bar and the acceleration of the car and trailer when
4   $M = 1200$, $m = 300$, $P = 2000$, $R_1 = 200$, $R_2 = 100$.
5   $M = 1000$, $m = 400$, $P = 1800$, $R_1 = 120$, $R_2 = 80$.
6   $M = 2000$, $m = 800$, $P = 2400$, $R_1 = 300$, $R_2 = 100$.

Figure 5.19 applies to questions 7–12.
The particle A of mass $m$ is on a smooth plane inclined at an angle $\alpha$ to the horizontal. It is connected to a particle B of mass $M$ by a light inextensible string which passes over a smooth fixed pulley C at the top of the plane. From A to C the string is parallel to a line of greatest slope of the plane and from C to B the

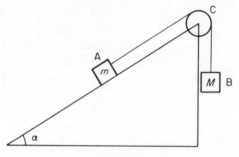

Fig. 5.19

string is vertical. The system is released from rest. Find
(a) the acceleration of A
(b) the tension in the string
(c) the resultant force exerted by the string on the pulley C
when

7  $m = 2$ kg, $M = 3$ kg,  $\alpha = 30°$.
8  $m = 1$ kg, $M = 2$ kg, $\sin \alpha = \frac{1}{3}$.
9  $m = 2$ kg, $M = 2$ kg, $\sin \alpha = \frac{1}{4}$.

If the plane in Fig. 5.19 is rough and the coefficient of friction between the plane and the particle is $\mu$, find
(a) the acceleration of B
(b) the tension in the string
when

10  $m = 2$ kg, $M = 4$ kg,  $\alpha = 30°, \mu = \frac{1}{2}$.
11  $m = 1$ kg, $M = 3$ kg, $\sin \alpha = \frac{1}{4}$,  $\mu = \frac{1}{3}$.
12  $m = 1$ kg, $M = 2$ kg, $\sin \alpha = \frac{1}{3}$,  $\mu = \frac{1}{4}$.

13  A car of mass 1500 kg tows a caravan of mass 500 kg up a road inclined at an angle $\sin^{-1} (1/10)$ to the horizontal. The engine produces a constant tractive force of magnitude 3500 N. There is a resistance of magnitude 0.3 N per kg opposing the motion of both the car and the caravan. Find
(a) the acceleration of the car and caravan
(b) the tension in the tow-bar.
(Take $g$ to be $10 \text{ m s}^{-2}$.)

## 5.3  Variable forces

In all the preceding examples using the equation $F = ma$, the force $\mathbf{F}$ was constant. Consequently the acceleration $\mathbf{a}$ was constant and so the equations for uniformly accelerated motion could be used. In the following examples the magnitude of $\mathbf{F}$ is not constant and so the magnitude of $\mathbf{a}$ is not constant. Consequently the equations for uniformly accelerated motion *cannot be used*. Instead the methods of Sections 3.3 and 3.4 are used in conjunction with the equation $F = ma$.

*Example 1*

A particle of mass $m$ moves along the $x$-axis and at time $t$ is at a distance $x$ from the origin O. The particle is acted upon by a force of magnitude $\lambda(2 + 6t^2)$ in the direction O$x$, where $\lambda$ is a constant. When $t = 0$, the particle is at rest at O. Find the speed of the particle when $t = 2$ and the distance travelled in the interval $2 \leqslant t \leqslant 4$.

Using $F = ma$ gives $\quad \lambda(2 + 6t^2) = ma$

and putting $a = \dfrac{dv}{dt}$ gives $\quad \dfrac{dv}{dt} = \dfrac{\lambda}{m}(2 + 6t^2).$

Integrating $\qquad\qquad\qquad v = \dfrac{\lambda}{m}(2t + 2t^3) + C_1$

and $v = 0$ when $t = 0$, $\therefore C_1 = 0$,

giving $\qquad\qquad\qquad v = \dfrac{\lambda}{m}(2t + 2t^3).$

When $t = 2$ $\qquad\qquad v = \dfrac{\lambda}{m}(4 + 16) = 20\lambda/m,$

i.e. when $t = 2$ the speed of the particle is $20\lambda/m$.

Since $v = \dfrac{dx}{dt}$, $\qquad \dfrac{dx}{dt} = \dfrac{\lambda}{m}(2t + 2t^3).$

Integrating $\qquad\qquad\quad x = \dfrac{\lambda}{m}(t^2 + \tfrac{1}{2}t^4) + C_2$

and $x = 0$ when $t = 0$, $\therefore C_2 = 0$,

giving $\qquad\qquad\qquad x = \dfrac{\lambda}{m}(t^2 + \tfrac{1}{2}t^4).$

When $t = 4$, $\qquad\qquad x = \dfrac{\lambda}{m}(16 + 128) = \dfrac{144\lambda}{m},$

When $t = 2$, $\qquad\qquad x = \dfrac{\lambda}{m}(4 + 8) \quad = \dfrac{12\lambda}{m}.$

$\therefore$ Distance travelled in the interval $2 \leqslant t \leqslant 4$ is $\dfrac{\lambda}{m}(144 - 12) = 132\lambda/m.$


*Example 2*

A particle of mass $m$ kg starts from rest and moves in a straight line under the action of a constant force of magnitude $X$ newtons. There is a resistance to motion of $ks^2$ newtons, where $k$ is a constant and $s$ metres is the distance the particle has moved. Find the speed of the particle when it has moved a distance of 6 metres.

Using $F = ma$    $X - ks^2 = ma$   and since   $a = v\dfrac{dv}{ds}$

$$X - ks^2 = mv\dfrac{dv}{ds}.$$

Integrating with respect to $s$   $\displaystyle\int (X - ks^2)ds = \int mv\dfrac{dv}{ds}ds = \int mv\,dv.$

$\therefore$    $Xs - \tfrac{1}{3}ks^3 = \tfrac{1}{2}mv^2 + C.$

$v = 0$ when $s = 0$, $\therefore$ $C = 0$,

giving    $$v^2 = \dfrac{6Xs - 2ks^3}{3m}.$$

When $s = 6$    $$v^2 = \dfrac{36X - 432k}{3m},$$

i.e. when the particle has moved 6 metres its speed is $\left[\dfrac{12(X - 12k)}{m}\right]^{1/2}.$

*Example 3*

The tractive force produced by the engine of a car of mass 800 kg varies, and its values at speeds from 0 to 30 m s$^{-1}$ are given in the following table. The car is moving on a straight level road and there is a constant resistance to motion of magnitude 300 N. Obtain an estimate, to 3 significant figures, of the time taken by the car to reach a speed of 30 m s$^{-1}$.

| Speed (m s$^{-1}$) | 0 | 5 | 10 | 15 | 20 | 25 | 30 |
|---|---|---|---|---|---|---|---|
| Tractive force (N) | 2700 | 2650 | 2500 | 2340 | 2180 | 1980 | 1500 |

The forces acting on the car *at the start* are shown in Fig. 5.20(a). Figure 5.20(b) shows the acceleration.

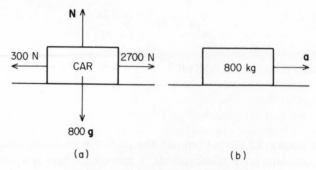

Fig. 5.20 (a) Forces (b) Acceleration

Using $F = ma$, $2700 - 300 = 800a$
giving $a = 3$,
i.e. the initial acceleration is $3 \text{ m s}^{-2}$.

Similarly the acceleration at each of the other given speeds can be calculated. The reciprocal of the acceleration for each speed can then be calculated. The values of the seven accelerations and their reciprocals are given in the following table:

| Speed $(\text{m s}^{-1})$ | 0 | 5 | 10 | 15 | 20 | 25 | 30 |
|---|---|---|---|---|---|---|---|
| $a$ $(\text{m s}^{-2})$ | 3.00 | 2.94 | 2.75 | 2.55 | 2.35 | 2.10 | 1.50 |
| $1/a$ | 0.33 | 0.34 | 0.36 | 0.39 | 0.43 | 0.48 | 0.67 |

From Section 3.4 it is known that, if the graph of $1/a$ is plotted against $v$, the 'area under the curve' between $v = 0$ and $v = 30$, the time taken by the car to reach a speed of $30 \text{ m s}^{-1}$, will be obtained. Evaluating this area by means of Simpson's rule gives the required time as approximately 12.4 seconds.

## Exercises 5.3

1  A particle of mass $\frac{1}{2}$ kg moves along the $x$-axis under the action of a force of magnitude $F$ N in the direction $Ox$. After moving for $t$ seconds the particle is at a distance $x$ metres from O and has velocity $v \text{ m s}^{-1}$. Given that $v = 6$ and $x = 4$ when $t = 0$, find the velocity and position of the particle when $t = 5$ for
(a) $F = 2t$        (b) $F = 6t^2 + 3$        (c) $F = 4 \sin t$.

2  A particle of mass 2 kg starts from a point O with speed $3 \text{ m s}^{-1}$. The particle moves in a straight line under the action of a force of magnitude $F$ N, given by $F = 2s + 6$, where $s$ metres is the distance of the particle from O. Find
(a) the speed of the particle when $s = 8$
(b) the distance of the particle from O after 6 seconds.

3  A particle of mass 1.5 kg starts from rest and moves in a straight line under the action of a force of magnitude $F$ N, given by $F = 6 \cos 2s$, where $s$ metres is the distance of the particle from its starting point. Find the velocity of the particle when $s = \pi/6$.

4  A particle of mass 3 kg starts from rest and after moving in a straight line for $t$ seconds it has velocity $v \text{ m s}^{-1}$. The particle is acted upon by a variable force of magnitude $(6v + 3)$ N. Find
(a) the speed of the particle when $t = 4$
(b) the distance moved by the particle in the interval $0 \leqslant t \leqslant 4$.

5  A particle of mass 3 kg moves in a straight line under the action of a variable force of magnitude $F$, given by $F = 3v^2 + 6$, where $v \text{ m s}^{-1}$ is the speed of the particle. Given that the particle starts from rest, find
(a) the distance travelled by the particle as $v$ increases from 0 to 10
(b) the speed of the particle after travelling 30 m.

6  A particle of mass 2 kg initially at rest is acted upon by a variable force

whose magnitude at one second intervals is given in the following table.

| Time (s) | 0 | 1 | 2 | 3 | 4 | 5 | 6 |
|---|---|---|---|---|---|---|---|
| Force (N) | 2 | 6 | 18 | 38 | 66 | 102 | 146 |

Given that the direction of the force is constant estimate graphically
(a) the speeds of the particle after 1, 2, 3, 4, 5, 6 seconds
(b) the distance moved by the particle in the six seconds.

7  A particle of mass $\frac{1}{2}$ kg starts with an initial speed of 4 m s$^{-1}$. It moves in a straight line under the action of a variable force whose magnitudes at intervals of $\pi/6$ seconds are given in the following table.

| Time (s) | 0 | $\pi/6$ | $\pi/3$ | $\pi/2$ |
|---|---|---|---|---|
| Force (N) | 2.5 | 2.73 | 2.37 | 1.5 |

Estimate graphically
(a) the speeds of the particle after $\pi/6$, $\pi/3$ and $\pi/2$ seconds
(b) the distance moved by the particle in $\pi/2$ seconds.

8  A particle of mass 4 kg starts with speed 2 m s$^{-1}$ and moves in a straight line under the action of a force which has magnitude $F$ N when the particle has moved $s$ metres. Corresponding values of $s$ and $F$ are given in the following table.

| $s$ | 0 | 10 | 20 | 30 | 40 |
|---|---|---|---|---|---|
| $F$ | 12 | 92 | 172 | 252 | 332 |

Estimate graphically
(a) the speeds of the particle when $s = 10, 20, 30, 40$
(b) the time taken by the particle in moving 40 m.

9  A particle of mass 20 kg moves in a straight line under the action of a variable force whose magnitude at different speeds is given in the following table.

| Speed (m s$^{-1}$) | 0 | 5 | 10 | 15 | 20 |
|---|---|---|---|---|---|
| Force (N) | 400 | 350 | 300 | 250 | 200 |

Given that there is a constant resistance to the motion of the particle of magnitude 100 N, find, graphically or otherwise, the time taken by the particle to reach a speed of 20 m s$^{-1}$ from rest.

10  The engine of a vehicle of mass 2 tonnes exerts a constant tractive force of a magnitude 2000 N against a variable resistance. The magnitudes of this resistance at different speeds are given in the following table.

| Speed (m s$^{-1}$) | 0 | 6 | 12 | 18 | 24 | 30 |
|---|---|---|---|---|---|---|
| Resistance (N) | 100 | 118 | 172 | 262 | 388 | 550 |

Estimate graphically the time taken by the vehicle in reaching a speed of 30 m s$^{-1}$ from rest on a straight level road.

## Miscellaneous exercises 5

1  A train is travelling on a straight level track at a speed of 108 km h$^{-1}$. The

brakes are applied to produce a constant resisting force of magnitude $R$ newtons per tonne. The train is brought to rest in a distance of 1 km. Assuming that other resistances are negligible, find $R$.

When the train is travelling at a speed of $108 \text{ km h}^{-1}$ down a track inclined at an angle $\alpha$ to the horizontal, it is brought to rest in a distance of 1.5 km. Assuming that the brakes produce the same constant resisting force and that other resistances are negligible, find

(a) the time taken in stopping the train

(b) $\sin \alpha$.

(Take $g$ to be $10 \text{ m s}^{-2}$.)

2 A cyclist pedals to produce a tractive force whose magnitude exceeds the magnitude of the non-gravitational resistance to his motion by a constant 20 N. The mass of the man and his cycle is 90 kg. Find the time it would take for him to reach from rest a speed of $72 \text{ km h}^{-1}$ down a road inclined at an angle $\sin^{-1}(1/40)$ to the horizontal. Find also the distance he would travel in this time.

Given that he starts moving up this road when his speed is $54 \text{ km h}^{-1}$, find how far he would travel before coming to rest, assuming that he continues to produce the same net tractive force.

(Take $g$ to be $10 \text{ m s}^{-2}$.)

3 The engine of a car of mass 1800 kg produces a constant tractive force of magnitude 4300 N against a constant resistance. Given that the car reaches a speed of $96 \text{ km h}^{-1}$ from rest in 12 seconds on a level road, find the magnitude of this resistance.

Given that the pull and the resistance remain unchanged, find the distance travelled by the car in reaching a speed of $96 \text{ km h}^{-1}$ up a road inclined at an angle $\sin^{-1}(1/15)$ to the horizontal.

(Take $g$ to be $10 \text{ m s}^{-2}$.)

4 A car of mass 800 kg runs at constant speed with the engine switched off down a road which makes an angle $\sin^{-1}(1/40)$ with the horizontal. Find the non-gravitational resistance to the motion of the car.

With the engine switched on, the car can reach a speed of $30 \text{ m s}^{-1}$ in 40 seconds up this road. Assuming that the pull of the engine is constant and that the non-gravitational resistance remains unchanged, find the pull of the engine.

When the car is moving up this road at a speed of $30 \text{ m s}^{-1}$, the engine is switched off. Find the distance travelled by the car in coming to rest when the brakes are not applied. When the brakes are applied the car comes to rest in a distance of 100 m. Find the resisting force produced by the brakes assuming this to be constant.

(Take $g$ to be $10 \text{ m s}^{-2}$.)

5 A parcel lies on the floor of a lorry which is travelling along a straight level road. Given that the lorry accelerates uniformly from a speed of $10 \text{ m s}^{-1}$ to a speed of $25 \text{ m s}^{-1}$ in a distance of 100 m and that the parcel does not

slide on the floor of the lorry, find the least possible value of $\mu$, the coefficient of friction, between the parcel and the floor of the lorry.

If $\mu = \frac{1}{4}$, find the speed of the parcel relative to the lorry after sliding 4 m relative to the lorry.

(Take $g$ to be $10 \text{ m s}^{-2}$.)

6  A car of mass 1200 kg pulls a trailer of mass 300 kg. There is a total non-gravitational resistance of 500 N and this is divided between the car and the trailer in the ratio of their masses. The engine of the car produces a constant pull and the car and trailer accelerate from a speed of 12 m s$^{-1}$ to a speed of 24 m s$^{-1}$ in a distance of 162 m. Find the magnitude of
(a) the pull of the engine   (b) the tension in the tow-bar
when the motion takes place
(i) on level ground
(ii) up a road inclined at an angle $\sin^{-1}(1/30)$
(iii) down this road.
(Take $g$ to be $10 \text{ m s}^{-2}$.)

7  A particle of mass $m$ moving vertically upwards with speed $u$ enters a fixed horizontal layer of material of thickness $a$ which resists its motion with a constant force of magnitude $R$. Show that it will pass through the layer if $u > u_0$, where

$$u_0^2 = 2a\left(\frac{R}{m} + g\right).$$

The particle emerges from the layer and moves freely under gravity until it re-enters the layer, which resists its motion as before. Show that the particle will pass completely through the layer in its downward motion only if $u^2 \geqslant 4aR/m$.                                              [JMB]

8  A particle moves along a line of greatest slope of a rough plane inclined at an angle $\beta$ to the horizontal, where $\tan \beta = \frac{3}{4}$. The particle passes through a point A when moving upwards with speed $u$, comes momentarily to rest at a point C and subsequently passes through a point B when moving downwards at the same speed $u$. Given that the coefficient of friction between the particle and the plane is $\frac{1}{4}$, find AC and the speed of the particle on its return to A. Show that AB = AC.                                      [L]

9  A lift travels vertically a distance of 22 m from rest at the basement to rest at the top floor. Initially the lift moves with constant acceleration $x$ m s$^{-2}$ for a distance of 5 m; it continues with constant speed $u$ m s$^{-1}$ for 14 m; it is then brought to rest at the top floor by a constant retardation $y$ m s$^{-2}$. State, in terms of $u$ only, the times taken by the lift to cover the three stages of its journey.

If the total time that the lift is moving is 6 seconds, calculate
(i) the value of $u$
(ii) the values of $x$ and $y$.

If a box of mass 30 kg is standing on the floor of the lift, calculate the reaction between the box and the floor in each of the three stages of the journey.

(Take the acceleration due to gravity to be $10 \text{ m s}^{-2}$.)  [AEB]

10  A particle of mass $m$ hangs freely from a light inextensible string, the other end of which is attached to a lift which moves upwards with a constant acceleration $f$, and the tension in the string is $T$. Express $T$ in terms of $f$, $g$ and $m$.

In a lift at rest at the bottom of a vertical shaft a particle of mass $m$ is suspended from a light inextensible string. The lift then rises with constant acceleration for a time $t_1$ and then slows down with constant retardation (less than $g$) to rest after a further time $t_2$. During the period of acceleration the tension in the string is $T_1$ and during the period of retardation it is $T_2$. Show that

$$mg(t_1 + t_2) = T_1 t_1 + T_2 t_2.$$  [JMB]

11  A light inextensible string passes over a fixed light smooth pulley and carries scale pans A and B, each of mass 4 kg, at its ends. An inelastic ball of mass 2 kg is placed in the scale pan A. The system is released from rest with the string taut and the straight parts of the string vertical. Find

(a) the tension, in newtons, in the string

(b) the acceleration, in $\text{m s}^{-2}$, of the ball

(c) the force, in newtons, exerted by the ball on the scale pan A

(d) the force, in newtons, exerted by the string on the pulley.  [L]

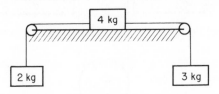

Fig. 5.21

12  Figure 5.21 shows a mass of 4 kg on a rough horizontal table, and two light inextensible strings attached to it passing over smooth pulleys and supporting masses of 2 kg and 3 kg which hang freely. The whole system is in the vertical plane containing the two pulleys. If the system is in limiting equilibrium, show that the coefficient of friction between the 4 kg mass and the table is $\frac{1}{4}$.

The 3 kg mass is replaced by another mass of 4 kg, and the system is released. Find the acceleration of the 2 kg mass and the tensions in the strings. (Take $g$ as $10 \text{ m s}^{-2}$.)

The system is moving at $5 \text{ m s}^{-1}$ when the 4 kg mass which is falling vertically hits the ground (from which it does not rebound). Find the time that elapses before the other two masses become instantaneously at rest. (Assume throughout that no mass hits a pulley.)  [C]

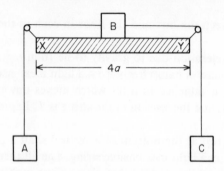

Fig. 5.22

13  Figure 5.22 shows a particle B on a horizontal bench XY of length $4a$. The particle B is attached by two inextensible light strings to two particles A and C, which hang over frictionless small pulleys at opposite ends of the bench. The two strings are each longer than $4a$ and the particle B is initially held at X. The particles A, B and C have masses $m$, $3m$ and $5m$, respectively. If the system is released from rest and friction between the particle B and the bench is negligible, calculate

(i) the common acceleration of the particles A, B and C
(ii) the speed of particle B at the instant it reaches Y
(iii) the time taken by the particle B to reach Y from X.       [AEB]

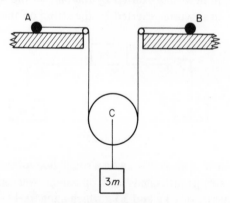

Fig. 5.23

14  Particles A and B rest on smooth horizontal planes and are connected by a light inextensible string which passes over smooth pulleys at the edges of the planes and under a smooth light pulley C which carries a mass $3m$ suspended from its axis (Fig. 5.23). When the system is released from rest the distances moved by A and B in a given time are $x$ and $y$ respectively. Express the distance fallen by C in the same time in terms of $x$ and $y$. If the particles A and B are of mass $m$ and $\lambda m$ respectively, find the acceleration of C and the tension in the string when (i) $\lambda = 1$ (ii) $\lambda = 2$.       [L]

**15** Two particles A and B of masses $m$ and $nm$ respectively, are connected by a light inextensible string. Particle A is placed on a rough fixed plane inclined at an angle $\beta$ to the horizontal. The coefficient of friction between A and the inclined plane is $\frac{1}{3} \tan \beta$. The string lies along a line of greatest slope of the inclined plane and after passing over a small smooth pulley P at the top of the plane supports B which hangs freely. The system is released from rest with the hanging part PB vertical. Prove that B will not descend unless $n > (4 \sin \beta)/3$.

Find the acceleration of each particle when $n = 3$ and show that the tension in the string is then

$$mg(3 + 4 \sin \beta)/4.$$

Also, if $\beta = \pi/6$, show that the force exerted by the string on the pulley is of magnitude $(5 \, mg\sqrt{3})/4$. [L]

**16** A particle of mass $M$ rests on a rough plane inclined at an angle $\alpha$ to the horizontal, where $\sin \alpha = 0.8$. The particle is attached to one end of a light inelastic string which lies along a line of greatest slope of the plane, and passes over a small smooth pulley at the top of the plane. Hanging freely from the other end of the string is a particle of mass $m$. If the largest value of $m$ for which equilibrium is possible is $M$, find the smallest such value.

Find the acceleration of the particles and the tension in the string when $m = 5M$. [L]

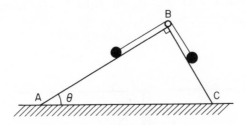

Fig. 5.24

**17** In Fig. 5.24, ABC is the right section of a prism; the angle BAC is $\theta$ ($< 45°$) and the angle ABC is $90°$. Two particles, each of mass $m$, are on the smooth sloping faces of the prism and are connected by a light inextensible string which passes over a smooth pulley in the top edge of the prism. The prism stands on a horizontal plane which is rough enough to prevent the prism moving. The system is released from rest when the string is in the plane ABC. Find the acceleration of the particles and the tension in the string when the particles are moving freely.

If the prism is of mass $M$, find the vertical component of the reaction between the prism and the horizontal plane. [L]

**18** A particle of mass $m$ is at rest on a smooth horizontal table at a distance $d$ from a small hole in the table. A light inextensible string has one end attached to the particle and passes through the hole carrying at its other end a second particle of mass $3m$. The system is released from rest when both parts of the string are taut. Find the time taken by the first particle to reach the hole.

The second particle is replaced by a light pulley and a light inextensible string is hung over this pulley. To one end of this string is attached a particle of mass $2m$ and a particle of mass $m$ is attached to the other end. Show that, when the system is released from rest as before, the acceleration of the particle on the table is $8g/11$, and find the tension in each string.

[L]

19 A light inextensible string passes over a smooth fixed pulley and has a particle of mass $5m$ attached to one end and a second smooth pulley of mass $m$ attached to the other end. Another light inextensible string passes over this second pulley and carries a mass $3m$ at one end and a mass $m$ at the other end. If the system moves freely under gravity, find the acceleration of the heaviest particle and the tension in each string.      [L]

20 In the pulley system shown in Fig. 5.25 the pulleys A, B and C are smooth. The pulley C is attached to one end of a light inextensible string which passes round the pulleys B and A and has its other end attached at a fixed point D. The straight portions of the string are vertical and the masses of the pulleys A and C are $4m$ and $m$ respectively. Find the vertical accelerations of A and C and also the tension in the string when the system is moving freely.

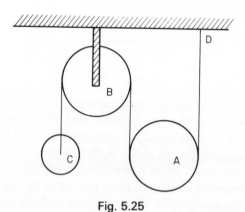

Fig. 5.25

A second string is now passed over the pulley C and carries masses $m$ and $2m$ at its ends. Calculate the vertical acceleration of A and the tension in each string when this new system is moving freely.      [AEB]

21 One end of a light inextensible string is attached to a ceiling. The string passes under a smooth light pulley carrying a weight C and then over a fixed smooth light pulley. To the free end of the string is attached a light scale pan in which two weights A and B are placed with A on top of B as shown in Fig. 5.26. The portions of the string not in contact with the pulleys are vertical. Each of the weights A and B has mass $m$ and the weight C has mass $km$. If the system is released from rest, find the accelerations of the movable pulley and of the scale pan and show that the scale pan will ascend if $k > 4$.

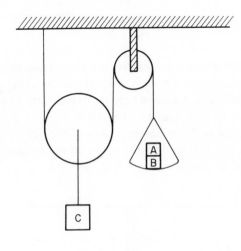

Fig. 5.26

When the system is moving freely find
(a) the tension in the string
(b) the reaction between the weights A and B. [L]

22 Figure 5.27 shows three smooth light pulleys, two of which, A and C, are fixed. The third pulley B carries a mass $km$ and is free to move. A light inextensible string passes over both fixed pulleys and under the movable pulley, and has masses $m$ and $3m$ attached at the ends P and Q, as shown. The portions of the string not in contact with the pulleys are all vertical. The system is released from rest, and moves freely under gravity.
(a) If P remains at rest find the value of $k$ and the accelerations of B and Q.
(b) If B remains at rest find the value of $k$ and the accelerations of P and Q. [L]

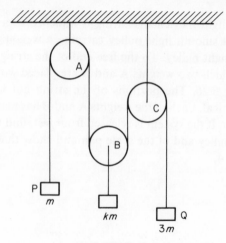

Fig. 5.27

**23** A light inextensible string passes over a smooth fixed pulley A. To one end
of the string is attached a mass of 8 kg and to the other end is attached a
light smooth pulley B. A second light inextensible string passes over B and
has attached at its ends masses 3 kg and $M$ kg (Fig. 5.28). The system is
released from rest with the hanging parts of the strings vertical.
(i) If $M = 3$, find the acceleration of B and the tension in each string.
(ii) If the mass $M$ remains at rest, find the value of $M$.           [L]

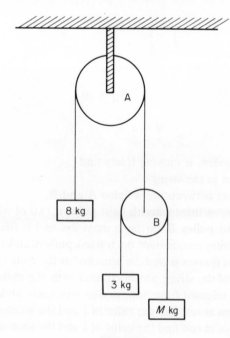

Fig. 5.28

**24**  As a car of mass 1000 kg travels along a level road its motion is opposed by a constant resistance of 100 newtons. The values of the pull of the engine for given values of the time $t$ seconds are as follows:

| Time $t$ (s) | 0 | 1 | 2 | 3 | 4 | 5 | 6 |
|---|---|---|---|---|---|---|---|
| Pull (N) | 1100 | 1200 | 1350 | 1550 | 1850 | 2300 | 3100 |

Find the acceleration of the car at each of the given values of $t$. When $t = 0$ the speed of the car is 50 km h$^{-1}$. Find graphically the speed of the car in km h$^{-1}$ when $t = 6$.

State how you would find graphically the distance travelled by the car in the period from $t = 0$ to $t = 6$. [AEB]

**25**  Explain briefly how to find the velocity from an acceleration–distance graph.

A car of mass 800 kg starts from rest and moves along a level road. It is subjected to a constant resistance of 400 N and the variable pull exerted by the engine is given at different distances travelled in the following table.

| Distance moved (metres) | 0 | 20 | 50 | 80 | 120 | 150 |
|---|---|---|---|---|---|---|
| Pull of engine (newtons) | 1200 | 1520 | 2000 | 2480 | 3120 | 3600 |

Show that the acceleration–distance graph of this motion is a straight line. Hence find, graphically or otherwise, the speed attained by the car after travelling 150 m.

Find also the speed the car would reach in moving 150 m under the same conditions up a road inclined at an angle $\theta$ to the horizontal, where $\sin \theta = \dfrac{1}{20}$.

(Take $g$ as 10 m s$^{-2}$.) [L]

**26**  A vehicle of mass 3000 kg starts from rest on a level road and accelerates to a speed of 30 m s$^{-1}$ against a constant resistance of 600 N. The pull of the engine varies and its values at speeds between 0 and 30 m s$^{-1}$ are as follows:

| Speed (m s$^{-1}$) | 0 | 10 | 15 | 25 | 30 |
|---|---|---|---|---|---|
| Pull (N) | 15 600 | 9600 | 8100 | 6225 | 5600 |

By drawing a suitable straight line graph, or otherwise, find the time taken to reach the speed of 30 m s$^{-1}$ from rest.

By using your graph to express the vehicle's acceleration in terms of its speed, or otherwise, find the distance travelled in reaching the speed of 30 m s$^{-1}$. [L]

**27**  A car of mass 1000 kg starts from rest and moves on level ground against a constant resistance of 1000 N. The pull of the engine increases uniformly with the distance travelled, starting at 1500 N and increasing to 4500 N when the car has travelled 150 m. Sketch the acceleration–distance graph. Hence, or otherwise, calculate the speed of the car when it has travelled 150 m. [L]

**28** A particle of mass $m$ is initially at rest at a point O on a rough horizontal table. The coefficient of friction between the particle and the table is $\frac{3}{4}$. A variable force $\lambda mg$ is applied to the particle in the direction of a fixed horizontal straight line OA, where $\lambda$ is given in terms of the time $t$ by the equations

$$\lambda = \begin{cases} \dfrac{t+1}{2} & (0 \leqslant t \leqslant 3) \\[3mm] \dfrac{9-2t}{4} & (t > 3). \end{cases}$$

Find

(i) the intervals of time during which the particle is in motion

(ii) the velocity of the particle when $t = 8$.                          [JMB]

**29** A car of mass $m$ moves along a straight level road with its engine switched off. The resistance to its motion at speed $v$ is proportional to $v^2 + U^2$, where $U$ is a constant. It comes to rest from speed $U$ in time $T$. Show that the resistance at zero speed is $\frac{1}{4}m\pi U/T$.

The car now starts from rest and moves, under the same law of resistance, with its engine exerting a constant force of magnitude $m\pi U/T$. Show that it reaches speed $U$ after travelling a distance $\dfrac{2UT}{\pi}\log_e\left(\dfrac{3}{2}\right)$. State the limiting speed under this force.                          [JMB]

**30** (a) A particle moves on the positive $x$-axis under the action of a force directed away from the origin. The magnitude of the force is proportional to the square root of the time. The particle starts from rest at time $t = 0$. Prove that the time taken to accelerate from speed $u$ to speed $2u$ is a little less than three-fifths of the time taken to accelerate from rest to speed $u$.

Find, to two significant figures, the ratio of the corresponding distances travelled by the particle.

(b) A second particle also moves on the positive $x$-axis under the action of a single force directed away from the origin, but in this case the magnitude of the force is proportional to the square root of the distance, $x$, of the particle from the origin. This particle has speed $u$ when $x = a$, and speed $2u$ when $x = 4a$. Find its speed when $x = 9a$.                          [C]

# 6 Work, power and energy

## 6.1 Work

When a force moves its point of application it is said to do work. The work done by a constant force is defined as the product of the magnitude of the force and the distance moved by the point of application in the direction of the force.

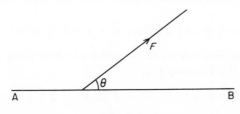

Fig. 6.1

Suppose a force **F** moves its point of application from A to B (Fig. 6.1). The magnitude of the force is $|\mathbf{F}|$ and the distance moved by the point of application in the direction of the force is AC (Fig. 6.2).

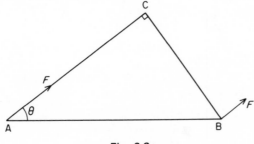

Fig. 6.2

The work done by the force $\mathbf{F} = |\mathbf{F}| \, (\text{AC})$
$$= |\mathbf{F}| \, (\text{AB} \cos \theta)$$
$$= |\mathbf{F}| \, |\overrightarrow{\text{AB}}| \cos \theta$$
$$= \mathbf{F} \cdot \overrightarrow{\text{AB}},$$

i.e. the work done is the scalar product of the force and the displacement of the point of application.

The same result may also be obtained by regarding **F** as equivalent to $\mathbf{F} \cos \theta$ along AB and $\mathbf{F} \sin \theta$ perpendicular to AB. Then the force $\mathbf{F} \cos \theta$ does work

$|\mathbf{F} \cos \theta|$ (AB) and the force $\mathbf{F} \sin \theta$ does no work, since its point of application does not move perpendicular to AB, i.e. the work done is $|\mathbf{F}|$ (cos $\theta$)(AB) or $\mathbf{F} \cdot \overrightarrow{AB}$.

It should be noted that work is a scalar quantity.

The unit of work in the SI system is the *joule* which is the work done when a force of 1 newton moves its point of application 1 metre in the direction of the force. The abbreviation for joule is J.

*Example 1*

A force of magnitude 20 N acts in the plane of coordinate axes Ox, Oy. It moves its point of application from O to A $(a, b)$. Find the work done if
(i) $a = 8$ m, $b = 0$ and the force is in the direction Ox
(ii) $a = 8$ m, $b = 0$ and the force acts in a direction which makes an angle of 60° with Ox
(iii) $a = 8$ m, $b = 6$ m and the force acts in the direction Ox.

(i) The point of application moves a distance of 8 m along Ox.
The force acts in the direction Ox.

$\therefore$          Work done $= 20 \times 8$ joules $= 160$ joules.

(ii) The point of application moves a distance of 8 m along Ox.
The force acts at an angle of 60° to Ox.

$\therefore$ The distance moved by the point of application *in the direction of the force* $= OB = 8 \cos 60° = 4$ metres (Fig. 6.3).

$\therefore$          Work done $= 20 \times 4$ joules $= 80$ joules.

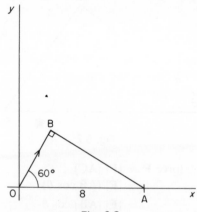

Fig. 6.3

(iii) The point of application moves from O to A (Fig. 6.4).
The force acts in the direction Ox.
$\therefore$ The distance moved by the point of application *in the direction of the force* $= OB = 8$ metres.

$$\therefore \quad \text{Work done} = 20 \times 8 \text{ joules} = 160 \text{ joules}.$$

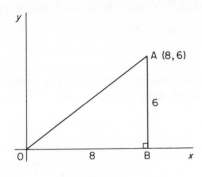

Fig. 6.4

*Example 2*

Find the work done when a force **F**, where **F** = (3**i** + 4**j** + 5**k**), moves its point of application from A to B which have position vectors (**i** − **j** + **k**) and (2**i** − 3**j** + 4**k**) respectively the units of distance and force being the metre and the newton.

The displacement of the point of application of the force **F** is $\overrightarrow{AB}$

and
$$\overrightarrow{AB} = (2\mathbf{i} - 3\mathbf{j} + 4\mathbf{k}) - (\mathbf{i} - \mathbf{j} + \mathbf{k})$$
$$= \mathbf{i} - 2\mathbf{j} + 3\mathbf{k}.$$

$$\therefore \quad \text{Work done by the force } \mathbf{F} = \mathbf{F} \cdot \overrightarrow{AB}$$
$$= (3\mathbf{i} + 4\mathbf{j} + 5\mathbf{k}) \cdot (\mathbf{i} - 2\mathbf{j} + 3\mathbf{k})$$
$$= 3 - 8 + 15 = 10,$$

i.e. the work done by the force **F** is 10 joules.

*Example 3*

A light rope is attached to a small wagon which runs on a straight level track. A man pulls the wagon slowly along the track by exerting a pull of magnitude 200 N on the rope. Find the work he does in pulling the wagon 50 m along the track if the rope is horizontal and makes an angle of 60° with the track.

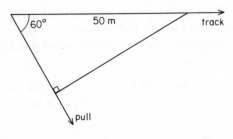

Fig. 6.5

As the wagon moves 50 m the point of application of the pull has moved 50 m along the track but only 50 cos 60° m in the direction of the pull (Fig. 6.5).

∴ Work done = 200 × 50 cos 60° = 500 joules.

*Example 4*

A particle of mass $m$ slides a distance $d$ down a plane inclined at an angle $\alpha$ to the horizontal. Find the work done by the weight of the particle.

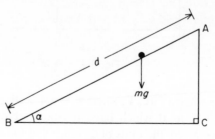

Fig. 6.6

The point of application of the weight, $m\mathbf{g}$, moves from A to B (Fig. 6.6). The distance moved in the direction of the weight is AC, i.e. $d \sin \alpha$.

∴ Work done by the weight = $mg$ AC = $mgd \sin \alpha$.

This result could also be obtained by resolving the weight $m\mathbf{g}$ into two components, of magnitude $mg \sin \alpha$ down the plane and $mg \cos \alpha$ perpendicular to the plane (Fig. 6.7).

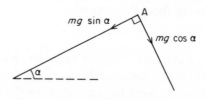

Fig. 6.7

The component of magnitude $mg \cos \alpha$ does no work since the point of application moves at right angles to this force.

The component of magnitude $mg \sin \alpha$ moves its point of application a distance $d$ in the direction of the force.

∴ Work done by the component $mg \sin \alpha = (mg \sin \alpha)d$
$$= mgd \sin \alpha \quad \text{as before.}$$

**Exercises 6.1**

In questions 1–15 the units of distance and force are the metre and newton.

A force **F** acts in the plane of coordinate axes O*x*, O*y*. Find the work done as the force moves its point of application from O to A when

1  **F** is in the direction O*x*, $F = 40$ and A has coordinates $(12, 0)$
2  **F** is in the direction O*x*, $F = 50$ and A has coordinates $(0, 12)$
3  **F** is in the direction O*y*, $F = 30$ and A has coordinates $(0, 7)$
4  **F** is in the direction O*y*, $F = 20$ and A has coordinates $(8, 0)$
5  **F** is in the direction O*x*, $F = 30$ and A has coordinates $(10, 8)$
6  **F** is in the direction O*y*, $F = 40$ and A has coordinates $(10, 8)$
7  **F** is in the direction O*x*, $F = 8$ and A has coordinates $(6, 8)$
8  **F** is in the direction O*y*, $F = 12$ and A has coordinates $(9, 15)$.

A force of magnitude 50 N acts at an angle $\alpha$ to O*x*, in the plane of coordinate axes O*x*, O*y*. Find the work done when the force moves its point of application from O to A when

9   $\alpha = 60°$ and the coordinates of A are $(10, 0)$.
10  $\alpha = 60°$ and the coordinates of A are $(0, 10)$.
11  $\tan \alpha = \frac{3}{4}$ and the coordinates of A are $(15, 0)$.
12  $\tan \alpha = \frac{3}{4}$ and the coordinates of A are $(0, 15)$.

Find the work done by a force **F** which moves its point of application from point A with position vector $\mathbf{r}_A$ to point B with position vector $\mathbf{r}_B$ when

13  $\mathbf{F} = (\mathbf{i} + 2\mathbf{j} + 3\mathbf{k})$, $\mathbf{r}_A = (3\mathbf{i} + 4\mathbf{j} + 5\mathbf{k})$, $\mathbf{r}_B = (7\mathbf{i} + 6\mathbf{j} + 5\mathbf{k})$.
14  $\mathbf{F} = (2\mathbf{i} - \mathbf{j} + 2\mathbf{k})$, $\mathbf{r}_A = (\mathbf{i} - 2\mathbf{k})$, $\mathbf{r}_B = (3\mathbf{j} + 4\mathbf{k})$.
15  $\mathbf{F} = (\mathbf{i} - 2\mathbf{j} - 4\mathbf{k})$, $\mathbf{r}_A = (3\mathbf{j} - 4\mathbf{k})$, $\mathbf{r}_B = (-\mathbf{i} + 6\mathbf{k})$.

In questions 16–18, ABCD is a rectangle in which AB $= 8$ m, BC $= 6$ m and M is the mid-point of BC.

16  Find the work done by a force of magnitude 20 N in the direction $\overrightarrow{AD}$ moving its point of application from
(a) A to B  (b) B to C  (c) C to B  (d) A to C.

17  Find the work done by a force of magnitude 30 N in the direction $\overrightarrow{AC}$ moving its point of application from
(a) A to C  (b) A to B  (c) A to D  (d) A to M.

18  Find the work done by a force of magnitude 12 N in the direction $\overrightarrow{BD}$ moving its point of application from
(a) B to C  (b) B to A  (c) A to C.

19  The engine of a car exerts a constant pull of magnitude 400 N. Find the work done by this force as the car travels 2 km along a road.

20  A barge is towed a distance of 200 m along a canal by a horse walking along the canal bank. The horse exerts a constant horizontal pull of magnitude $F$ newtons. The rope connecting the barge to the horse makes an angle $\alpha$ with the canal bank. Find the work done by this force when
(a) $F = 300$ and   $\alpha = 30°$
(b) $F = 400$ and   $\alpha = 45°$
(c) $F = 350$ and $\tan \alpha = \frac{3}{4}$.

**21** A man drags a sledge a distance of 80 m across a field by means of a rope which is attached to the sledge and which passes over the man's shoulder. The man exerts a constant pull of magnitude $F$ newtons and the rope makes an angle $\alpha$ with the ground. Given that the paths of the man and the sledge are in the same line, find the work done by this force when
  (a) $F = 80$ and $\alpha = 30°$
  (b) $F = 120$ and $\alpha = 60°$
  (c) $F = 60$ and $\tan \alpha = 5/12$.

**22** Find the work done by the weight of a parcel of mass 5 kg falling a distance of 8 m
  (a) vertically
  (b) down a slope at 45° to the horizontal
  (c) down a slope at an angle $\tan^{-1} 2.4$ to the horizontal.
  (Take $g$ to be $10 \text{ m s}^{-2}$.)

## 6.2 Power

Power is defined as rate of working. The unit of power is the watt which is a rate of working of 1 joule per second. This is a small unit and so the kilowatt, which is 1000 watts, is often used. The abbreviations for watt and kilowatt are W and kW respectively.

Consider a constant force of magnitude $F$ newtons acting in the line AB and moving a particle from A to C, a distance of $s$ metres in $t$ seconds (Fig. 6.8).

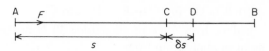

Fig. 6.8

The work done by the force in this time $= Fs$ joules

and        the average rate of working $= \dfrac{Fs}{t}$ joules per second,

i.e. the average power over this period of $t$ seconds is $\dfrac{Fs}{t}$ watts.

If the particle moves with constant speed $v$, then $v = s/t$ and the rate of working, i.e. the power, will be constant and equal to $F(s/t)$ or $Fv$ joules per second, i.e. $Fv$ watts.

Suppose that the force moves the particle from C to D, a distance $\delta s$ metres in the next $\delta t$ seconds. Over this period of time the average power is

$$\frac{F\delta s}{\delta t} \text{ watts.}$$

The power at the instant when the particle is at C, i.e. at time $t$, is defined as

$$\lim_{\delta t \to 0} \frac{F\delta s}{\delta t} = F \lim_{\delta t \to 0} \frac{\delta s}{\delta t} = F\frac{ds}{dt} = Fv,$$

where $v$ is the velocity of the particle at time $t$.

*Example 1*
A train of mass 400 tonnes is travelling at a constant speed of 96 km h$^{-1}$ up a track inclined at $\sin^{-1}$ (1/200) to the horizontal. There is a constant non-gravitational resistance to motion of magnitude 16 000 N. Find the power exerted by the engine. (Take $g$ to be 10 m s$^{-2}$.)

$$96 \text{ km h}^{-1} = \frac{96 \times 1000}{60 \times 60} \text{ m s}^{-1} = \frac{80}{3} \text{ m s}^{-1}.$$

Let the pull of the engine at this speed have magnitude $F$ newtons.
The forces in newtons acting on the train are shown in Fig. 6.9.

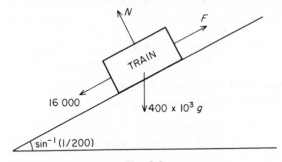

Fig. 6.9

Since the speed is constant the train has no acceleration, i.e. there is no resultant force on the train.

Thus, resolving up the track $\quad F - 16\,000 - 400 \times 10^3 \, g\left(\frac{1}{200}\right) = 0$

giving $F = 36\,000$.

Using $\qquad\qquad\qquad$ pull $\times$ speed = power

$$36\,000 \times \left(\frac{80}{3}\right) = \text{power in watts,}$$

i.e. the power of the engine = 960 000 watts = 960 kilowatts.

*Example 2*

The engine of a car works at a constant rate of 16 kW. The car has mass 1200 kg and on a straight level road there is a constant resistance to motion of magnitude 400 N. Find

(a) the maximum speed of the car

(b) the acceleration of the car when its speed is 25 m s$^{-1}$.

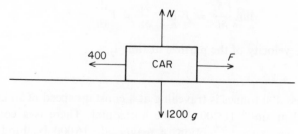

Fig. 6.10

Figure 6.10 shows diagrammatically the forces, in newtons, acting on the car.

(a) Let the pull of the engine have magnitude $F$ newtons at the maximum speed $v$ m s$^{-1}$.

At the maximum speed the car has no acceleration

$$\therefore \qquad F - 400 = 0,$$

i.e. the pull of the engine is 400 N.

Using

$$\text{pull} \times \text{speed} = \text{power}$$
$$400v = 16\,000 \quad \text{giving} \quad v = 40,$$

i.e. the maximum speed of the car is 40 m s$^{-1}$.

(b) Let the pull of the engine have magnitude $F_1$ newtons at speed 25 m s$^{-1}$.

Using

$$\text{pull} \times \text{speed} = \text{power}$$
$$F_1 \times 25 = 16\,000 \quad \text{giving} \quad F_1 = 640.$$

Now using $F = ma$, $640 - 400 = 1200a$ giving $a = 0.2$,

i.e. at a speed of 25 m s$^{-1}$ the car has an acceleration of 0.2 m s$^{-2}$.

*Example 3*

The engine of a vehicle of mass 1500 kg works at a constant rate of 20 kW. On a straight level road there is a constant resistance to motion of magnitude 500 N. Find

(a) how long the vehicle takes to accelerate from 10 m s$^{-1}$ to 30 m s$^{-1}$

(b) the distance travelled in this time.

Let the pull of the engine at a speed of $v$ m s$^{-1}$ be $F$ newtons and let the acceleration at this speed be $a$ m s$^{-2}$. Figure 6.11(a) shows the forces, in newtons, acting on the vehicle and Fig. 6.11(b) shows the acceleration.

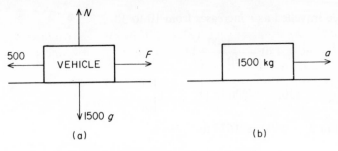

Fig. 6.11 (a) Forces  (b) Acceleration

Using $F = ma$ gives      $F - 500 = 1500a$ ...(1)
Using pull × speed = power, $Fv = 20\,000$ ...(2)

(a) Putting $F = \dfrac{20\,000}{v}$ from (2) and $a = \dfrac{dv}{dt}$ in (1)

$$\frac{20\,000}{v} - 500 = 1500\frac{dv}{dt},$$

i.e. $$3\frac{dv}{dt} = \frac{40 - v}{v}.$$ ...(3)

Separating the variables gives $\displaystyle\int \frac{3v}{40 - v}\,dv = \int dt$

or $$\int \left( -3 + \frac{120}{40 - v} \right) dv = \int dt.$$

∴ time taken as $v$ increases from 10 to 30

$$= \int_{10}^{30} \left( -3 + \frac{120}{40 - v} \right) dv$$

$$= \left[ -3v - 120 \ln (40 - v) \right]_{10}^{30}$$

$$= 120 \ln 3 - 60 \approx 71.8 \text{ seconds.}$$

(b) Putting $a = v\dfrac{dv}{ds}$, equation (3) becomes

$$3v\frac{dv}{ds} = \frac{40 - v}{v}$$

i.e. $$\int \frac{3v^2}{40 - v}\,dv = \int ds$$

or $$\int \left( -3v - 120 + \frac{4800}{40 - v} \right) dv = \int ds.$$

∴ distance travelled as $v$ increases from 10 to 30

$$= \int_{10}^{30} \left( -3v - 120 + \frac{4800}{40 - v} \right) dv$$

$$= \left[ -\frac{3v^2}{2} - 120v - 4800 \ln (40 - v) \right]_{10}^{30}$$

$$= 4800 \ln 3 - 3600 \approx 1673 \text{ m}.$$

## Exercises 6.2

Where required, take $g = 10 \text{ m s}^{-2}$.

In questions 1–4 a car of mass $M$ kg travels at a constant speed of $v$ m s$^{-1}$ up a road inclined at an angle $\alpha$ to the horizontal. Given that there is a non-gravitational resistance to motion of magnitude $X$ N, find the power exerted by the engine of the car when

1   $M = 1000, v = 30, \quad \alpha = 0, \quad X = 400.$
2   $M = 1200, v = 25, \sin \alpha = 1/10, X = 500.$
3   $M = 2000, v = 35, \sin \alpha = 1/20, X = 400.$
4   $M = 1500, v = 40, \sin \alpha = 1/8, \quad X = 300.$

In questions 5–10 the engine of a car of mass $M$ kg works at a constant rate of $H$ kW. There is a constant non-gravitational resistance to motion of magnitude $X$ N and the road rises at an angle $\alpha$ to the horizontal. Find
(a) the maximum speed of the car
(b) the acceleration of the car when its speed is $v$ m s$^{-1}$, if

5   $M = 1000, v = 10, \quad \alpha = 0, \quad X = 400, H = 10.$
6   $M = 1500, v = 15, \quad \alpha = 0, \quad X = 500, H = 15.$
7   $M = 1200, v = 20, \quad \alpha = 0, \quad X = 300, H = 12.$
8   $M = 1000, v = 10, \sin \alpha = 1/10, X = 500, H = 24.$
9   $M = 1200, v = 12, \sin \alpha = 1/5, \quad X = 300, H = 54.$
10   $M = 2000, v = 16, \sin \alpha = 1/20, X = 400, H = 28.$
11   The engine of a car of mass 1200 kg provides constant power 16 kW. The car is moving on a level road and there is a constant resistance to motion of magnitude 400 N. Find
     (a) the time taken to increase the speed from 10 m s$^{-1}$ to 20 m s$^{-1}$
     (b) the distance travelled in this time.
12   A car of mass 1600 kg travels up a road inclined at an angle $\sin^{-1} (1/20)$ to the horizontal. The engine works at a constant rate of 26 kW and there is a constant non-gravitational resistance to motion of magnitude 500 N. Find
     (a) the time taken for the car to increase its speed from 6 m s$^{-1}$ to 12 m s$^{-1}$
     (b) the distance travelled in this time.

## 6.3   Energy

Energy is the capacity to do work, and, like work, it is a scalar quantity.

In the SI system the unit of energy is the joule, which is also the unit for work.

Energy can take a variety of forms such as heat, light and chemical energy. Here our concern is only with mechanical energy. There are two types of mechanical energy which a particle may possess:

(a) kinetic energy, the capacity of a particle to do work by virtue of its speed,

(b) potential energy, the capacity of a particle to do work by virtue of its position.

Consider a particle of mass $m$ moving with speed $u$ in a particular direction. Let this particle be brought to rest by a constant force of magnitude $F$ acting in the line of motion.

From $F = ma$, the constant retardation produced will be of magnitude $F/m$. Using the relation $v^2 = u^2 + 2as$ with initial speed $u$ and final speed zero, (since the retardation is uniform)

$$0 = u^2 + 2(-F/m)s,$$

i.e.
$$Fs = \tfrac{1}{2}mu^2.$$

$Fs$ is the work done against the force which brought the particle to rest. Therefore in coming to rest from speed $u$ the particle has done work of magnitude $Fs$, which equals $\tfrac{1}{2}mu^2$. Thus by virtue of its speed $u$ the particle had the capacity to do work of magnitude $\tfrac{1}{2}mu^2$, i.e. the kinetic energy of a particle of mass $m$ travelling at speed $u$ is $\tfrac{1}{2}mu^2$.

Next consider a particle of mass $m$ which is released from rest, at a height $h$ above a table, and which falls freely under gravity. Using the relation $v^2 = u^2 + 2as$ with $u = 0$, $a = g$ and $s = h$

$$v^2 = 2gh.$$

Therefore the kinetic energy of the particle when it reaches the table is given by

$$\tfrac{1}{2}mv^2 = mgh.$$

Thus in falling through a vertical distance $h$ the particle acquires kinetic energy $mgh$, so that when the particle was released from rest it possessed the capacity to do work of magnitude $mgh$ due to its height above the table.

When the particle is at a height $h$ above the table, the particle is said to possess potential energy $mgh$ with reference to the level of the table.

Potential energy has no absolute value, only a value relative to some origin or level of reference.

The potential energy of a particle of mass $m$ at a height $h$ above some level of reference is $mgh$ referred to that level.

The total mechanical energy of a particle of mass $m$ equals the sum of its kinetic energy and its potential energy, where

(a) kinetic energy $= \tfrac{1}{2}mv^2$, $v$ being the speed of the particle,

(b) potential energy $= mgh$, $h$ being the height of the particle above the chosen level of reference.

When the work done by a force or against a force in moving a particle from a point A to a point B is independent of the path taken from A to B, the force is said to be a conservative force.

An example of a conservative force is the weight of the particle. When the particle is moved from a point A to a lower point B the work done by the weight depends on the difference in level between A and B, and is the same whatever path is taken.

The 'Principle of Conservation of Mechanical Energy' states that the sum of the kinetic energy and the potential energy of a system of particles under the action of conservative forces is constant.

When a projectile moves freely under gravity with no air resistance, it is moving under the action of just one force, its own weight. This is a conservative force, and the sum of the kinetic energy and the potential energy of the projectile is constant.

The same is true when a particle slides down a smooth inclined plane. However, when the plane is rough, part of the energy of the particle is used to do work against the force of friction and is converted into heat. In this case the total mechanical energy of the particle is not constant.

*Example 1*

A particle is projected with speed $u$ at an angle of elevation $\alpha$. Find the speed of the particle when it is at a height $h$ above the level of the point of projection.

Let the horizontal plane through the point of projection be taken as the level of reference for potential energy. Let $m$ be the mass of the particle.
At the point of projection,

$$\text{kinetic energy} = \tfrac{1}{2}mu^2,$$
$$\text{potential energy} = 0.$$

At height $h$,

$$\text{kinetic energy} = \tfrac{1}{2}mv^2,$$

where $v$ is the speed at height $h$,

$$\text{potential energy} = mgh.$$

By the principle of conservation of mechanical energy,

$$\tfrac{1}{2}mu^2 + 0 = \tfrac{1}{2}mv^2 + mgh,$$

i.e.
$$v^2 = u^2 - 2gh.$$

Therefore at height $h$ the speed of the particle is $\sqrt{(u^2 - 2gh)}$.

*Example 2*

One end of a light inextensible string of length $a$ is fixed and a particle of mass $m$ is attached to the other end. The particle is released from rest when the string is taut and horizontal. Find the speed of the particle when the string is (i) vertical (ii) at an angle $\theta$ to the vertical.

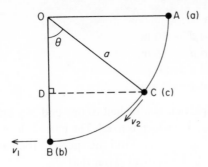

Fig. 6.12

Figure 6.12 shows the three positions of the string and particle. Position (a) is the initial position, position (b) is when the string is vertical and position (c) shows the string when it is at an angle $\theta$ to the vertical.

Only two forces act on the particle, its own weight $mg$ and the tension in the string. Since the particle always moves perpendicular to the string the tension does no work and so the law of the conservation of mechanical energy may be applied.

Take the level of the lowest position of the particle as the reference level for potential energy. The speeds of the particle when the string is vertical and when it is at an angle $\theta$ to the vertical are $v_1$ and $v_2$ respectively

(i) potential energy in position (a) = $mga$
    kinetic energy in position (a) = 0
    potential energy in position (b) = 0
    kinetic energy in position (b) = $\frac{1}{2}mv_1{}^2$.

Using conservation of mechanical energy

$$mga + 0 = 0 + \tfrac{1}{2}mv_1{}^2 \quad \text{giving} \quad v_1{}^2 = 2ga,$$

i.e. the speed of the particle when the string is vertical is $\sqrt{(2ga)}$.

(ii) potential energy in position (c) = $mg(\text{BD}) = mg(a - a\cos\theta)$
    kinetic energy in position (c) = $\frac{1}{2}mv_2{}^2$.

Using the conservation of mechanical energy with positions (a) and (c)

$$mga + 0 = mga(1 - \cos\theta) + \tfrac{1}{2}mv_2{}^2$$

giving
$$v_2{}^2 = 2ga\cos\theta,$$

i.e. the speed of the particle when the string makes an angle $\theta$ with the vertical is $\sqrt{(2ga\cos\theta)}$.

Putting $\theta = 0$ gives $\sqrt{(2ga)}$, the speed when $\theta = 0$, i.e. when the string is vertical.

**Exercises** 6.3

Where required, take $g$ to be $10\ \mathrm{m\,s^{-2}}$.

In questions 1–4 a particle slides from rest down a smooth plane inclined at an angle $\alpha$ to the horizontal. Find the speed of the particle when it has moved a distance $d$ given that

1    $\alpha = 30°$ and $d = 10$ m.
2    $\alpha = 60°$ and $d = 15$ m.
3    $\tan \alpha = \frac{3}{4}$  and $d = 20$ m.
4    $\tan \alpha = 2.4$ and $d = 2.6$ m.

In questions 5–10 a particle is attached to one end of a light inextensible string of length $l$, the other end of which is fixed. The particle is released from rest when the string is taut and horizontal. Find the speed of the particle when the string makes an angle $\theta$ with the vertical given that

5    $l = 2$ m,          $\theta = 0.$
6    $l = 50$ cm,        $\theta = 0.$
7    $l = 3$ m,          $\theta = 30°.$
8    $l = 80$ cm,        $\theta = 60°.$
9    $l = 120$ cm,       $\theta = 45°.$
10   $l = 60$ cm,  $\tan \theta = 0.75.$

In questions 11–14 a particle is projected with speed $u$ from a point A up a line of greatest slope of a smooth plane inclined at an angle $\alpha$ to the horizontal. The particle comes momentarily to rest at the point B and C is the mid-point of AB. Find

(a) the distance AB
(b) the speed at C, when

11   $u = 10$ m s$^{-1}$,      $\alpha = 30°.$
12   $u = 20$ m s$^{-1}$,      $\alpha = 60°.$
13   $u = 24$ m s$^{-1}$, $\tan \alpha = \frac{3}{4}.$
14   $u = 30$ m s$^{-1}$, $\tan \alpha = 2.4.$

In questions 15–18 a particle hangs at rest attached to one end of a light inextensible string of length $l$. The particle is then given a speed $u$ in a horizontal direction and the particle comes to instantaneous rest when the string makes an angle $\theta$ with the vertical.

15   Find $u$ when $l = 3$ m and $\theta = 90°.$
16   Find $u$ when $l = 2$ m and $\theta = 60°.$
17   Find $\theta$ when $l = 5$ m and $u = 10$ m s$^{-1}.$
18   Find $\theta$ when $l = 160$ cm and $u = 4$ m s$^{-1}.$

## 6.4   The equation of energy

When a particle of mass $m$ falls freely under gravity, the sum of its kinetic energy and its potential energy remains constant. This is expressed by the equation of energy

$$\tfrac{1}{2}mv^2 + mgh = \text{a constant},$$

where $v$ is the speed of the particle when it is at a vertical distance $h$ above the level of reference.

Consider a particle of mass $m$ at rest on a smooth horizontal table. Let the particle be acted on by a constant horizontal force of magnitude $F$. When the particle has moved through a distance $s$ let its speed be $v$. The work done by the force is $Fs$ and this equals the kinetic energy given to the particle. The energy equation for the particle is

$$Fs = \tfrac{1}{2}mv^2.$$

Suppose now that the table is rough, the coefficient of friction between the particle and the table being $\mu$. Let the particle acquire a speed $v$ in moving through a distance $s$. The work done against the force of friction is $(\mu mg)s$ and the energy equation for the particle is

$$Fs = \mu mgs + \tfrac{1}{2}mv^2.$$

In general, the equation of energy can be stated as follows:

(work done by external forces) = (work done against resistances)
$\quad\quad\quad$ + (increase in mechanical energy of the particle).

*Example 1*
A pump working at 5 kW lifts water to a height of 50 m and delivers it with a speed of 12 m s$^{-1}$. Assuming that the resistances are negligible, find the amount of water raised per minute. (Take $g$ to be 10 m s$^{-2}$.)

Work done by pump per second = 5000 J.
Let the mass of water raised per second be $x$ kg.

Gain in potential energy per second = $xgh = x(10)(50) = 500x$ J.

The water emerges at the top with speed 12 m s$^{-1}$.

Gain in kinetic energy per second = $\tfrac{1}{2}xv^2 = \tfrac{1}{2}x(144) = 72x$ J.

The energy equation is thus

$$5000 = 500x + 72x = 572x,$$

giving $\quad\quad\quad\quad\quad x \approx 8.74.$

Mass of water raised per minute = $60x \approx 524$ kg.

*Example 2*
A particle of mass $m$ slides from rest a distance $d$ down a rough plane inclined at an angle $\alpha$ to the horizontal. If the coefficient of friction between the particle and the plane is $\mu$, find the speed acquired by the particle.

The forces acting on the particle are its own weight, the normal reaction of magnitude $N$, and the frictional force of magnitude $F$ (Fig. 6.13). The motion takes place at right angles to the normal reaction and so this force does no work.

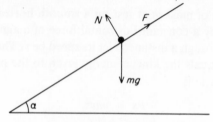

Fig. 6.13

Resolving perpendicular to the plane $N = mg \cos \alpha$.
The particle is in motion and so $F = \mu N = \mu mg \cos \alpha$.

Let the final speed of the particle be $v$.
The particle will have fallen a vertical distance $d \sin \alpha$.

$\therefore$          the loss of potential energy $= mg(d \sin \alpha)$

and          the gain in kinetic energy $= \frac{1}{2}mv^2$.

The point of application of the frictional force moves a distance $d$

         work done against friction $= Fd = (\mu mg \cos \alpha)d$.

The energy equation is

$$mgd \sin \alpha = \tfrac{1}{2}mv^2 + (\mu mg \cos \alpha)d$$

giving   $v^2 = 2gd \sin \alpha - 2\mu gd \cos \alpha$, i.e. $v = \sqrt{[2gd(\sin \alpha - \mu \cos \alpha)]}$.

*Example 3*
A particle of mass 2 kg is pushed up a rough plane inclined at an angle $\tan^{-1}\frac{3}{4}$ to the horizontal by a horizontal force of magnitude 40 N. The particle starts from rest at point A and is moved up a line of greatest slope of the plane to the point B, where AB is of length 15 m. The coefficient of friction between the particle and the plane is $\frac{1}{4}$. Find the speed of the particle at B. (Take $g$ to be $10 \text{ m s}^{-2}$.)

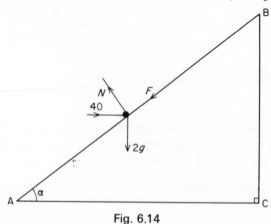

Fig. 6.14

Figure 6.14 shows the forces in newtons acting on the particle.
There is no motion perpendicular to the plane and so the normal reaction does no work and

$$N = 2g \cos \alpha + 40 \sin \alpha = 16 + 24 = 40.$$

The particle is in motion and so

$$F = \mu N = \tfrac{1}{4}(40) = 10.$$

The force of 40 N moves its point of application a distance AC which equals 15 cos $\alpha$, i.e. 12 m.
Work done on the particle by this external force $= 40(12) = 480$ joules.
Gain in kinetic energy $= \tfrac{1}{2}(2)v^2$, where $v$ m s$^{-1}$ is the speed at B.
Gain in potential energy $= 2g(\text{BC}) = 2(10)15 \sin\alpha = 180$ joules.
Work done against friction $= F(\text{AB}) = 10(15) = 150$ joules.
The energy equation is

$$480 = v^2 + 180 + 150 \quad \text{giving} \quad v^2 = 150,$$

i.e. the speed at B is $5\sqrt{6}$ m s$^{-1}$.

## Exercises 6.4
Where required, take $g$ to be 10 m s$^{-2}$.
In questions 1 and 2 a particle slides from rest a distance $d$ m down a rough plane inclined at an angle $\alpha$ to the horizontal. The coefficient of friction between the particle and the plane is $\mu$. Find the speed acquired by the particle when
1   $d = 10$ m, $\sin \alpha = 3/5$, $\mu = \tfrac{1}{2}$.
2   $d = 20$ m, $\tan \alpha = 12/5$, $\mu = \tfrac{1}{3}$.
3   A pump working at 6 kW pumps 400 kg of water per minute to a height of 30 m. Assuming that friction is negligible, find the speed with which the water is delivered at the top.
4   A pump raises 600 litres of water per minute to a height of 20 m and delivers it with a speed of 12 m s$^{-1}$. Assuming the frictional forces to be negligible, calculate the rate of working of the pump.
5   A pump working at 4 kW raises $x$ litres of water per hour to a height of 40 m and delivers it with a speed of 10 m s$^{-1}$. Assuming that friction may be neglected, find the value of $x$.

In questions 6–8 a particle of mass $m$ kg is at rest at a point A on a rough plane inclined at an angle $\alpha$ to the horizontal. The coefficient of friction between the particle and the plane is $\mu$. The particle is pushed up the plane by a force of magnitude $P$ N which acts along a line of greatest slope of the plane. Find the speed of the particle when it reaches the point B, where AB $= d$ m, given that
6       $\alpha = 30°$, $\mu = \tfrac{1}{4}$, $m = 3$, $P = 40$, $d = 10$.
7   $\tan \alpha = \tfrac{3}{4}$,   $\mu = \tfrac{1}{3}$, $m = 2$,   $P = 50$, $d = 8$.
8   $\tan \alpha = 12/5$, $\mu \doteqdot \tfrac{1}{2}$, $m = 0.5$, $P = 30$, $d = 12$.

**9** The engine of a car of mass 1200 kg exerts a constant tractive force of magnitude 1000 N when the car is travelling on a road inclined at an angle $\sin^{-1} (1/20)$ to the horizontal. There is a constant non-gravitational resistance of magnitude 300 N. Find the speed of the car when it has moved 200 m from rest (a) up the slope (b) down the slope.

## 6.5   Energy in a stretched elastic string

A light elastic string, of natural length $l$ and modulus of elasticity $\lambda$, is given an extension $x$. The work done in producing this extension will now be found.

When the extension is $s$ the tension in the string is $\dfrac{\lambda}{l} s$.

Let $\delta W$ denote the work done in stretching the string a further distance $\delta s$.

Then
$$\frac{\lambda}{l} s\,\delta s < \delta W < \frac{\lambda}{l}(s + \delta s)\delta s.$$

$\therefore$
$$\frac{\lambda}{l} s < \frac{\delta W}{\delta s} < \frac{\lambda}{l}(s + \delta s).$$

In the limit when $\delta s \to 0$, this gives $\dfrac{dW}{ds} = \dfrac{\lambda}{l} s$.

Therefore work done in producing an extension $x$ is given by

$$W = \int_0^x \frac{\lambda}{l} s\,ds = \left[\frac{\lambda s^2}{2l}\right]_0^x$$

$$= \frac{1}{2}\left(\frac{\lambda}{l}\right)x^2.$$

This is the potential energy stored in the stretched string since this is the work which can be done as the string returns to its natural unstretched length,

i.e. the potential energy stored in a stretched string $= \dfrac{1}{2}\left(\dfrac{\lambda}{l}\right)x^2$,

where $\lambda$ is the modulus of elasticity, $l$ is the natural length and $x$ is the extension.

*Example 1*
A light elastic string AB, of natural length $a$ and modulus of elasticity $4mg$, has its end A fixed and a particle of mass $m$ is attached to the other end B. The particle is released from rest at A. Find how far it is from A when it comes momentarily to rest.

The mechanical energy of the system remains constant.
Let the particle come momentarily to rest at a distance $(a + x)$ below A, i.e. when the string has an extension $x$. Take the level of this lowest position of the particle as the level of reference for potential energy.

At A          the potential energy $= mg(a + x)$
the kinetic energy $= 0$,
the energy stored in the string $= 0$, since the string is unstretched.

At the lowest point          the potential energy $= 0$,
the kinetic energy $= 0$,
the energy stored in the string $= \dfrac{1}{2}\left(\dfrac{4mg}{a}\right)x^2$.

The energy equation is

$$mg(a + x) = \dfrac{1}{2}\left(\dfrac{4mg}{a}\right)x^2$$

giving          $2x^2 - ax - a^2 = 0$    or    $(2x + a)(x - a) = 0$.

Thus                    $x = a$   or   $-\tfrac{1}{2}a$.

The root with the minus sign can be discarded since $x$ must clearly be positive. Thus the particle comes momentarily to rest at a distance $2a$ below A.

*Example 2*
A particle of mass 2 kg hangs at rest on the end of a light elastic string of natural length 1.5 m and modulus of elasticity 120 N. The particle is pulled vertically downwards a distance of 25 cm and released. Find how high the particle will rise. (Take $g$ to be $10\ \mathrm{m\,s}^{-2}$.)

Initially the particle is at rest and so the tension in the string is equal to the weight, $2g$ N, of the particle.

Using $T = \dfrac{\lambda}{l}x$,    $2g = \dfrac{120}{1.5}x$   giving   $x = \tfrac{1}{4}$ m or 25 cm,

i.e. in the initial position of equilibrium the extension in the string is 25 cm. The particle is then pulled down 25 cm so that the extension is then 50 cm or 0.5 m. Taking the level of this lowest point as the level of reference for potential energy,

at lowest point          the potential energy $= 0$
the kinetic energy $= 0$
the energy stored in the string $= \dfrac{1}{2}\left(\dfrac{120}{1.5}\right)(0.5)^2$.

Let $h$ m be the height above the lowest point to which the particle rises.

At the highest point          the potential energy $= 2gh$
the kinetic energy $= 0$
the energy stored in the string $= \dfrac{1}{2}\left(\dfrac{120}{1.5}\right)(0.5 - h)^2$

when    $h \leqslant 0.5$.

Using the energy equation

$$\frac{1}{2}\left(\frac{120}{1.5}\right)(0.5)^2 = 2gh + \frac{1}{2}\left(\frac{120}{1.5}\right)(0.5 - h)^2 \quad \text{when} \quad h \leqslant 0.5$$

giving $\qquad\qquad 20h = 40h^2 \quad \text{and} \quad h = 0.5 \text{ or } 0.$

∴ the particle rises 0.5 m or 50 cm above the lowest point which was at a distance 1.5 m + 0.5 m (natural length + extension in initial position). Thus the particle rises to a point 1.5 m below the fixed end of the string.

It should be noted that this working requires $h \leqslant 0.5$. If $h$ were to exceed 0.5, the string would go slack before the particle reached its highest point. This would happen if the energy stored in the string at the lowest point were to exceed the potential energy of the particle at the instant when the string goes slack. In that case the energy equation becomes

(energy stored in string at lowest point) = (potential energy of particle at highest point).

### Exercises 6.5
Find the energy stored in a natural string of natural length $l$, modulus of elasticity $\lambda$ and extension $x$ when
1  $l = 2$ m, $\quad \lambda = 100$ N, $x = 0.2$ m.
2  $l = 3$ m, $\quad \lambda = 200$ N, $x = 30$ cm.
3  $l = 2.5$ m, $\quad \lambda = 250$ N, $x = 10$ cm.
4  $l = 120$ cm, $\lambda = 300$ N, $x = 40$ cm.

In questions 5–7 a particle of mass $m$ is attached to one end of a light elastic string, the other end of which is fixed at a point A. The string has natural length $l$ and modulus of elasticity $kmg$. The particle is released from rest at A. Find the distance below A at which the particle comes momentarily to rest when
5  $k = 4$ and $l = 3$ m.
6  $k = 12$ and $l = 4$ m.
7  $k = 5$ and $l = 2$ m.

In questions 8 and 9 a particle of mass $m$ hangs at rest suspended from a beam by a light elastic string of natural length $l$ and modulus of elasticity $kmg$. The particle is pulled vertically downwards a distance $d$ and released. Find the distance below the beam of the point at which the particle comes momentarily to rest when
8  $l = 3$ m, $k = 6$, $d = 50$ cm.
9  $l = 2$ m, $k = 8$, $d = 40$ cm.

### Miscellaneous exercises 6
1  A hollow hemispherical bowl, of internal radius $r$, is fixed with its rim horizontal and uppermost. A particle is released from rest at a point on the rim. It slides down the inner surface of the bowl and has speed $v$ when it

reaches the lowest point. Given that the bowl is smooth, find $v$.

On a rough but otherwise identical bowl the particle has speed $\frac{1}{2}v$ at the lowest point. In this case find the work done against friction.

2   A particle P starts with speed $v$ in a horizontal direction from the lowest point A on the smooth inner surface of a hollow sphere, of centre O and internal radius $a$. The particle comes to rest when the angle AOP $= \theta$. Find $v$ when

(i) $\theta = \pi/6$              (ii) $\theta = \pi/3$              (iii) $\theta = \pi/2$.

3   A piston of mass 4 kg is moving to and fro in a cylinder which is at rest. At time $t$ seconds the distance, $x$ metres, of the piston from one end of the cylinder is given by

$$5x = 3 + 2 \cos 3t.$$

Express its acceleration and its kinetic energy in terms of $x$. Sketch graphs to show how the acceleration and the kinetic energy vary with $x$.     [L]

4   A particle is released from rest at a point on a plane inclined at an angle $\tan^{-1}\frac{3}{4}$ to the horizontal. Find the speed of the particle when it has moved 10 m down the plane when

(a) the plane is smooth

(b) the plane is rough and the coefficient of friction between the plane and the particle is $\frac{1}{3}$.

(Take $g$ to be $10 \text{ m s}^{-2}$.)

5   A particle is projected with speed $20 \text{ m s}^{-1}$ up a line of greatest slope of a plane inclined at an angle $\tan^{-1}(5/12)$ to the horizontal. Find

(a) how far the particle travels before coming to instantaneous rest

(b) the speed with which the particle returns to its starting point

when (i) the plane is smooth

        (ii) the plane is rough and the coefficient of friction between the particle and the plane is $\frac{1}{4}$.

(Take $g$ to be $10 \text{ m s}^{-2}$.)

6   (i) A pump raises 100 kg of water per second from a depth of 30 m. The water is delivered at a speed of $30 \text{ m s}^{-1}$. Find, in joules, (a) the potential energy (b) the kinetic energy gained by the water delivered each second. Neglecting frictional losses calculate, in kW, the rate at which the pump is working.

(Take $g$ as $9.8 \text{ m s}^{-2}$.)

(ii) A car of mass 1000 kg moves along a horizontal road with acceleration proportional to the cube root of the time $t$ seconds after starting from rest. When $t = 8$, the speed of the car is $8 \text{ m s}^{-1}$. Neglecting frictional resistances, calculate, in kW, the rate at which the engine driving the car is working when $t = 27$.     [L]

7   A pump which works at an effective rate of 17.6 kW raises water from a depth of 24 m and delivers it with a speed of $20 \text{ m s}^{-1}$. Find, in kg, the mass of water raised per hour. (Take $g$ to be $10 \text{ m s}^{-2}$.)

**8** A pump raises 80 kg of water per second from a depth of 15 m and delivers it at ground level with a speed of 40 m s$^{-1}$. Find the effective power of the pump.

   If the water emerges from the ground as a jet making an angle of 45° with the horizontal, find the height above ground level to which the water will rise.

   (Take $g$ to be 10 m s$^{-2}$.)

**9** A pump raises 108 000 kg of water per hour from a depth of 30 m and delivers it through a pipe of 15 cm$^2$ cross-sectional area. Find, in kW, the effective rate of working of the pump.

   The water emerges as a jet from a point A on level ground. Find the greatest distance from A at which the water can hit the ground.

   (Take $g$ to be 10 m s$^{-2}$.)

**10** The mass of a car is 1000 kg and the total non-gravitational resistance to its motion is constant and equal to a force of 400 N.

   (i) Find, in kW, the rate of working of the engine of the car when it is moving along a level road at a constant speed of 20 m s$^{-1}$.

   (ii) What is the acceleration, in m s$^{-2}$, when the car is moving along a level road at 20 m s$^{-1}$ with the engine working at 20 kW?

   (iii) When the car is being driven at 15 m s$^{-1}$ up a hill inclined at sin$^{-1}$ $(1/n)$ to the horizontal with the engine working at 10 kW, the acceleration is $1/5$ m s$^{-2}$. Calculate the value of $n$.

   (Take the acceleration due to gravity to be 10 m s$^{-2}$.)       [AEB]

**11** At the instant a car of mass 840 kg passes a sign post on a level road its speed is 90 km h$^{-1}$ and its engine is working at 70 kW. If the total resistance is constant and equal to 2100 N, find the acceleration of the car in m s$^{-2}$ at the instant it passes the sign post. Calculate the maximum speed in km h$^{-1}$ at which this car could travel up an incline of sin$^{-1}$ $(1/10)$ against the same resistance with the engine working at the same rate.

   (Take $g$ to be 9.8 m s$^{-2}$.)       [AEB]

**12** A car is moving along a level road at a steady speed of 72 km h$^{-1}$ against constant resistances which total 1250 N. Calculate the rate, in kW, at which the engine of the car is working. The car climbs a hill whose inclination to the horizontal is sin$^{-1}$ $(1/12)$ against the same total non-gravitational resistances. The mass of the car is 1500 kg and its engine is working steadily at 30 kW. Calculate

   (i) the maximum steady speed that is possible in km h$^{-1}$.

   (ii) the acceleration of the car, in m s$^{-2}$, at the instant when the car is moving at 36 km h$^{-1}$.

   (Take the acceleration due to gravity to be 10 m s$^{-2}$.)       [AEB]

**13** A railway train of mass 10$^5$ kg has a maximum speed of 25 m s$^{-1}$ on the level. When the train is climbing an incline of sin$^{-1}$ $(1/50)$ the maximum speed is 15 m s$^{-1}$. If the resistance to motion other than gravity is constant, find the maximum rate of working of the engine in kilowatts.

Find the maximum acceleration when the train is travelling at 20 m s$^{-1}$ down this incline.

(Take $g$ to be 9.8 m s$^{-2}$.)                                                                 [AEB]

14  A car of mass 1000 kg has a maximum speed of 15 m s$^{-1}$ up a slope inclined at an angle $\theta$ to the horizontal where $\sin \theta = 0.2$. There is a constant frictional resistance equal to one-tenth of the weight of the car. Find the maximum speed of the car on a level road.

If the car descends the same slope with its engine working at half its maximum power, find the acceleration of the car at the moment when its speed is 30 m s$^{-1}$.

(Take $g$ to be 10 m s$^{-2}$.)                                                                 [L]

15  A car of weight $W$ has maximum power $H$. In all circumstances there is a constant resistance $R$ due to friction. When the car is moving up a slope of 1 in $n$ ($\sin^{-1}(1/n)$) its maximum speed is $v$ and when it is moving down the same slope its maximum speed is $2v$. Find $R$ in terms of $W$ and $n$.

The maximum speed of the car on level road is $u$. Find the maximum acceleration of the car when it is moving with speed $\frac{1}{2}u$ up the given slope.
                                                                                                 [AEB]

16  The engine of a car, of mass $M$ kg, works at a constant rate of $H$ kW. The non-gravitational resistance to the motion of the car is constant. The maximum speed on level ground is $V$ m s$^{-1}$. Find, in terms of $M$, $V$, $H$, $\alpha$ and $g$, expressions for the accelerations of the car when it is travelling at speed $\frac{1}{2}V$ m s$^{-1}$ (a) directly up a road of inclination $\alpha$ (b) directly down this road.

Given that the acceleration in case (b) is twice that in case (a), find $\sin \alpha$ in terms of $M$, $V$, $H$ and $g$. Find also, in terms of $V$ alone, the greatest steady speed which the car can maintain when travelling directly up the road.                                                                 [L]

17  The engine of a car of mass $M$ kg develops constant power $H$ watts when the car is in motion. The resistance to the motion of the car is constant. The maximum speed of the car on a level road is $v$ m s$^{-1}$. Find the maximum speed of the car (a) directly up a road inclined at an angle $\phi$ to the horizontal, where $\sin \phi = 1/n$ (b) directly down the same road. If the maximum speed down the road is twice the maximum speed up the road, show that $3Mgv = Hn$. When the speed of the car is $\frac{1}{2}v$ m s$^{-1}$, show that the acceleration of the car on a level road is $H/(Mv)$ m s$^{-2}$.                                                                 [L]

18  The resistance to the motion of a car of mass $M$ is constant and equal to $R$. Throughout the motion the power developed by the engine of the car remains constant. The maximum speeds of the car up and down a slope of inclination $\theta$ to the horizontal are $2u$ and $3u$ respectively. Find an expression for $R$ and the greatest speed of the car on level ground. Show that the acceleration of the car when travelling on level ground at speed $u$ is $7g \sin \theta$.                                                                 [L]

**19** The magnitude of the force resisting the motion of a lorry is proportional to the speed of the lorry. When the lorry moves along a horizontal road at a steady speed of 36 km h$^{-1}$, its engine is working at 48 kW. Show that the magnitude of the force resisting motion is 4800 N at this speed.

The lorry, whose mass is 7000 kg, descends an incline of sin$^{-1}$ (1/20) to the horizontal at a steady speed of 60 km h$^{-1}$. Calculate the rate at which the engine is now working.

Find, in m s$^{-2}$, the immediate retardation of the lorry if the rate of working of its engine is suddenly decreased by 10 kW when it is descending the incline at 60 km h$^{-1}$.

(Take the acceleration due to gravity to be 10 m s$^{-2}$.)          [AEB]

**20** A lorry of mass 10 000 kg has a maximum speed of 24 km h$^{-1}$ up a slope of 1 in 10 against a resistance of 1200 newtons. Find the effective power of the engine in kilowatts.

If the resistance varies as the square of the speed, find the maximum speed on the level to the nearest km h$^{-1}$.

(Take $g$ as 9.8 m s$^{-2}$.)          [L]

**21** The non-gravitational resistance to the motion of a car of mass 1000 kg is $(A + Bv^2)$ newtons, where $v$ is the speed of the car in m s$^{-1}$. When the engine of the car is working at 45.9 kW the car can travel on a horizontal road at a steady 15 m s$^{-1}$. When the engine of the car is working at 16 kW the car can ascend an incline of sin$^{-1}$ (1/20) to the horizontal at a steady 5 m s$^{-1}$. Calculate the values of the constants $A$ and $B$.

Using these values of $A$ and $B$, find the rate at which the engine of the car is working when it is moving along a horizontal road at a steady 20 m s$^{-1}$.

(Take the acceleration due to gravity to be 10 m s$^{-2}$.)          [AEB]

**22** A car is moving along a horizontal straight road at a steady speed of $v$ m s$^{-1}$ when the engine of the car is working at $H$ kilowatts. In a series of test runs the car produced the results shown in the table.

| $v$ | 10 | 20 | 40 |
|---|---|---|---|
| $H$ | 6.2 | 13.6 | 36.8 |

Show that these results are consistent with the assumption that the forces resisting the motion of the car can be expressed in the form $(A + Bv^2)$ newtons, where $A$ and $B$ are constants.

The mass of the car is 1200 kg. The car climbs a hill of inclination sin$^{-1}$ (1/40) to the horizontal at a steady speed of 30 m s$^{-1}$.

(i) Calculate the rate at which the engine of the car is working.

(ii) If this rate of working is suddenly reduced to 23.4 kW, find the initial retardation of the car.

(Take the acceleration due to gravity to be 10 m s$^{-2}$.)          [AEB]

**23** A train is travelling along a straight level track at a speed of 36 km h$^{-1}$. If the power exerted by the engine is 500 kW find, in newtons, the tractive force exerted by the engine. If the train has total mass 6 × 10$^5$ kg and has an acceleration of $0.05 \text{ m s}^{-2}$, calculate the total resistance to the motion.

The mass of the engine alone is $10^5$ kg. Assuming that the resistance to the motion of any part of the train is proportional to its mass, find the tension in the coupling between the engine and the rest of the train. [L]

24 The maximum rate of working of the engine of a car is $S$ kW. Against a constant resistance, the car can attain a maximum speed of $u \text{ m s}^{-1}$ on level ground and a maximum speed of $\frac{1}{2}u \text{ m s}^{-1}$ directly up a slope of inclination $\alpha$ where $\sin \alpha = \frac{1}{16}$. Calculate the maximum speed of the car up a slope of inclination $\beta$, where $\sin \beta = \frac{1}{8}$, assuming that the resistance remains unchanged.

Given that $S = 15$ and that $u = 20$, find the maximum acceleration that can be attained when the car is towing a trailer of mass 300 kg at $10 \text{ m s}^{-1}$ on level ground. It may be assumed that the resistance to the motion of the car remains unchanged and that the resistance to the motion of the trailer can be neglected.

(Take $g$ as $10 \text{ m s}^{-2}$.) [L]

25 A car of mass 1000 kg whose maximum power is constant at all speeds experiences a constant resistance $R$ newtons. If the maximum speed of the car on the horizontal is $120 \text{ km h}^{-1}$ and the maximum speed up a slope of angle $\theta$ where $\sin \theta = 1/100$ is $60 \text{ km h}^{-1}$, calculate the power of the car.

Calculate also the maximum speed of the car (a) on the horizontal and (b) up the slope when it is pulling a caravan of mass 1000 kg if the total resistance to the motion of the car and the caravan is $3R$ newtons.

(Take $g$ to be $9.81 \text{ m s}^{-2}$.) [L]

26 A car of mass 960 kg is towing a trailer of mass 240 kg attached to the rear of the car by means of a tow-bar. The non-gravitational resistances opposing the motions of the car and the trailer are constant and of magnitudes 384 N and 48 N respectively. Calculate the maximum speed of the car, in $\text{km h}^{-1}$, when it is travelling along a horizontal road with the engine working at 7.2 kW.

With the car travelling at this speed the power of the engine is suddenly increased to 12 kW. Calculate the initial acceleration of the car, in $\text{m s}^{-2}$, and the tension in the tow-bar at this instant.

The car and trailer climb a hill of road length 1 km and vertical height 100 m. Calculate the total work done by the engine of the car if the speed during the ascent decreases from $20 \text{ m s}^{-1}$ to $10 \text{ m s}^{-1}$, giving your answer in megajoules to three significant figures.

(Take the acceleration due to gravity to be $10 \text{ m s}^{-2}$.) [AEB]

27 State Hooke's law for a stretched elastic string.

By integration prove that the work done in stretching an elastic string of natural length $c$ and modulus of elasticity $\lambda$ to a length $(c + x)$ is $\lambda x^2/(2c)$.

A particle of mass $m$ is attached at one end of an elastic string of natural length $a$ and modulus of elasticity $4mg$. The other end of the string is attached at a fixed point O. The particle is initially held at O and then allowed to fall vertically. Using the conservation of energy principle, or otherwise, show that the particle falls a distance $2a$ before coming to instantaneous rest. Find the greatest speed of the particle during this fall, stating the distance from O at which it occurs. [AEB]

28  Prove that the work done in stretching a light elastic string from its natural length $a$ to a length $(a + x)$ is proportional to $x^2$.

One end of this string is fastened to a fixed point A, and at the other end a particle of mass $m$ is attached. The particle is released from rest at A, and first comes to rest when it has fallen a distance $3a$. Show that at the lowest point of its path the acceleration of the particle is $2g$ upwards.

Find in terms of $g$ and $a$ the speed of the particle at the instants when the magnitude of its acceleration is $\frac{1}{2}g$. [L]

29  A particle of mass $m$ is suspended from a ceiling by a light elastic string, of natural length $a$ and modulus $12mg$. When the particle hangs at rest, find the extension in the string. The particle is then pulled down vertically a distance $x$ and released. If the particle just reaches the ceiling, find
(a) the value of $x$
(b) the maximum speed and the maximum acceleration during the motion. [L]

30  Prove that the work done in stretching an elastic string of natural length $l$ and modulus $\lambda$ from a length $(l + x_0)$ to a length $(l + x_1)$ is $\lambda(x_1^2 - x_0^2)/(2l)$.

A particle of mass $m$ is attached to one end of a light elastic string of natural length $a$ and modulus $3mg$. The other end of the string is fastened to a fixed point O. The particle is released from rest at a point C vertically below O, where $OC = \frac{1}{2}a$. Find how far the particle falls before it comes instantaneously to rest.

Find also the greatest speed of the particle during the motion. [C]

31  State Hooke's law and obtain an expression for the potential energy stored in a light elastic string of natural length $a$ and modulus $\lambda$ stretched to a length $(a + b)$.

The ends of this string are fixed to two points A and B on the same level at a distance $3a$ apart. A particle of mass $m$ is fixed to the mid-point of the string and is released from rest at the mid-point of AB. If the particle first comes to rest after falling through a distance $2a$, find its acceleration at this instant. [L]

# 7 Moments

## 7.1 Moment of a force

The effect of a force on a particle can be calculated when the mass of the particle and the magnitude and direction of the force are known. In order to find the effect of a force on a rigid body, it is essential to know at which point of the body the force is applied.

Consider a force acting on a uniform sphere with centre O. If the line of action of the force passes through O, the sphere will accelerate without rotating. If the line of action of the force does not pass through O, the sphere will rotate while accelerating. A rigid body can be treated as a particle only when the question of rotation does not arise.

The turning effect of a force is measured by its moment. In Fig. 7.1 the line AN is at right angles to the force **F**. The moment of **F** about the point A is defined as the product of the distance AN and the magnitude of the force. The unit of moment is the newton-metre, or N m. When the turning effect is anticlockwise as in Fig. 7.1(a), the moment is taken to be positive. When the turning effect is clockwise, as in Fig. 7.1(b), the moment is taken to be negative.

(a)                    (b)

Fig. 7.1

When all the forces involved are in one plane it is sufficient to consider their moments about a point in that plane. However, when the forces are not coplanar, it becomes necessary to take moments about an axis.

Consider a straight line drawn through the point A at right angles to the plane containing the line AN and the force **F**. The moment of the force **F** about this line as an axis of rotation is again defined as the product of the distance AN and the magnitude of the force.

*Example 1*
In the rectangle ABCD, AB = 4 m and BC = 3 m. A force F of magnitude 10 N acts at B towards D. Find the moment of F about A, about C and about D.

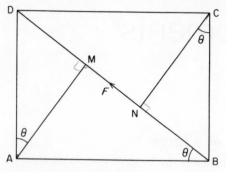

Fig. 7.2

Let AM and CN be perpendicular to BD.
Then $\angle BCN = \angle DAM = \angle ABD = \theta$.
Since $AB = 4$ and $AD = 3$, $BD = 5$ and $\cos \theta = 0.8$.
Hence $AM = CN = 3 \cos \theta = 2.4$.
Moment of $F$ about $A = F \times AM = 10 \times 2.4 = 24$ N m.
Moment of $F$ about $C = F \times CN = 24$ N m.
Since the turning effect of the force about C is clockwise, the moment of $F$ about
C is taken to be $-24$ N m.
The distance of D from the line of action of $F$ is zero. Therefore the moment of $F$
about D is zero.

*Example 2*
A force $F$ of magnitude 7 N acts at the point $P(2.5, 4)$ towards the point $Q(1, 2)$.
Find the moment of $F$ about the point $A(4, 1)$.

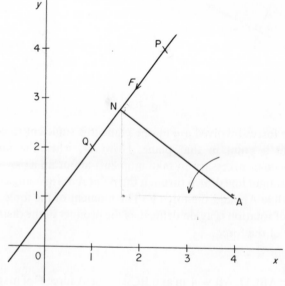

Fig. 7.3

The equation of the line PQ is $4x - 3y + 2 = 0$.
The perpendicular distance AN from A to this line equals

$$\frac{4 \times 4 - 3 \times 1 + 2}{\sqrt{(4^2 + 3^2)}},$$

i.e. AN $= (16 - 3 + 2)/5 = 3$.
Moment of **F** about A $=$ AN $\times 7 = 21$ N m anticlockwise.

In order to find the moment of a force about a point it is often convenient to resolve the force into components. Let the force **P** act at the point O and let AN be the perpendicular from A to the line of action of **P**. Let $\angle OAN = \theta$ (Fig. 7.4).

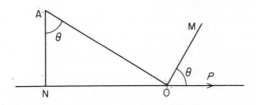

Fig. 7.4

$$\text{Moment of } \mathbf{P} \text{ about A} = P \times \text{AN}$$
$$= P \times \text{AO} \cos \theta$$
$$= (P \cos \theta) \times \text{AO},$$

i.e. the moment about A of the force **P** acting at O equals the product of AO and the component of **P** at right angles to AO.

Now let another force **Q** act at O. The moment of **Q** about A will equal the product of AO and the component of **Q** at right angles to AO.

If **R** is the resultant of **P** and **Q**, the component of **R** at right angles to AO equals the sum of the components of **P** and **Q** at right angles to AO. Therefore the moment of **R** about A equals the sum of the moments of **P** and **Q** about A. This result is particularly useful when a force **R** is resolved into two components at right angles to one another.

### Example 3
The force **R** acts at the point A$(a, b)$ and the components of **R** parallel to the $x$ and $y$ axes are $X$ and $Y$ respectively. Find the moment of **R** about the origin O.

Moment of the force **X** about O $= Xb$ clockwise (Fig. 7.5).
Moment of the force **Y** about O $= Ya$ anticlockwise.
Moment of **R** about O equals the sum of the moments of **X** and **Y** about O.
Moment of **R** about O $= (Ya - Xb)$ anticlockwise.

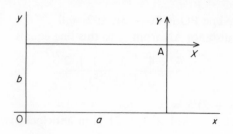

Fig. 7.5

## Exercises 7.1

1  The sides of a square ABCD are of length 4 m and M is the mid-point of AB. A force of 10 N acts at M in the direction $\overrightarrow{MC}$. Find the moment of this force about D and about B.

2  Each side of the triangle ABC is of length $2a$. Forces each of magnitude $P$ act along AB, along BC and along CA. Find the sum of the moments of these forces (a) about A (b) about the mid-point of BC (c) about the centroid of the triangle.

3  A force of magnitude 10 acts along the line $4x = 3y$ in the direction of $x$ increasing. Find its moment about  (a) the point $(5, 1)$  (b) the point $(1, 5)$  (c) the point $(-6, -8)$.

4  A force of known magnitude acts through the point $(2, 1)$. Find the equation of its line of action if the moment of this force about the origin is to be a maximum.

5  A force has a clockwise moment of magnitude 6 about the point $A(5, 2)$ and an anticlockwise moment of magnitude 4 about the point $B(1, 4)$. The force acts at right angles to AB. Find the equation of its line of action and its magnitude.

6  The force $\mathbf{F}$ acts at the point P with position vector $4\mathbf{i} + 3\mathbf{j}$. Find the moment of $\mathbf{F}$ about the point A with position vector $\mathbf{a}$ in the following cases.
   (a) $\mathbf{F} = 6\mathbf{i} + 5\mathbf{j}, \quad \mathbf{a} = \mathbf{i} - \mathbf{j}$.
   (b) $\mathbf{F} = 2\mathbf{i} - 3\mathbf{j}, \quad \mathbf{a} = -2\mathbf{i} + 2\mathbf{j}$.
   (c) $\mathbf{F} = -3\mathbf{i} + 2\mathbf{j}, \quad \mathbf{a} = 6\mathbf{i} + \mathbf{j}$.
   (d) $\mathbf{F} = -5\mathbf{i} + \mathbf{j}, \quad \mathbf{a} = 4\mathbf{i} - \mathbf{j}$.

7  A force $6\mathbf{i} + 5\mathbf{j}$ acts at the point with position vector $\mathbf{i} + 2\mathbf{j}$, and a force $-3\mathbf{i} - 2\mathbf{j}$ acts at the point with position vector $-2\mathbf{i} + 3\mathbf{j}$. Find the sum of their moments about the origin O. Find also the perpendicular distance from O to the line of action of their resultant.

8  A force of magnitude 10 N acts at the origin towards the point $(12, 5)$. Find the moment of the force  (a) about the point $(10, 2)$  (b) about the point $(-3, 2)$. The unit of length is the metre.

9  The sides of a regular hexagon ABCDEF are tangents to a circle of radius 10 cm. A force of 2 N acts along AB, a force of 3 N acts along BC and a

force of 4 N acts along CD. Find the sum of the moments of these forces (a) about D (b) about E (c) about F.

10 A force $2i + 2j$ acts at the point with position vector $i$, and a force $i - j$ acts at the point with position vector $2j$. Show that the sum of the moments of the two forces about the point with position vector $3ti + tj$ is zero for all values of $t$. Find the vector equation of the line of action of the resultant of the two forces.

## 7.2 The principle of moments

In the previous section it was shown that the sum of movements of two forces about a point equals the moment of their resultant about that point. It will be seen later that the sum of the moments of any number of forces about a point equals the moment of their resultant about that point. If the forces are in equilibrium their resultant is zero. Therefore when a system of forces is in equilibrium, the sum of their moments about any point will be zero. This is the *Principle of Moments*.

### Example 1

A uniform beam AB of length 120 cm and weight 50 N rests on a fulcrum at its mid-point M. A force of 20 N acts vertically downward at B, a force of 30 N acts vertically downward at C, the mid-point of AM, and a force of magnitude $x$ N acts vertically downward at A. Find $x$ if the beam is in equilibrium. Find also the force exerted by the fulcrum on the beam.

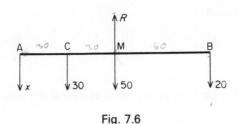

Fig. 7.6

Let $R$ newtons be the magnitude of the force exerted by the fulcrum.
The sum of the moments of the forces about any point will be zero.
Taking moments about M

$$(x \times 60) + (30 \times 30) - (20 \times 60) = 0$$
$$60x = 300$$
$$x = 5.$$

Taking moments about A

$$(R \times 60) - (30 \times 30) - (50 \times 60) - (20 \times 120) = 0$$
$$60R = 6300$$
$$R = 105.$$

Note that the sum of the downward forces is 105 N, giving a check on the value found for $R$.

*Example 2*
A uniform bar AB of length 80 cm and weight 17 N is supported in a horizontal position by two vertical strings, one attached at a point C where AC = 20 cm and the other at B. The bar carries a load of weight 5 N suspended from A and a load of weight 10 N suspended from D where DB = 30 cm. Find the tension in each string.

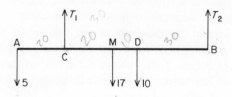

Fig. 7.7

Let $T_1$ be the tension in the string at C and let $T_2$ be the tension in the string at B. By taking moments about B, an equation not involving $T_2$ will be obtained.

$$(5 \times 80) - (T_1 \times 60) + (17 \times 40) + (10 \times 30) = 0$$
$$60T_1 = 1380$$
$$T_1 = 23 \text{ N}.$$

Taking moments about C

$$(5 \times 20) - (17 \times 20) - (10 \times 30) + (T_2 \times 60) = 0$$
$$60T_2 = 540$$
$$T_2 = 9 \text{ N}.$$

As a check, note that the sum of the tensions equals the sum of the downward forces.

**Exercises 7.2**
1   A uniform rod AB of length 30 cm and weight 12 N rests on a knife edge at a point C where AC = 10 cm. A vertical force applied at B keeps the rod in equilibrium when horizontal. Find the magnitude of this force.
2   A uniform bar AB of length 50 cm and weight 12 N can turn freely in a vertical plane about a pivot at a point C, where AC = 10 cm. The bar is kept in equilibrium in a horizontal position by a string attached to the bar at B. Find the tension in the string when the angle between the bar and the string is (a) 30° (b) 45° (c) 60° (d) 90°.
3   Four forces acting along the sides of a square are in equilibrium. Show that the forces are all of the same magnitude.
4   In the equilateral triangle ABC, M and N are the mid-points of AC and AB

respectively. A force of 10 N acts along MN and three unknown forces act along the sides of the triangle ABC. Given that the four forces form a system in equilibrium, find the unknown forces by taking moments about A, B and C.

5   A uniform rod AB of length $2a$ and weight $W$ can turn freely in a vertical plane about a pivot at A. The rod is kept in equilibrium inclined at 60° to the horizontal by a string attached to its highest point B. Find the tension in the string if the string is horizontal. Find the least possible tension in the string if its direction is varied.

6   A uniform square trapdoor ABCD of side 80 cm and weight 200 N can turn about hinges on the side AD, which is horizontal. P and Q are points on AB and CD such that AP = 2PB and DQ = 2QC. Find the magnitude of two equal vertical forces acting at P and Q which keep the trapdoor at rest in a horizontal position.

7   A uniform rod AB of length 80 cm and weight 10 N is supported in a horizontal position by vertical strings attached at its ends. A downward vertical force of 4 N is applied to the rod at a point C, where BC = 10 cm. Find the tension in each string.

8   A uniform bar AB of length 1.2 m and weight 3 N rests in a horizontal position on supports at A and B. A load of weight 9 N is suspended from a point C of the bar. If the reaction at each support must not exceed 9 N, find the shortest possible length of AC.

9   A uniform beam AB of weight 60 N and length 90 cm can turn about a pivot at a point C of its length. When a load of weight 40 N is suspended from A and a load of weight 80 N is suspended from B, the beam can remain in equilibrium when horizontal. Find the length of BC.

10  A uniform rod AB of length 30 cm and weight 16 N can turn freely in a vertical plane about a pivot at a point C, where AC = 20 cm. A load of weight 5 N is suspended from the lower end A and a horizontal force **F** applied at B maintains equilibrium with the rod inclined at an angle $\tan^{-1} 2$ to the horizontal. Find the value of $F$.

## 7.3   Resultant of two parallel forces

Let the parallel forces **P** and **Q** act at the points A and B respectively, where AB is perpendicular to the lines of action of the forces.

(a) Let **P** and **Q** act in the same direction.

In Fig. 7.8, the vector $\overrightarrow{AC}$ represents **P** and $\overrightarrow{BD}$ represents **Q**.

The magnitude $R$ of the resultant **R** of **P** and **Q** equals $P + Q$, and the direction of **R** is the direction of **P** and **Q**.

Let the line of action of **R** meet AB at E, and let the vector $\overrightarrow{EF}$ represent **R**. Then the sum of the moments of **P** and **Q** about E will be zero. This gives

$$P \times AE = Q \times EB,$$

i.e.                               $$AE/EB = Q/P.$$

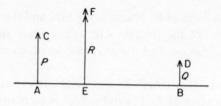

Fig. 7.8

This shows that the point E divides AB internally in the ratio $Q:P$. When $P > Q$, as in Fig. 7.8, E will be nearer to A than to B. When $P < Q$, E will be nearer to B than to A.

The moment of **P** about B will equal the moment of **R** about B.

$$P \times AB = R \times EB$$

$$EB = (P/R)AB = \frac{P}{P + Q}AB.$$

The moment of **Q** about A will equal the moment of **R** about A.

$$Q \times AB = R \times AE$$

$$AE = (Q/R)AB = \frac{Q}{P + Q}AB.$$

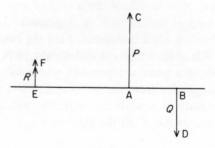

Fig. 7.9

(b) Let **P** and **Q** act in opposite directions.
The case when $P > Q$ is shown in Fig. 7.9.

Let the forces **P** and **Q** be represented by the vectors $\overrightarrow{AC}$ and $\overrightarrow{BD}$ respectively.

When $P > Q$, the resultant **R** of **P** and **Q** is in the same direction as **P**, and its magnitude $R$ equals $P - Q$.

In Fig. 7.9 let the vector $\overrightarrow{EF}$ represent **R**. The sum of the moments of **P** and **Q** about the point E will be zero. This gives

$$P \times EA = Q \times EB.$$

Since $P$ is greater than $Q$, this shows that EB is longer than EA and that

$$EA/EB = Q/P.$$

Thus the point E divides AB externally in the ratio $Q:P$. When $P > Q$, as in Fig. 7.9, the lines of action of **Q** and **R** lie on opposite sides of the line of action of **P**.

*Example*
Two vertical forces **P** and **Q** act at the points A and B respectively, where AB is horizontal and AB = 12 cm. $P = 6$ N and $Q = 3$ N. The line of action of **R**, the resultant of **P** and **Q**, meets AB at C. Find the distance of C from A
(a) when **P** and **Q** act upwards
(b) when **P** acts upwards and **Q** acts downwards.

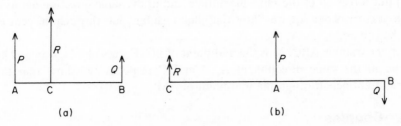

Fig. 7.10

(a) When **P** and **Q** act upwards,

$$R = P + Q = (6 + 3) \text{ N} = 9 \text{ N}.$$

In Fig. 7.10(a), the moment of **R** about any point will equal the sum of the moments of **P** and **Q** about that point.
Taking moments about A

$$R \times AC = Q \times AB$$
$$AC = (Q/R)AB = 4 \text{ cm}.$$

(b) In this case, $R$ equals $P - Q$, i.e. 3 N. Since the moment of **R** about A will equal the moment of **Q** about A, the line of action of **R** is as shown in Fig. 7.10(b). Then

$$R \times CA = Q \times AB$$
$$CA = (Q/R)AB = 12 \text{ cm}.$$

### Exercises 7.3
1  Two forces **P** and **Q** act in the same direction through the points $A(a, 0)$ and $B(b, 0)$ respectively. Find the $x$-coordinate of the point C on the $x$-axis through which their resultant acts.
(a) $P = 8$ N, $Q = 4$ N, $a = 1$, $b = 4$.
(b) $P = 2$ N, $Q = 6$ N, $a = 3$, $b = 23$.
(c) $P = 6$ N, $Q = 8$ N, $a = 2$, $b = 9$.
(d) $P = 7$ N, $Q = 4$ N, $a = -2$, $b = 20$.

**2**  Two forces **P** and **Q** act in opposite directions through the points A($a$, 0) and B($b$, 0) respectively. Find the $x$-coordinate of the point C on the $x$-axis through which their resultant acts.

(a) $P = 3$ N, $Q = 5$ N, $a = 3$, $b = 5$.
(b) $P = 4$ N, $Q = 3$ N, $a = 4$, $b = 5$.
(c) $P = 6$ N, $Q = 8$ N, $a = 6$, $b = 5$.
(d) $P = 7$ N, $Q = 4$ N, $a = 6$, $b = 9$.

**3**  Five forces represented by the vectors 3i, 6i, $-2$i, 4i and $-3$i act at the points with position vectors 2j, 3j, 6j, 7j and 8j respectively. Their resultant is represented by the vector $p$i acting at the point with position vector $q$j. Find $p$ and $q$.

**4**  Four forces all of the same magnitude and in the same direction act at the four corners of a square. Show that their resultant acts through the centre of the square.

**5**  In the triangle ABC, M is the mid-point of BC. Forces of 8 N, 4 N and 4 N act in the same direction at A, B and C respectively. Show that their resultant acts through the mid-point of AM.

## 7.4   Couples

Two parallel forces of the same magnitude acting in opposite directions (but not in the same line) form a couple. The moment of a couple about a point is the sum of the moments of the two parallel forces.

(1) A couple has the same moment about every point in its plane.

Consider the two parallel forces each of magnitude $P$ represented in Fig. 7.11 by $\overrightarrow{AC}$ and $\overrightarrow{BD}$ at right angles to AB. Let L, M and N be points on AB.

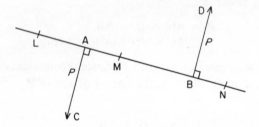

Fig. 7.11

Moment of the couple about L $= P \times$ LB $- P \times$ LA $= P \times$ AB.
Moment of the couple about M $= P \times$ AM $+ P \times$ BM $= P \times$ AB.
Moment of the couple about N $= P \times$ AN $- P \times$ BN $= P \times$ AB.

This shows that the moment of the couple about any point in the plane is the product $P \times$ AB. The distance AB between the two forces is called the arm of the couple.

(2) A couple can be replaced by any couple having the same moment, i.e. of the same magnitude in the same sense.

(3) Two couples can be combined to form a single couple. The combined moment of two couples is the sum of their moments.

*Example 1*
In the rectangle ABCD, AB = 1.5 m and BC = 0.8 m. A force of 10 N acts along AB, a force of 4 N acts along AD, a force of 4 N acts along CB and a force of 10 N acts along CD. Show that this system of forces is equivalent to a couple of moment 2 N m.

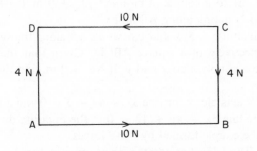

Fig. 7.12

The given forces along AB and CD form an anticlockwise couple of moment (10 × 0.8) N m, i.e. 8 N m.
The given forces along CB and AD form a clockwise couple of moment (4 × 1.5) N m, i.e. 6 N m.
Therefore the four forces are equivalent to a couple of anticlockwise moment 2 N m.

*Example 2*
Three forces are represented in magnitude, direction and line of action by the vectors $\overrightarrow{AB}$, $\overrightarrow{BC}$ and $\overrightarrow{CA}$. Show that this system of forces is equivalent to a couple.

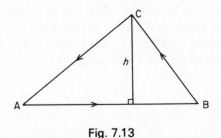

Fig. 7.13

By vector addition, the resultant of the forces represented by $\overrightarrow{CA}$ and $\overrightarrow{BC}$ is a force acting at C which is represented in magnitude and direction by $\overrightarrow{BA}$. Now the force represented by $\overrightarrow{AB}$ acting at A and the force represented by $\overrightarrow{BA}$ acting at C form a couple of moment AB $\times$ $h$, where $h$ is the perpendicular distance from C to AB.

Thus the three forces are equivalent to a couple, and the magnitude of the moment of this couple is given by twice the area of the triangle ABC.

**Exercises 7.4**

1　Six forces each of magnitude 4 N act along the sides AB, BC, CD, DE, EF and FA of a regular hexagon of side 6 m. Show that this system of forces is equivalent (a) to a couple of moment $72\sqrt{3}$ N m (b) to forces each of magnitude 24 N acting along AB and CF.

2　Forces of magnitude 4, 5, $x$ and $y$ newtons act along the sides AB, BC, CD and DA respectively of a square ABCD. Given that the four forces are equivalent to a couple, find $x$ and $y$. If AB $=$ 2 m, find the moment of the couple.

3　A force of 4 N acts along the line $3x - 4y - 3 = 0$ and a force also of 4 N acts along the line $3x - 4y + 12 = 0$ in the opposite direction. Find the moment of the couple formed by these forces.

4　Forces $3\mathbf{i}$, $-7\mathbf{i}$ and $4\mathbf{i}$ act at points with position vectors $\mathbf{i} + \mathbf{j}$, $2\mathbf{i} + 3\mathbf{j}$ and $4\mathbf{i} + 2\mathbf{j}$ respectively. Show that they are equivalent to a couple and find its moment.

5　In the triangle ABC, AB $=$ 3 m, BC $=$ 4 m and CA $=$ 5 m. A force of 6 N acts along AB, a force of 8 N acts along BC and a force of 10 N acts along CA. Show that these forces are equivalent to a couple of moment 24 N m in the sense ABC.

6　In the triangle ABC, AB $=$ BC $=$ 4 m and the angle ABC is a right angle. Forces acting along the sides of this triangle are equivalent to a couple of moment 20 N m in the sense ABC. Find the forces.

## 7.5　A force and a couple

A force and a couple in the same plane can always be reduced to a single force.

Let the force $\mathbf{F}$ act at the point A and let $C$ be the moment of the couple. The couple can be represented by a force $-\mathbf{F}$ acting at A and a force $\mathbf{F}$ acting at B, where AB is at right angles to $\mathbf{F}$ and $F \times \text{AB} = C$.

The two forces acting at A have a zero resultant, and so the force and the couple are equivalent to the force $\mathbf{F}$ acting at B.

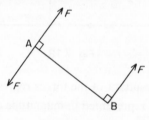

Fig. 7.14

Conversely, a force **F** acting at a point B is equivalent to a force **F** acting at A together with a couple of moment $F \times AB$.

*Example*
A force of 3 N acts at the point P(7, 1) towards the point Q(10, 5). Show that this force is equivalent to a force of the same magnitude acting at the origin together with a couple of magnitude 15 N m.

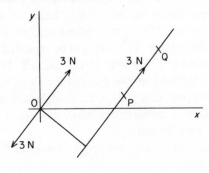

Fig. 7.15

The equation of the line PQ is $4x - 3y - 25 = 0$. The perpendicular distance from the origin O to this line is 5.
Let two forces each of magnitude 3 N act at O, both parallel to PQ but in opposite directions. The force of 3 N at P and the force of 3 N at O in the direction $\overrightarrow{QP}$ form a couple of moment 15 N m. This couple together with the force of 3 N acting at O in the direction $\overrightarrow{PQ}$ are equivalent to the given force of 3 N acting at P.

**Exercises 7.5**
1. Show that the resultant of a force 3**i** acting through the point with position vector 4**j** and an anticlockwise couple of moment 12 is a force of 3**i** acting at the origin.
2. A force of 3 N acts at the point (4, 1) towards the point (5, 2). Show that the resultant of this force and a clockwise couple of moment $6\sqrt{2}$ N m is a force of 3 N acting at the point (2, 3).
3. A force of 4 N acts in the direction of the vector 3**i** + 4**j** at the point with position vector 5**i** + 6**j**. Show that the resultant of this force and a clockwise couple of moment 20 N m is a force of 4 N in the same direction acting at the point with position vector **i** + 9**j**.
4. A force 4**i** acts at the point with position vector 3**i** + 2**j** and a force 3**j** acts at the point with position vector **i** + 4**j**. Show that these two forces are equivalent to a force 4**i** + 3**j** acting at the origin together with a clockwise couple of magnitude 5.

**5** A force **P** acting along the line $x - y\sqrt{3} = 4$ has an anticlockwise moment about the origin of magnitude 10. Find the equation of the line of action of the force which is equivalent to

(a) the force **P** and a clockwise couple of moment 10

(b) the force **P** and an anticlockwise couple of moment 20.

## 7.6 Reduction of a system of coplanar forces

Any system of coplanar forces can be reduced either to a single force or to a couple. For any two forces of the system whose lines of action intersect can be combined to form a single force. This step can be repeated until either only one force remains or a number of parallel forces are left. In the latter case, the parallel forces can be reduced to two parallel forces in opposite directions. If these two forces are unequal, they will have a single resultant. If they are equal in magnitude, they will form a couple except when they act in the same line. In that case, the system is in equilibrium.

A more practical procedure for reducing a system of forces is based on the following two principles:

(a) the component of the resultant in any direction equals the sum of the components of all the forces of the system in that direction

(b) the moment of the resultant about any point equals the sum of the moments about that point of all the forces of the system.

*Example 1*

Forces of magnitude $4P$, $3P$, $2P$ and $P$ act along AB, BC, CD and DA, where ABCD is a square of side $2a$. Find the resultant of this system. If the line of action of the resultant cuts AB produced at E, find the length of BE.

A couple is added to the system and the line of action of the resultant of the enlarged system passes through B. Find the moment and sense of the couple.

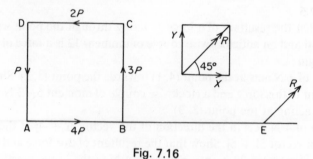

Fig. 7.16

Let $X$ and $Y$ be the components of the resultant in the directions $\overrightarrow{AB}$ and $\overrightarrow{AD}$ respectively (Fig. 7.16).

Resolving along AB

$$X = 4P - 2P = 2P.$$

Resolving along AD, $\qquad\qquad Y = 3P - P = 2P.$

The magnitude $R$ of the resultant is given by

$$R^2 = X^2 + Y^2 = 8P^2$$
$$R = 2\sqrt{2}P.$$

Since $Y = X$, the resultant makes an angle of $45°$ with AB produced.

To find the length of BE, take moments about B.

The moment of the resultant about B is $2P \times$ BE, and hence

$$2P \times BE = 2P \times CB + P \times AB = 6Pa$$
$$BE = 3a.$$

Alternatively, taking moments about E, and making use of the fact that the resultant has no moment about E, gives

$$2P \times 2a + P(2a + BE) - 3P \times BE = 0$$
$$6Pa - 2P \times BE = 0$$
$$BE = 3a.$$

The couple required will consist of a force equal to the resultant **R** acting at B with a force equal to $-$**R** acting at E, as in Fig. 7.17. The magnitude of this couple is the product of $R$ and BE/$\sqrt{2}$, i.e. $6Pa$. From Fig. 7.17 it can be seen that the sense of the couple is clockwise.

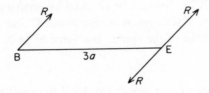

Fig. 7.17

*Example 2*
ABCDEF is a regular hexagon of side $2a$. Forces of 4, 8, 10 and 4 N act along BA, BC, CD and DE respectively. Find the magnitude and the line of action of the resultant of this system.

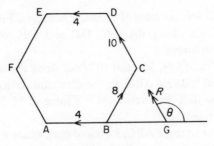

Fig. 7.18

Let $X$ and $Y$ be the components of the resultant in the directions $\overrightarrow{AB}$ and $\overrightarrow{AE}$ respectively (Fig. 7.18).

Resolving along AB

$$X = -4 + 8 \cos 60° - 10 \cos 60° - 4 = -9.$$

Resolving along AE

$$Y = 8 \cos 30° + 10 \cos 30° = 9\sqrt{3}.$$

The magnitude $R$ of the resultant is given by

$$R^2 = X^2 + Y^2 = 81 + 243 = 324$$
$$R = 18.$$

Let the line of action of the resultant cut AB produced at G. The moment of the resultant about B equals $Y \times BG$.

Taking moments about B

$$Y \times BG = 10(2a \cos 30°) + 4(4a \cos 30°)$$
$$(9\sqrt{3})BG = 18\sqrt{3}a$$
$$BG = 2a.$$

Let the resultant make an angle $\theta$ with AB produced.

Then $\tan \theta$ equals $Y/X$, i.e. $-\sqrt{3}$. Hence $\theta = 120°$.

Now BG = BC and the angle GBC is 60°. It follows that the line of action of the resultant is the line GCD. This can be checked by verifying that the sum of the moments of the forces of the system is zero about C and also about D.

Therefore the resultant of the system is a force of 18 N acting along CD.

### Exercises 7.6

1  ABCD is a square of side one metre. Find the resultants of the following systems of forces:

(a) 2 N along BA, 2 N along BC and 2 N along DC

(b) 2 N along AB, 4 N along CB, 2 N along CD and 4 N along AD

(c) 3 N along AB, 6 N along BC, 6 N along DC and 3 N along AD.

2  In the rectangle ABCD, AB = $3a$ and BC = $a$. Forces of magnitude $3P$, $P$, $5P$ and $P$ act along AB, CB, CD and AD respectively. Find the resultant of the system.

3  AB, BC, CD and DE are sides of a regular hexagon. Forces of magnitude 4, 12, 8 and 4 N act along BA, BC, DC and DE respectively. Find the resultant of these forces.

4  Forces of magnitude 5 N, 5 N and 10 N act along the sides BA, AC and BC of the equilateral triangle ABC, in the direction indicated by the order of the letters. Show that their resultant is a force of 15 N acting through the centroid of the triangle.

5  In the equilateral triangle ABC, forces of magnitude 6 N, 4 N and 2 N act along AB, CA and BC respectively. Show that the resultant of this system

of forces is of magnitude $2\sqrt{3}$ N and that its line of action is perpendicular to AC. Show that the distance of its line of action from B equals AB.

6  In the rectangle ABCD, AB $= 2a$ and BC $= a$. Forces of magnitude $2P$, $5P$, $4P$ and $3P$ act along AB, BC, CD and DA respectively. Show that the magnitude of the resultant of these forces is $2\sqrt{2}P$. The line of action of the resultant cuts AB produced at E and BC produced at F. Find the lengths of BE and CF.

7  ABCD is a rectangle in which AB $= 8a$ and BC $= 6a$. Show that the resultant of forces 6 N along AB, 6 N along BC, 15 N along CA and 9 N along AD is a force of $6\sqrt{2}$ N acting along the bisector of the angle ABC. This system of forces is equivalent to a force along CD and a force along AP, where P is a point on CD. Find these forces and the length of CP.

8  Four coplanar forces are represented in magnitude, direction and line of action by the vectors $\overrightarrow{AB}$, $\overrightarrow{BC}$, $\overrightarrow{CD}$ and $\overrightarrow{DA}$. Show that this system of forces can be reduced to a couple.

9  In the square ABCD, forces **P**, **Q** and **R** act along AC, DB and AB respectively and forces of 3, 4 and 5 N act along CB, AD and CD respectively. This system of forces is in equilibrium. Find the magnitudes of **P**, **Q** and **R**.

10  Forces **P**, **Q** and **R** acting along the sides BC, ED and FA of a regular hexagon ABCDEF are equivalent to a force of 6 N acting along AB. Find $P$, $Q$ and $R$.

## 7.7 Centre of mass

Let a particle of mass $m_1$ be situated at the point $P_1$ with position vector $\mathbf{r}_1$ and let a particle of mass $m_2$ be situated at the point $P_2$ with position vector $\mathbf{r}_2$. The centre of mass of the two particles is defined to be the point G such that

$$m_1\overrightarrow{GP_1} + m_2\overrightarrow{GP_2} = \mathbf{0}.$$

The line $P_1P_2$ is divided by G in the ratio $m_2{:}m_1$.
If $m_1 > m_2$, G is nearer to $P_1$ than to $P_2$.
If $m_1 = m_2$, G will be the mid-point of $P_1P_2$.
Let **r** be the position vector of the point G. Then

$$m_1(\mathbf{r}_1 - \mathbf{r}) + m_2(\mathbf{r}_2 - \mathbf{r}) = 0,$$

i.e.
$$\mathbf{r} = \frac{m_1\mathbf{r}_1 + m_2\mathbf{r}_2}{m_1 + m_2}.$$

If $\mathbf{r} = \bar{x}\mathbf{i} + \bar{y}\mathbf{j}$, $\mathbf{r}_1 = x_1\mathbf{i} + y_1\mathbf{j}$, $\mathbf{r}_2 = x_2\mathbf{i} + y_2\mathbf{j}$, this gives

$$\bar{x} = \frac{m_1 x_1 + m_2 x_2}{m_1 + m_2}, \quad \bar{y} = \frac{m_1 y_1 + m_2 y_2}{m_1 + m_2}.$$

*Example 1*
Find the centre of mass of two particles of masses $2m$ and $3m$ placed at the points $P_1(3, 1)$ and $P_2(8, 6)$ respectively.

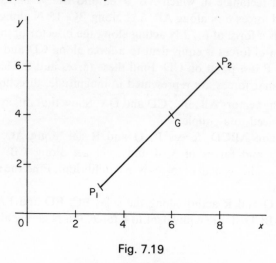

Fig. 7.19

The centre of mass G is the point $(\bar{x}, \bar{y})$ where

$$\bar{x} = \frac{2 \times 3 + 3 \times 8}{2 + 3} = 6, \quad \bar{y} = \frac{2 \times 1 + 3 \times 6}{2 + 3} = 4.$$

In Fig. 7.19 note that the point G divides $P_1 P_2$ in the ratio 3:2.

For three particles of masses $m_1, m_2, m_3$ situated at the points $P_1, P_2, P_3$ with position vectors $\mathbf{r}_1, \mathbf{r}_2, \mathbf{r}_3$ respectively, the centre of mass G is defined by the equation

$$m_1 \overrightarrow{GP_1} + m_2 \overrightarrow{GP_2} + m_3 \overrightarrow{GP_3} = \mathbf{0}.$$

Let $\mathbf{r}$ be the position vector of G. Then

$$m_1(\mathbf{r}_1 - \mathbf{r}) + m_2(\mathbf{r}_2 - \mathbf{r}) + m_3(\mathbf{r}_3 - \mathbf{r}) = \mathbf{0},$$

i.e.

$$\mathbf{r} = \frac{m_1 \mathbf{r}_1 + m_2 \mathbf{r}_2 + m_3 \mathbf{r}_3}{m_1 + m_2 + m_3}.$$

*Example 2*
Three particles each of mass $m$ are placed at the points A, B, C with position vectors $\mathbf{r}_1, \mathbf{r}_2, \mathbf{r}_3$ where $\mathbf{r}_1 = x_1\mathbf{i} + y_1\mathbf{j}, \mathbf{r}_2 = x_2\mathbf{i} + y_2\mathbf{j}, \mathbf{r}_3 = x_3\mathbf{i} + y_3\mathbf{j}$. Find their centre of mass.

The position vector $\mathbf{r}$ of the centre of mass G is given by

$$\mathbf{r} = \frac{m\mathbf{r}_1 + m\mathbf{r}_2 + m\mathbf{r}_3}{m + m + m} = (\mathbf{r}_1 + \mathbf{r}_2 + \mathbf{r}_3)/3.$$

Let $\mathbf{r} = \bar{x}\mathbf{i} + \bar{y}\mathbf{j}$. Then

$$\bar{x} = (x_1 + x_2 + x_3)/3, \quad \bar{y} = (y_1 + y_2 + y_3)/3.$$

This shows that the centre of mass of the three particles of equal mass is at the centroid of the triangle ABC.

Let $n$ particles of masses $m_1, m_2, \ldots, m_n$ be situated at points with position vectors $\mathbf{r}_1, \mathbf{r}_2, \ldots, \mathbf{r}_n$ respectively. Let the position vector $\mathbf{r}$ of their centre of mass be $\bar{x}\mathbf{i} + \bar{y}\mathbf{j}$, and let $M$ be the sum of their masses. Then

$$\bar{x} = \sum_{i=1}^{n} m_i x_i / M, \quad \bar{y} = \sum_{i=1}^{n} m_i y_i / M,$$

where $\mathbf{r}_i = x_i\mathbf{i} + y_i\mathbf{j}$ for each value of $i$.

*Example 3*
Find the centre of mass of five particles of masses $2m$, $6m$, $3m$, $7m$ and $2m$ situated at the points $(1, 4)$, $(2, 3)$, $(2, 4)$, $(3, -1)$ and $(-3, 7)$ respectively.

Sum of the five masses $= 20m$

$$(20m)\bar{x} = 2m \times 1 + 6m \times 2 + 3m \times 2 + 7m \times 3 + 2m(-3)$$
$$\bar{x} = 35/20 = 7/4.$$

$$(20m)\bar{y} = 2m \times 4 + 6m \times 3 + 3m \times 4 + 7m(-1) + 2m \times 7$$
$$\bar{y} = 45/20 = 9/4.$$

Hence the centre of mass is the point $(7/4, 9/4)$.

In order to find the centre of mass of a rod or a lamina (a thin plane sheet) it is possible in some cases to use ideas of symmetry.

A uniform rod can be regarded as composed of pairs of particles of equal mass equidistant from the mid-point of the rod. Therefore the mid-point is the centre of mass of the rod. A uniform rectangular lamina can be regarded as composed of pairs of particles of equal mass, a typical pair being situated at points $P_1$ and $P_2$ such that the mid-point of $P_1P_2$ is the centre of the rectangle. Therefore this point is the centre of mass of the lamina.

It is important to be able to find the centre of mass of a set of particles formed by combining two sets $S_1$ and $S_2$.

Let the set $S_1$ of total mass $M_1$ have its centre of mass at the point $(\bar{x}_1, \bar{y}_1)$ and let the set $S_2$ of total mass $M_2$ have its centre of mass at the point $(\bar{x}_2, \bar{y}_2)$. The summation $\sum m_i x_i$ taken over the two sets is equal to $M_1\bar{x}_1 + M_2\bar{x}_2$. The summation $\sum m_i y_i$ taken over the two sets is equal to $M_1\bar{y}_1 + M_2\bar{y}_2$.

It follows that the coordinates $(\bar{x}, \bar{y})$ of the centre of mass of the set formed by combining the sets $S_1$ and $S_2$ are given by

$$\bar{x} = \frac{M_1\bar{x}_1 + M_2\bar{x}_2}{M_1 + M_2}, \quad \bar{y} = \frac{M_1\bar{y}_1 + M_2\bar{y}_2}{M_1 + M_2}.$$

In effect, the set $S_1$ is treated as a particle of mass $M_1$ at the point $(\bar{x}_1, \bar{y}_1)$ and the set $S_2$ is treated as a particle of mass $M_2$ at the point $(\bar{x}_2, \bar{y}_2)$.

*Example 4*
Show that the centre of mass of a uniform triangular lamina is at the point of intersection of the medians.

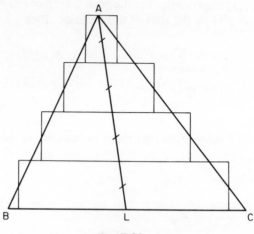

Fig. 7.20

In the triangle ABC let L be the mid-point of BC. Let the median AL be divided into $n$ equal parts by lines parallel to BC. Let $n$ rectangles be constructed having these lines as horizontal sides, while the mid-points of their vertical sides lie on AB and AC. Figure 7.20 shows the case $n = 4$.

The centre of mass of each rectangle lies on AL. Therefore the centre of mass of the set of rectangles lies on AL.

As $n$ increases and tends to infinity, the difference between the triangle and the $n$ rectangles tends to zero, but for all values of $n$ the centre of mass of the set of rectangles lies on AL.

Therefore the centre of mass of the triangular lamina lies on AL. It must also lie on each of the other medians, and so it is the point of intersection of the three medians. Thus it coincides with the centre of mass of three equal particles placed at the vertices of the triangle.

*Example 5*
Find the centre of mass of a uniform lamina ABCD in the shape of a trapezium in which AB is parallel to CD, AB = 8 m, CD = 2 m and AD = BC = 5 m.

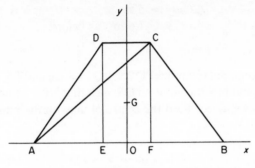

Fig. 7.21

Since AD = 5 m and AE = 3 m, DE = 4 m.
Since BC = 5 m and BF = 3 m, CF = 4 m.
*Method 1*
Take axes as in Fig. 7.21. By symmetry the centre of mass lies on the $y$-axis.
Divide the trapezium into two triangles ADE and BCF and the rectangle
CDEF. Let the mass per unit area be $m$.

| Shape | Area | Mass $m_i$ | $y_i$ | $m_i y_i$ |
|-------|------|------------|-------|-----------|
| ADE | 6 | 6$m$ | 4/3 | 8$m$ |
| CDEF | 8 | 8$m$ | 2 | 16$m$ |
| BCF | 6 | 6$m$ | 4/3 | 8$m$ |
| ABCD | 20 | 20$m$ | $\bar{y}$ | 32$m$ |

$$(20m)\bar{y} = 32m$$
$$\bar{y} = 1.6 \text{ m.}$$

Hence the distance of the centre of mass from AB is 1.6 m.
*Method 2*
Divide the trapezium into two triangles ACD and ABC.
Area of $\triangle ABC = \frac{1}{2}AB \times CF = 16$.
Replace $\triangle ABC$ by particles of mass 16$m$/3 at A, B and C.
Area of $\triangle ACD = \frac{1}{2}CD \times DE = 4$.
Replace $\triangle ACD$ by particles of mass 4$m$/3 at A, C and D.

| Vertex | Mass $m_i$ | $y_i$ | $m_i y_i$ |
|--------|------------|-------|-----------|
| A | 16$m$/3 + 4$m$/3 | 0 | 0 |
| B | 16$m$/3 | 0 | 0 |
| C | 16$m$/3 + 4$m$/3 | 4 | 80$m$/3 |
| D | 4$m$/3 | 4 | 16$m$/3 |
| G | 20$m$ | $\bar{y}$ | 32$m$ |

$$(20m)\bar{y} = 32m$$
$$\bar{y} = 1.6 \text{ m} \quad \text{as before.}$$

*Example 6*
In the uniform rectangular lamina PQRS, PQ = 8 cm and PS = 7 cm. M is the mid-point of RS and N is the point on QR such that NR = 3 cm. The triangular portion NRM is removed. Find the centre of mass of the remainder.

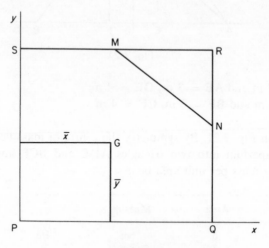

Fig. 7.22

Take axes as in Fig. 7.22.
Let the mass per cm² of the lamina be $m$. The centre of mass of the triangular lamina NRM is at a distance 4/3 cm from QR and at a distance 1 cm from RS. Let the point G($\bar{x}$, $\bar{y}$) be the centre of mass of the remainder.

| Shape | Mass $m_i$ | Distance of centre of mass | | Product | |
|-------|-----------|---------------------------|---------------|---------|---------|
| | | from PS | from PQ | $m_i x_i$ | $m_i y_i$ |
| NRM | 6m | 20/3 | 6 | 40m | 36m |
| PQNMS | 50m | $\bar{x}$ | $\bar{y}$ | 50m$\bar{x}$ | 50m$\bar{y}$ |
| PQRS | 56m | 4 | 3.5 | 224m | 196m |

The distance of G from PS is given by

$$40 + 50\bar{x} = 224$$
$$\bar{x} = 3.68 \text{ cm.}$$

The distance of G from PQ is given by

$$36 + 50\bar{y} = 196$$
$$\bar{y} = 3.2 \text{ cm.}$$

## Exercises 7.7

1 Find the distance from the origin of the centres of mass of the following sets of particles on the $x$-axis.

(a) Particles of mass 5, 8 and 3 units at the points $x = 3$, 4 and 11 respectively

(b) Particles of mass 4, 3 and 5 units at the points $x = 5$, 6 and 2 respectively.

(c) Particles of mass 6, 3, 7 and 4 units at the points $x = 3$, $-2$, 4 and 5 respectively.

(d) Particles of mass 5, 2, 4, 6 and 3 units at the points $x = 4$, $-2$, 5, $-3$ and $-6$ respectively.

2 Particles of mass 4, 3, 6, 2 and 3 kg are placed at the points (4, 2), (2, 4), (3, 4), (4, 5) and (2, 6) respectively. Find the coordinates of their centre of mass.

3 Find the position vector of the centre of mass of particles of masses $m_1$ and $m_2$ situated at points with position vectors $r_1$ and $r_2$ in the following cases:

(a) $m_1 = 4$, $m_2 = 4$, $r_1 = 2i - 3j$, $r_2 = 4i + 5j$

(b) $m_1 = 3$, $m_2 = 7$, $r_1 = 7i + 3j$, $r_2 = 2i + 8j$

(c) $m_1 = 2$, $m_2 = 3$, $r_1 = 6i - j$, $r_2 = i + 4j$

(d) $m_1 = 5$, $m_2 = 3$, $r_1 = 9i - 3j$, $r_2 = i + 5j$.

4 Find the position vector of the centre of mass of particles of mass 4, 5 and 3 kg placed at points with position vectors $5i + 2j$, $-6i + 4j$ and $6i - 8j$ respectively.

5 A particle of mass 4 kg is at the point A($-6, 9$) and a particle of mass 6 kg is at a point B. The centre of mass of the two particles is at the origin. Find the coordinates of B.

6 Particles of mass 4, 3, 6 and 7 kg are placed at the corners A, B, C and D respectively of a square ABCD of side 10 cm. Find the distance of their centre of mass from AB and from AD.

7 Particles of mass 4 kg and 5 kg are placed at the points A(2, 3) and B(2, $-6$) respectively. A third particle of mass 6 kg is placed at a point C. Given that the centre of mass of the three particles is the origin, find the coordinates of C.

8 A uniform lamina PQRS is in the form of a trapezium in which PQ and RS are parallel, PQ = 24 cm, RS = 8 cm and PS = QR = 10 cm. Find the distance of the centre of mass of the lamina from PQ.

9 A uniform lamina is in the form of a square ABCD of side 30 cm with a triangle CED on the side CD, the angle CED being a right angle. Find the distance of the centre of mass of the lamina from AB.

10 A uniform lamina OPQR has its vertices at the origin O and at the points P(3, $-4$), Q(6, 0) and R(6, 4). Find the coordinates of the centre of mass of the lamina.

11 A uniform lamina in the form of a regular hexagon ABCDEF of side $a$ is cut in half by the diameter AD. Find the distance from AD of the centre of mass

of each half of the hexagon.

12   In the uniform rectangular lamina ABCD, AB = 14 cm and AD = 11 cm.
P is a point on BC such that BP = 2 cm and Q is a point on CD such that
CQ = 12 cm. The triangular portion PCQ is cut away. Find the distance
of the centre of mass of the remainder from AB and from AD.

13   A uniform rectangular lamina PQRS has PQ = 17 cm and QR = 4 cm. M
is a point on QR such that MR = 3 cm and N is a point on RS such that
RN = 12 cm. The triangular lamina MRN is removed. Find the distance of
the centre of mass of the remainder from PS and from PQ.

14   The sides AB and BC of a uniform rectangular lamina ABCD are of
length 7 cm and 11 cm respectively. A point X is taken on AB such that
AX = 1 cm and a point Y is taken on BC such that YC = 2 cm. Find the
distances from AD and AB of the centre of mass of the portion AXYCD of
the lamina.

15   Three particles of equal mass are placed at the mid-points of the sides of a
triangle. Show that their centre of mass is at the centroid of the triangle.

## 7.8   Centroids

### Centroid of a plane region

The region ABCD in Fig. 7.23 is defined by the inequalities

$$a \leqslant x \leqslant b, \quad 0 \leqslant y \leqslant f(x).$$

AD is the line $x = a$ and BC is the line $x = b$.

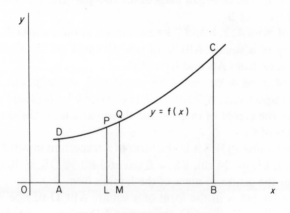

Fig. 7.23

It was shown in Section 10.6 of *Advanced mathematics 1* that the area $A$ of this
region is given by

$$A = \int_a^b f(x)\,dx.$$

Let P and Q be the points $(x, y)$ and $(x + \delta x, y + \delta y)$ respectively on the arc DC, so that LM $= \delta x$.

(a) Let $\delta M$ denote the moment about the $y$-axis of the area of the strip LMQP. Then

$$x(y\,\delta x) < \delta M < (x + \delta x)(y + \delta y)\delta x$$
$$xy < \frac{\delta M}{\delta x} < (x + \delta x)(y + \delta y).$$

As $\delta x$ tends to zero, $\delta y$ tends to zero and in the limit this gives

$$\frac{\mathrm{d}M}{\mathrm{d}x} = xy.$$

It follows that

$$M = \int_a^b xy\,\mathrm{d}x.$$

This is known as the first moment of the area of the region about the $y$-axis.

(b) Let $\delta M$ denote the moment about the $x$-axis of the area of the strip LMQP. Since the $y$-coordinate of the centroid of this strip is between $\frac{1}{2}y$ and $\frac{1}{2}(y + \delta y)$,

$$(y\,\delta x)(\tfrac{1}{2}y) < \delta M < [(y + \delta y)\delta x][\tfrac{1}{2}(y + \delta y)]$$
$$\tfrac{1}{2}y^2 < \frac{\delta M}{\delta x} < \tfrac{1}{2}(y + \delta y)^2.$$

When $\delta x$ tends to zero, in the limit this gives

$$\frac{\mathrm{d}M}{\mathrm{d}x} = \tfrac{1}{2}y^2.$$

It follows that

$$M = \int_a^b \tfrac{1}{2}y^2\,\mathrm{d}x.$$

This is known as the first moment of the area of the region about the $x$-axis.
The centroid of the region is defined to be the point $G(\bar{x}, \bar{y})$ where

$$\bar{x} = \int_a^b xy\,\mathrm{d}x \bigg/ A, \quad \bar{y} = \int_a^b \tfrac{1}{2}y^2\,\mathrm{d}x \bigg/ A.$$

The same point G will be the centre of mass of a uniform lamina having the same boundary as the region ABCD.
    It was assumed above that $f(x)$ increases as $x$ increases as in Fig. 7.23. The results are true when $f(x)$ decreases as $x$ increases.

## Centroid of a solid of revolution

The region ABCD in Fig. 7.23 is defined by the inequalities

$$a \leqslant x \leqslant b, \quad 0 \leqslant y \leqslant f(x).$$

When this region is rotated completely about the x-axis, it sweeps out a solid of revolution. By symmetry, the centroid G of this solid lies on the x-axis.

It was shown in Section 10.9 of *Advanced mathematics 1* that the volume $V$ of this solid is given by

$$V = \int_a^b \pi y^2 \, dx.$$

Let P and Q be the points $(x, y)$ and $(x + \delta x, y + \delta y)$ respectively on the arc DC, so that LM $= \delta x$. The volume of the disc swept out by the strip LMQP lies between $\pi y^2 \, \delta x$ and $\pi (y + \delta y)^2 \, \delta x$.

For each element of volume $\delta V$ of this disc form the product $\delta V \times x$ and let $\delta M$ denote this product. Then

$$(\pi y^2 \, \delta x) \times x < \delta M < \pi (y + \delta y)^2 \, \delta x \times (x + \delta x)$$

$$\pi x y^2 < \frac{\delta M}{\delta x} < \pi (x + \delta x)(y + \delta y)^2.$$

In the limit when $\delta x$ tends to zero, this gives

$$\frac{dM}{dx} = \pi x y^2.$$

It has been assumed in the diagram that $f(x)$ increases as $x$ increases, but this result holds good when $f(x)$ decreases as $x$ increases.

It follows that $M$, the first moment of the volume of the solid about the y-axis, is given by

$$M = \int_a^b \pi x y^2 \, dx.$$

The distance OG is given by

$$\text{OG} \times V = M.$$

Therefore

$$\text{OG} = \int_a^b \pi x y^2 \, dx \bigg/ \int_a^b \pi y^2 \, dx.$$

The centre of mass of a uniform solid of revolution coincides with its centroid.

*Example 1*
Find the coordinates of the centroid of the region bounded by the arc of the curve $y = 6 - 3x^2$ for which $0 \leqslant x \leqslant 1$, the axes and the line $x = 1$.

$$\text{Area of the region} = \int_0^1 y \, dx$$

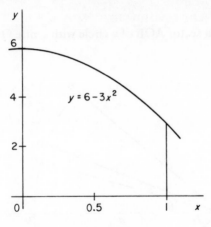

$y = 6 - 3x^2$

**Fig. 7.24**

$$= \int_0^1 (6 - 3x^2)\, dx$$

$$= \left[ 6x - x^3 \right]_0^1$$

$$= 5.$$

Moment of the area about y-axis $= \displaystyle\int_0^1 xy\, dx$

$$= \int_0^1 (6x - 3x^3)\, dx$$

$$= \left[ 3x^2 - 3x^4/4 \right]_0^1$$

$$= 2.25.$$

Therefore $\bar{x} = (2.25)/5 = 0.45.$

Moment of the area about x-axis $= \displaystyle\int_0^1 \tfrac{1}{2} y^2\, dx$

$$= \frac{1}{2} \int_0^1 (36 - 36x^2 + 9x^4)\, dx$$

$$= \frac{1}{2} \left[ 36x - 12x^3 + 9x^5/5 \right]_0^1$$

$$= 12.9.$$

Therefore $\bar{y} = (12.9)/5 = 2.58.$

*Example 2*

Find the centroid of a sector AOB of a circle with centre O and radius $a$, where $\angle AOB = 2\alpha$.

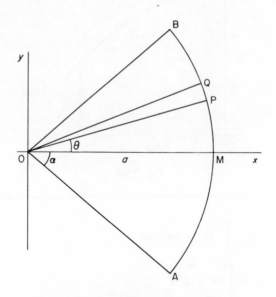

Fig. 7.25

Let the $x$-axis bisect $\angle AOB$. Then by symmetry $\bar{y} = 0$.

Let $\angle MOP = \theta$ and $\angle POQ = \delta\theta$. Area of the sector POQ $= \frac{1}{2}a^2\,\delta\theta$.

The distance of the centroid of the sector POQ from the $y$-axis is approximately equal to $\frac{2}{3}a\cos\theta$ (Fig. 7.25). The moment of the area of the sector POQ about the $y$-axis is given approximately by $(\frac{2}{3}a\cos\theta)(\frac{1}{2}a^2\,\delta\theta)$.

Moment of the area of the sector AOB about the $y$-axis

$$= \int_{-\alpha}^{\alpha} \tfrac{1}{3}a^3 \cos\theta\,d\theta$$

$$= \left[ \tfrac{1}{3}a^3 \sin\theta \right]_{-\alpha}^{\alpha}$$

$$= \tfrac{2}{3}a^3 \sin\alpha.$$

Area of the sector AOB $= a^2\alpha$.

Therefore

$$\bar{x} = (\tfrac{2}{3}a^3 \sin\alpha)/(a^2\alpha)$$
$$= (2a\sin\alpha)/(3\alpha).$$

When $\alpha = \pi/2$, the sector AOB becomes a semicircle. Hence the distance of the centroid of a semicircle of radius $a$ from its diameter is $4a/(3\pi)$.

*Example 3*

The chord AB subtends an angle $2\alpha$ at the centre O of a circle of radius $a$. Find the distance from O of the centroid of the minor segment of the circle cut off by AB.

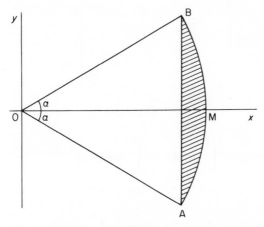

Fig. 7.26

| Shape | Area | Distance of centroid from $y$-axis | Moment about $y$-axis |
|---|---|---|---|
| triangle AOB | $\frac{1}{2}a^2 \sin 2\alpha$ | $\frac{2}{3}a \cos \alpha$ | $\frac{1}{3}a^3 \sin 2\alpha \cos \alpha$ |
| segment AMB | $\frac{1}{2}a^2(2\alpha - \sin 2\alpha)$ | $\bar{x}$ | $\frac{1}{2}\bar{x}a^2(2\alpha - \sin 2\alpha)$ |
| sector AOB | $a^2\alpha$ | $(\frac{2}{3}a \sin \alpha)/\alpha$ | $\frac{2}{3}a^3 \sin \alpha$ |

From the last column of the table,

$$\frac{1}{2}\bar{x}a^2(2\alpha - \sin 2\alpha) + \frac{1}{3}a^3 \sin 2\alpha \cos \alpha = \frac{2}{3}a^3 \sin \alpha$$
$$\frac{1}{2}x(2\alpha - \sin 2\alpha) = \frac{2}{3}a \sin \alpha - \frac{1}{3}a \sin 2\alpha \cos \alpha$$
$$= \frac{1}{3}a(2 \sin \alpha - 2 \sin \alpha \cos^2 \alpha)$$
$$= \frac{2}{3}a \sin \alpha(1 - \cos^2 \alpha)$$
$$= \frac{2}{3}a \sin^3 \alpha.$$

Therefore
$$\bar{x} = \frac{4a \sin^3 \alpha}{3(2\alpha - \sin 2\alpha)}.$$

*Example 4*

Show that the centroid of a right circular cone of height $h$ is at a distance $h/4$ from its base.

In Fig. 7.27, P and Q are the points $(h, 0)$ and $(h, mh)$ respectively, and the equation of the line OQ is $y = mx$. If the triangular region OPQ is rotated completely about the $x$-axis, it will sweep out a cone.

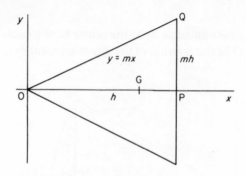

Fig. 7.27

Volume of the cone $= \int_0^h \pi y^2 \, dx = \int_0^h \pi m^2 x^2 \, dx$

$$= \pi m^2 h^3 / 3.$$

Moment of the volume about the $y$-axis $= \int_0^h \pi x y^2 \, dx$

$$= \int_0^h \pi m^2 x^3 \, dx$$

$$= \pi m^2 h^4 / 4.$$

If G is the centroid of the cone,

$$OG = (\pi m^2 h^4 / 4) / (\pi m^2 h^3 / 3) = 3h/4.$$

Therefore the distance of G from the centre of the base is $h/4$.

*Example 5*
Show that the centroid G of a hemisphere of radius $a$ is at a distance $3a/8$ from its plane face.

The hemisphere is swept out when the semicircle OAMB of radius $a$ and centre O is rotated completely about the $x$-axis (Fig. 7.28). Since the equation of the circle is $x^2 + y^2 = a^2$, on the arc MB

$$y = \sqrt{(a^2 - x^2)}.$$

Volume of hemisphere $= \int_0^a \pi y^2 \, dx$

$$= \int_0^a \pi (a^2 - x^2) \, dx$$

$$= \pi \left[ a^2 x - x^3 / 3 \right]_0^a = 2\pi a^3 / 3.$$

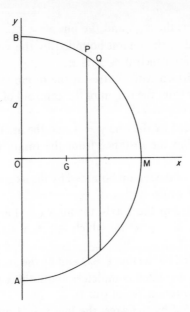

Fig. 7.28

Moment of the volume about the $y$-axis $= \displaystyle\int_0^a \pi x y^2 \, dx$

$$= \int_0^a \pi(a^2 x - x^3) \, dx$$

$$= \pi \left[ a^2 x^2/2 - x^4/4 \right]_0^a$$

$$= \pi a^4/4.$$

$$OG = \int_0^a \pi x y^2 \, dx \Big/ \int_0^a \pi y^2 \, dx$$

$$= (\pi a^4/4)/(2\pi a^3/3)$$

$$= 3a/8.$$

### Exercises 7.8

1 Find the centroid of the region in the first quadrant bounded by the axes and the arc of the curve $y = 1 - x^2$.

2 Find the centroid of the region bounded by the $x$-axis, the line $x = 1$ and the arc of the curve $y = x^2$ for which $0 \leqslant x \leqslant 1$.

3 Find the centroid of the region in the first quadrant bounded by the $x$-axis, the line $x = 2$ and the arc of the parabola $y^2 = 4x$ for which $0 \leqslant x \leqslant 2$.

4 A sector of angle $60°$ is removed from a circle of radius $a$. Find the distance of the centroid of the remainder from the centre of the circle.

5 Find the centroid of the region bounded by the arc of the curve $y = e^x$ for

which $0 \leqslant x \leqslant 1$, the axes and the line $x = 1$.

6   Find the centroid of the region bounded by the $x$-axis and the arc of the curve $y = \sin x$ for which $0 \leqslant x \leqslant \pi$.

7   Two circles are drawn with centres at the origin and with radii $a$ and $3a$. Find the distance from the origin of the centroid of the region between the circles for which $x \geqslant 0$.

8   A region is bounded by the line $y = 1$ and the arc of the curve $y = x^2$ for which $y \leqslant 1$. Find the distance from the origin of the centroid of this region.

9   Find the centroid of the region bounded by the axes and the arc of the curve $y = \cos(x/2)$ for which $0 \leqslant x \leqslant \pi$.

10   A region is bounded by the $x$-axis, the lines $x = 1$ and $x = -1$, and by the arc of the curve $y = 1 + x^2$ for which $-1 \leqslant x \leqslant 1$. Find the centroid of this region.

11   The region bounded by the line $x = 3$ and by the arc of the curve $y^2 = x$ for which $0 \leqslant x \leqslant 3$ is rotated completely about the $x$-axis. Find the centroid of the solid of revolution swept out.

12   The region bounded by the axes, the line $x = 1$ and the arc of the curve $y = 1 + x^2$ for which $0 \leqslant x \leqslant 1$ is rotated completely about the $x$-axis. Find the distance from the origin of the centroid of the solid of revolution swept out.

13   The region defined by the inequalities $1 \leqslant x \leqslant 2, 0 \leqslant y \leqslant 1/x$ is rotated completely about the $x$-axis. Find the distance from the origin of the centroid of the solid of revolution which is swept out.

14   Sketch the curve $y = 1/(1 + x)$. The region bounded by the axes, the line $x = 1$ and the arc of this curve for which $0 \leqslant x \leqslant 1$ is rotated completely about the $x$-axis. Find the distance from the origin of the centroid of the solid of revolution generated.

15   The region bounded by the axes, the line $x = 1$ and the arc of the curve $y = e^x$ for which $0 \leqslant x \leqslant 1$ is rotated completely about the $x$-axis. Find the distance from the origin of the centroid of the solid of revolution swept out.

## 7.9   Centre of gravity

Let $n$ particles of masses $m_1, m_2, \ldots, m_n$ be placed at the points $(x_1, y_1), (x_2, y_2), \ldots, (x_n, y_n)$ respectively. Let the $x$–$y$ plane be horizontal, so that the weights of the particles form a set of vertical forces. Let $M$ be the sum of the masses of the particles. Then the resultant of the weights of the particles is a vertical force of magnitude $Mg$.

The centre of gravity of the particles is defined to be the point of the $x$–$y$ plane through which this resultant acts. The sum of the moments of the weights of the particles about the $y$-axis

$$= \sum_{r=1}^{n} (m_r g) x_r$$

$$= g \times \sum_{r=1}^{n} m_r x_r$$
$$= Mg\bar{x},$$

where $\bar{x}$ is the $x$-coordinate of the centre of mass of the particles. Therefore the resultant weight acts at a distance $\bar{x}$ from the $y$-axis.

The same method shows that the resultant weight acts at a distance $\bar{y}$ from the $x$-axis.

This shows that the centre of gravity of the particles coincides with the point $G(\bar{x}, \bar{y})$, their centre of mass. However, this result is based on the assumption that the weights of the particles form a set of parallel forces. This will not be true if the distances between the particles are not negligible in comparison with the radius of the earth.

The centre of gravity of a solid body is defined to be the point through which the line of action of its weight always passes. If the body is small compared to the earth, its centre of gravity can be taken to coincide with its centre of mass.

*Example 1*
A thin uniform wire is bent into the shape of an arc AB of a circle of radius $a$. The arc subtends an angle $2\alpha$ at O, the centre of the circle. If G is the centre of gravity of the wire, show that $OG = (a \sin \alpha)/\alpha$.

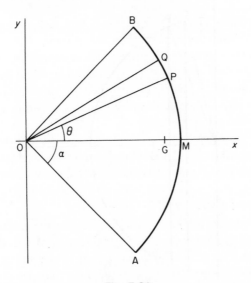

Fig. 7.29

By symmetry, G lies on the bisector of the angle AOB. Let the wire be in a horizontal plane, with axes Ox and Oy as shown in Fig. 7.29. Let P and Q be points on the arc such that OP makes an angle $\theta$ with Ox and $\angle POQ = \delta\theta$. Let the mass of the wire per unit length be $m$.
Mass of the arc PQ $= ma\,\delta\theta$.

Moment of the weight of the arc PQ about O$y \approx mga^2 \cos \theta \, \delta\theta$.

Moment of the weight of the wire about O$y$

$$= \int_{-\alpha}^{\alpha} mga^2 \cos \theta \, d\theta$$

$$= mga^2 \left[ \sin \theta \right]_{-\alpha}^{\alpha}$$

$$= 2mga^2 \sin \alpha.$$

Since the weight of the wire is $2mga\alpha$,

$$2mga\alpha \times \mathrm{OG} = 2mga^2 \sin \alpha$$

$$\mathrm{OG} = (a \sin \alpha)/\alpha.$$

For a semicircular arc, $\alpha = \pi/2$ and $\mathrm{OG} = 2a/\pi$.

*Example 2*

Find the distance of the centre of gravity of a hemispherical shell of radius $a$ from the centre of its rim.

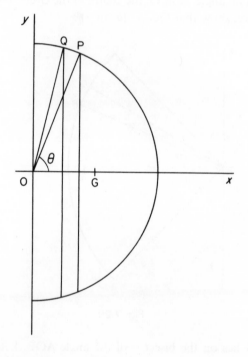

Fig. 7.30

The centre of gravity G will lie on the axis of symmetry. Take this axis as the $x$-axis, and let the $x$-axis and the $y$-axis through O be horizontal (Fig. 7.30).

If the radius OP makes an angle $\theta$ with Ox, P will lie on a circle of radius $a \sin \theta$ on the hemisphere. Let the radius OQ make an angle $\theta + \delta\theta$ with Ox. Then Q lies on a circle of radius $a \sin (\theta + \delta\theta)$ on the hemisphere.

The area of the zone of the hemisphere lying between these two circles is $(2\pi a \sin \theta)(a \, \delta\theta)$ approximately. Let $mg$ be the weight per unit area. The moment of the weight of this zone about the $y$-axis is

$$mg(2\pi a \sin \theta)(a \, \delta\theta) \times (a \cos \theta)$$

approximately. Therefore the moment of the weight of the hemispherical shell about the $y$-axis

$$= \pi a^3 mg \int_0^{\pi/2} 2 \sin \theta \cos \theta \, d\theta$$

$$= \pi a^3 mg \left[ \sin^2 \theta \right]_0^{\pi/2}$$

$$= \pi a^3 mg.$$

Since the surface area of the shell is $2\pi a^2$, its weight is $2\pi a^2 mg$ acting vertically downward at G. Therefore

$$2\pi a^2 mg \times OG = \pi a^3 mg$$

Hence $$OG = a/2.$$

*Example 3*
The radius of a uniform circular disc with centre O is $4a$. A circular hole with centre C and radius $2a$ is cut out. Given that $OC = a$, find the distance from O of the centre of gravity of the remainder.

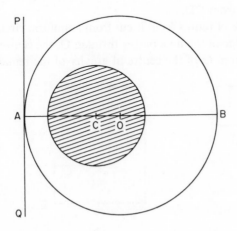

**Fig. 7.31**

By symmetry, the centre of gravity of the remainder will lie on AB, the diameter through O and C (Fig. 7.31). Let the disc be in a horizontal plane, and let PQ be a horizontal line through A at right angles to AB. The area of the disc is four times the area of the hole. Therefore, if the weight of the disc is $W$, the weight of the part removed is $W/4$ and the weight of the remainder is $3W/4$.

In order to find the moments of these weights about PQ, the table below is drawn up.

| Shape | Weight | Distance of centre of gravity from A | Moment about PQ |
|---|---|---|---|
| shaded part | $W/4$ | CA = $3a$ | $3Wa/4$ |
| remainder | $3W/4$ | GA = $\bar{x}$ | $3W\bar{x}/4$ |
| whole disc | $W$ | OA = $4a$ | $4Wa$ |

By equating the moments about PQ,

$$3Wa/4 + 3W\bar{x}/4 = 4Wa$$
$$\bar{x} = 13a/3.$$

Hence the distance of G from O is $a/3$.
Note that G does not lie on the remainder.

### Exercises 7.9

1   Three particles of equal weight are placed at the vertices of a triangle with sides 40 cm, 25 cm and 25 cm. Find the distance of their centre of gravity from each side.

2   M is the mid-point of the side AD of a square sheet of plywood ABCD of side 18 cm. A sawcut is made along the line BM and the triangular piece ABM is removed. Find the distance of the centre of gravity of the remainder from BC and from CD.

3   A circular hole of radius 6 cm is cut from a uniform circular disc of radius 12 cm. The edge of the hole passes through O, the centre of the disc. Find the distance from O of the centre of gravity of the remainder.

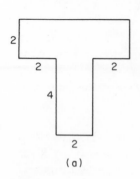

(a)

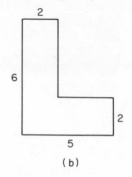

(b)

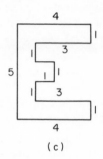

(c)

Fig. 7.32

**4** Three letters made of uniform sheet metal are shown in Fig. 7.32, the dimensions being in centimetres. All the angles are right angles. Find the distance of the centre of gravity of each letter (i) from its lower edge (ii) from its left-hand edge.

**5** ABCD is a square board of side 48 cm. A square hole of side 16 cm is cut in the board, its sides being parallel to the sides of the board and one of its corners being at the centre of the square. Find the distance of the centre of gravity of the remainder from the centre of the square.

**6** A uniform wire of length $\pi a$, bent into the shape of a semicircle of radius $a$, is joined to the curved edge of a uniform semicircular sheet of metal of the same radius. The mass of the wire is one-third of the mass of the metal sheet. Find the distance from the straight edge of the centre of gravity of the sheet and wire combined.

**7** Show that the centre of gravity of the curved surface of a hollow cone of height $h$ is at a distance $h/3$ from the centre of its base.

**8** The radii of the circular faces of the frustum of a uniform solid cone are $a$ and $2a$, and the distance between these faces is $4a$. Show that the centre of gravity of the frustum is at a distance $11a/7$ from the larger face.

**9** The distance of the vertex A of a uniform solid regular tetrahedron ABCD from the base BCD is $h$. Show by considering a thin section by planes parallel to the base that the centre of gravity of the tetrahedron is at a distance $h/4$ from the base.

**10** In the uniform rectangular lamina ABCD, $AB = 4a$ and $AD = 2a$. The point P on AB is such that $AP = 3a$, and the point Q on AD is such that $AQ = 4a/3$. The triangular lamina APQ is removed. Find the distance of the centre of gravity of the remainder from BC and from CD.

The triangular portion APQ is now joined to the remainder, with AP lying along CD and AQ lying along CB. Find the distance of the new centre of gravity from BC and from CD.

**11** The radius of the base of a cone is equal to the radius of a cylinder. The base of the cone is fastened to a plane face of the cylinder to form a uniform solid symmetrical about the axis of the cylinder. The centre of gravity of the solid is at the centre of the base of the cone. Find the ratio of the height of the cone to the length of the cylinder.

**12** The radii of the plane faces of the frustum of a cone are 2 cm and 3 cm, and the distance between the faces is 5 cm. A uniform paper collar is fitted to the curved surface of the frustum. Find the distance of the centre of gravity of the paper collar from the centre of the larger plane face of the frustum.

## Miscellaneous exercises 7

**1** ABCD is a uniform rectangular plate with $AB = 80$ cm, $AD = 60$ cm. A circular hole of radius 10 cm is cut from the plate, its centre being 20 cm from BC and 15 cm from CD. Find, to the nearest centimetre, the distances of the centre of gravity of the remainder from AD and AB.

2   In the triangle ABC, AB = 4 cm, BC = 3 cm and CA = 5 cm. Forces of magnitude 8, 9 and 10 N act along AB, BC and CA respectively. Find the resultant of this system of forces and the distance from A of its line of action.

3   Forces of magnitude 1, 4, 3 and 5 N act along the sides AB, BC, CD and DA respectively of a square ABCD of side 2 m. A force **P** acts at A, such that the system of five forces reduces to a couple. Find the magnitude of the force **P**, the angle which it makes with AB and the magnitude of the resultant couple.

4   ABCD is a uniform square metal plate of side 24 cm and E and F are the mid-points of AB and BC. The triangular portion BEF is bent over so that it lies flat with B coinciding with the centre of the square. Find the distance from AD of the new centre of gravity of the plate.

5   A uniform bar of weight 12 N and length one metre rests in a horizontal position on two supports 60 cm apart. The force exerted on the bar by one of the supports is 4 N. Find the distance of this support from the mid-point of the bar. A downward vertical force is now applied to one end of the bar so that the forces exerted by the supports become equal. Find the magnitude of this force.

6   Forces of magnitude 4, 2 and 2 N act along the sides AB, CB and AC of an equilateral triangle ABC. Show that this system of forces can be reduced to a single force of magnitude 6 N parallel to AB acting at the centroid of the triangle.

7   Forces $3P$, $4P$, $5P$ act along the sides AB, BC, CA of an equilateral triangle ABC of side $a$. Show that the line of action of their resultant passes through a point D on BA produced such that AD = $4a$.

When a force **Q** acting at A is added to the system, the system reduces to a couple. Find the magnitude of the force **Q** and the moment of the couple.

8   The radii of the plane faces of a frustum of a uniform solid cone are 5 cm and 15 cm, and the faces are 10 cm apart. Find the distance of the centre of gravity of the frustum from the centre of the larger plane face.

9   In the uniform rectangular lamina PQRS, PQ = 48 cm and QR = 24 cm. The points M and N on PQ and PS are such that PM = PN = 12 cm. The triangle PMN is turned through two right angles about MN. Find the distances of the new centre of gravity of the lamina from RQ and RS.

10   A region is bounded by the arc of the curve $xy = a^2$ for which $a \leqslant x \leqslant 2a$, the lines $x = a$, $x = 2a$ and the $x$-axis. Find the $x$-coordinate of the centroid of this region.

The region is rotated completely about the $x$-axis. Find the $x$-coordinate of the centroid of the volume swept out.

11   A triangle OAB is drawn outwards on the side AB of a square ABCD, with OA = OB. Find the ratio of OA to AB if the centroid of the whole figure lies on AB.

12   A rectangular lamina ABCD has AB = 12 cm, BC = 9 cm. P is a point on CB such that CP = 6 cm, and S is a point on DA such that DS = 3 cm. Q and R are points on CD such that CQ = DS and RD = CP. The triangles CPQ and DRS are removed. Find the distance of the centre of gravity of the remainder from AB and from AD.

13   The region defined by the inequalities $0 \leqslant x \leqslant a$, $y^2 \leqslant 4ax$ is rotated completely around the $x$-axis. Find the distance from the origin of the centroid of the volume swept out.

14   A quadrilateral ABCD has opposite sides AB and DC parallel. AB = 1, BC = 2, the angle ABC = 150° and the angle BAD = 60°. Forces of magnitude $2P$, $P$, $P$, $2P$ act along $\overrightarrow{AB}$, $\overrightarrow{BC}$, $\overrightarrow{CD}$, $\overrightarrow{AD}$ respectively in the directions indicated by the arrows. Prove that the resultant has magnitude $(8 + 3\sqrt{3})^{1/2}P$. Find the tangent of the angle it makes with AB and the distance from A of the point on AB through which it acts.          [L]

15   (i) A force **P** acts along the side BC of an equilateral triangle ABC in which M and N are the mid-points of AB and AC respectively. Find three forces acting along the sides of the triangle AMN such that their resultant is the force **P**.

   (ii) A particle at the centre O of a regular hexagon ABCDEF is acted on by forces of magnitude 1, 2, 3, 4, 5, 6 units acting along $\overrightarrow{OA}$, $\overrightarrow{OB}$, $\overrightarrow{OC}$, $\overrightarrow{OD}$, $\overrightarrow{OE}$, $\overrightarrow{OF}$ respectively. Find the magnitude, the direction and the line of action of the extra force required to keep the particle at rest.          [L]

16   A rectangular sheet of cardboard ABCD has AB = 48 cm, BC = 36 cm. The cardboard is folded so that the edge BC lies along CD. Find the distance of the centre of gravity of the cardboard thus folded from AD and from DC.

17   A horizontal uniform plank ABCD of length $6a$ and weight $W$ rests on two supports B and C, AB = BC = CD = 2a. A man of weight $4W$ wishes to stand on the plank at A, for which purpose he places a counterbalancing weight $Y$ at a point P between C and D at a distance $x$ from D. Show that the least value of $Y$ required is $7Wa/(4a - x)$.

   Show that, if the board is also to be in equilibrium when he is not standing on it, then the greatest value of $Y$ is $Wa/(2a - x)$.          [JMB]

18   Forces $P$, $2P$, $3P$, $kP$, $lP$ and $mP$ act along the sides AB, BC, CD, DE, EF and FA respectively of a regular hexagon ABCDEF, the directions of the forces being indicated by the order of the letters. If the system is in equilibrium, find the values of $k$, $l$ and $m$.

   The forces along DE, EF and FA are replaced by a coplanar force through O, the centre of the hexagon, and a coplanar couple. The resulting system is in equilibrium. If the length of each side of the hexagon is $2a$, find the couple and the force through O, giving the angle the line of action of this force makes with OA.          [AEB]

19 Forces of magnitude 3, 4 and 5 units act along the sides AB, BC, and CA respectively of an equilateral triangle ABC of side $2a$, and a coplanar couple of magnitude $6a\sqrt{3}$ acts in the sense CBA. Show that the resultant of this system of forces is a force of magnitude $\sqrt{3}$. The line of action of the resultant cuts CB produced at D and it cuts AB produced at E. Find the lengths of BD and BE.

20 In the rectangle ABCD, AB $= 2a$ and AD $= a$. Forces **P, Q, R, S** act along the sides AB, BC, CD, DA respectively in the sense indicated by the order of the letters. Find equations which must be satisfied by $P, Q, R, S$
(a) if the forces reduce to a couple
(b) if the forces are in equilibrium
(c) if the forces reduce to a resultant force acting through the mid-points of AB and CD.
  If the forces reduce to a single force $2P$ acting along AB, express $Q, R$ and $S$ in terms of $P$. [L]

21 The centre of gravity of a rod AB of length 90 cm and weight 8 N is at G where AG $= 40$ cm. The rod is supported in a horizontal position by two vertical strings CP and DQ, attached to the rod at C and D where AC $= 20$ cm and DB $= 10$ cm. A load of weight 4 N is hung from A. Find the vertical force which must be applied at B in order that the tensions in the strings may be equal.
  Each string can withstand a tension of 80 N. Show that if the force applied at B is gradually increased the rod will turn before either string breaks.

22 A canister made of thin uniform material consists of a right circular cylinder of radius 7 cm and length 14 cm, closed at one end by a plane base and surmounted at the other end by a hemispherical lid. Find the distance of the centre of gravity of the canister from the centre of the base.

23 A bowl with a flat base but no lid is made of uniform thin metal in the shape of a frustum of a cone. The diameter of the base of the bowl is 20 cm and the diameter of the rim is 30 cm. The height of the bowl is 12 cm. Find, to two significant figures, the distance of the centre of gravity of the bowl from its base.

24 In the rectangle ABCD, AB $= 4$ m and BC $= 3$ m. Forces of 16, 12, 4, 9 and 10 N act along AB, BC, CD, AD and DB respectively. Find the magnitude and direction of the resultant of this system and find the distance from A of the point on AB through which it acts.
  A couple of moment 24 N m in the sense ABCD is added to the system. Show that the resultant will now act through B.

25 A closed carton is made of thin uniform cardboard in the form of a frustum of a right square pyramid. The square faces have sides of length 10 cm and 20 cm and they are 12 cm apart. Find the distance of the centre of gravity of the carton from the larger square face.

**26** The diagonals AC and BD of the quadrilateral ABCD intersect at O. Given that $AO = a, CO = 2a, BO = DO = b$, find the position of the centroid G of the quadrilateral.

**27** The points A, B, C, D, E, F are the vertices of a regular hexagon. Forces each of 2 N act along AB and DC, and forces each of 1 newton act along BC and ED, in the directions indicated by the order of the letters. Forces of $P$ newtons and $Q$ newtons act along EF and AF respectively. Find $P$ and $Q$
(a) if the system reduces to a couple
(b) if the resultant of the system is a force acting along EB. [L]

**28** The plane face of a uniform solid hemisphere of radius $a$ coincides with the base of a uniform solid right circular cone of base radius $a$, semi-vertical angle $\pi/6$ and made of the same material. Show that the centre of mass of this body is at the centre of the base of the cone.

**29** Forces of magnitudes $5P$, $7P$, $mP$ and $nP$ act along AB, BC, CD and DA where ABCD is a square of side $a$. With A as origin and AB and AD as $x$-axis and $y$-axis respectively, the equation of the line of action of the resultant of this system of forces is $3x - 4y - 8a = 0$. Find the values of $m$ and $n$.

A couple is now added to the system and the line of action of the resultant of the enlarged system acts through B. Find the moment and state the sense of the couple. [AEB]

**30** With O as origin, the coordinates of A, B and C are $(5a, 0)$, $(8a, 4a)$ and $(0, 10a)$ respectively. Forces of magnitudes $14P$, $5P$, $15P$ and $P$ act along $\overrightarrow{OA}$, $\overrightarrow{AB}$, $\overrightarrow{BC}$ and $\overrightarrow{CO}$ respectively. Calculate the magnitude and the direction of the resultant of this system of forces and show that the line of action of the resultant cuts OA at E, where $OE:OA = 7:3$.

The force acting in $\overrightarrow{CO}$ is increased to $13P$ and the force acting in $\overrightarrow{OA}$ is decreased to $9P$, while the forces in BC and AB remain unchanged. Show that this new system reduces to a couple and find the moment and sense of this couple. [AEB]

**31** The centre of a regular hexagon ABCDEF of side $a$ is O. Forces of magnitude $P, 2P, 3P, 4P, mP$ and $nP$ act along $\overrightarrow{AB}$, $\overrightarrow{BC}$, $\overrightarrow{CD}$, $\overrightarrow{DE}$, $\overrightarrow{EF}$ and $\overrightarrow{FA}$ respectively. Given that the resultant of these six forces is of magnitude $3P$ acting in a direction parallel to $\overrightarrow{EF}$,
(i) determine the values of $m$ and $n$
(ii) show that the sum of the moments of the forces about O is $9Pa\sqrt{3}$.
The mid-point of EF is M.
(iii) Find the equation of the line of action of the resultant referred to OM as $x$-axis and OA as $y$-axis.

The forces $mP$ and $nP$ acting along $\overrightarrow{EF}$ and $\overrightarrow{FA}$ are removed from the system. The remaining four forces and an additional force **Q**, which acts through O, reduce to a couple. Calculate the magnitude of **Q** and the moment of the couple. [AEB]

**32** In a triangle ABC, AB $=$ AC $=$ 4 m and BC $=$ 6 m.

(i) Forces of magnitude 7 N, 12 N and 9 N act along $\overrightarrow{AB}$, $\overrightarrow{BC}$ and $\overrightarrow{CA}$ respectively. Show that the resultant of these forces acts along a line which is perpendicular to BC. Calculate the magnitude of the resultant and find the distance from B of the point where the line of action of the resultant cuts BC.

(ii) Forces of magnitude $P$ newtons, $Q$ newtons and 3 newtons act along $\overrightarrow{AB}$, $\overrightarrow{BC}$ and $\overrightarrow{CA}$ respectively. If this system of forces reduces to a couple, calculate the values of $P$ and $Q$ and the magnitude and sense of the moment of the couple. [AEB]

**33** Forces $\lambda\,\overrightarrow{OA}$ and $3\lambda\,\overrightarrow{OB}$ act at the point O. The position vectors of A and B relative to O are **a** and **b** respectively. Show that the position vector of the point C where the line of action of the resultant of these forces meets AB is $\frac{1}{4}(\mathbf{a} + 3\mathbf{b})$.

Forces of magnitude 11 N and $3\sqrt{13}$ N act along $\overrightarrow{OA}$ and $\overrightarrow{OB}$ respectively and a force of magnitude 28 N acts along $\overrightarrow{AB}$. Given that $\mathbf{a} = -9\mathbf{i} + 2\mathbf{j} + 6\mathbf{k}$ and $\mathbf{b} = 3\mathbf{i} - 2\mathbf{j}$, find in vector form the resultant of these three forces. Find also the position vector of the point where the line of action of this resultant meets AB. [L]

**34** (i) Forces $4P$, $4P$, $6P$ act respectively along $\overrightarrow{AB}$, $\overrightarrow{BC}$, $\overrightarrow{CA}$ where ABC is an equilateral triangle of side $d$. The line of action of the resultant **R** of these forces meets BC produced at D. Calculate the magnitude and direction of **R** and the distance BD.

(ii) A system of forces acts in the plane of an equilateral triangle LMN of side $a$. This system has moments $+K$, $+K$, 0 about L, M, N respectively ($+$ indicates the sense LMN). Show that the resultant of this system is parallel to ML and find its magnitude. [L]

**35** Weights of magnitude $W$, $5W$, $mW$, $2W$, $nW$, $3W$ are attached to points A, B, C, D, E, F respectively, equally spaced around the circumference of a uniform circular disc of radius $a$. The disc is free to rotate in a vertical plane about a fixed horizontal axis through its centre O. When E is vertically below O the disc is kept at rest by a couple $\frac{1}{2}Wa\sqrt{3}$ acting in the sense ABC in the plane of the disc. When D is vertically below O the disc is kept at rest by a similar couple of the same magnitude but acting in the sense CBA. Find the values of $m$ and $n$. [L]

**36** (i) Forces of magnitude $4P$, $3P$, $4P$, $3P$, $5P$ act along the sides AB, BC, CD, ED, FE respectively of a regular hexagon ABCDEF in the directions indicated by the order of the letters. Find the magnitude of the resultant of this system of forces and show that this resultant acts through A.

(ii) A non-uniform heavy beam ABCD rests horizontally on supports at B and C, where AB $=$ BC $=$ CD $=$ $a$. When a load of weight $W$ is hung from A, the beam is on the point of rotating about B. When an additional load of weight $5W$ is hung from D, the beam is on the point of rotating about C. Find the weight of the beam and show that the centre of mass of the beam is at a distance $a/4$ from B.    [L]

37    Find by integration the distance from its plane face of the centre of gravity of a uniform solid hemisphere of radius $a$.

A hollow sphere of external radius $R$ and internal radius $r$ is cut in half by a plane through its centre. Show that the distance of the centre of gravity of each half from the centre of the sphere is

$$\frac{3(R + r)(R^2 + r^2)}{8(R^2 + Rr + r^2)}.$$

Deduce that the centre of gravity of a hemispherical shell of radius $R$ is at a distance $R/2$ from its centre.    [L]

38    A uniform lamina is in the form of the quadrilateral whose vertices are the points $(4, 0)$, $(5, 0)$, $(0, 12)$, $(0, 3)$ referred to rectangular axes. Find the coordinates of the centre of mass of the lamina.

Find also the coordinates of the centre of mass of a uniform wire bent into the form of the perimeter of the quadrilateral.    [L]

39    A uniform solid body consists of a hemisphere, a cylinder and a right circular cone, the cone and the hemisphere having the ends of the cylinder as their bases. The radius of the cylinder is three times its length and the height of the cone is equal to the radius of its base. Show that the body can rest in equilibrium with any point of the curved surface of the hemisphere in contact with a horizontal table.

40    In the rectangle OABC, O is the origin, A is the point $(2a, 0)$ and B is the point $(2a, a)$. Forces of magnitude $P$, $Q$ and $R$ act respectively along OA, AB and BC in the directions indicated by the order of the letters. Their resultant lies along the line $x + 2y = 7a$. Find the magnitude of the resultant in terms of $P$.

Find also the moment of a couple which when added to the system would transfer the resultant to the line $x + 2y = 9a$.    [L]

41    A uniform solid body consists of a square pyramid mounted on a cube such that each of the sixteen edges of the solid is 4 cm long. Find the distance of the centre of gravity of the solid from the common interface.

42    From a uniform circular lamina of radius 30 cm a portion in the form of an equilateral triangle of side 20 cm with one vertex at the centre O of the circle, is removed. Find the distance from O of the centre of mass of the remainder.

43    Three forces are represented in magnitude, direction and line of action by the vectors $\overrightarrow{OA}$, $\overrightarrow{OB}$ and $\overrightarrow{OC}$. Show that the three forces are equivalent to a force represented in magnitude, direction and line of action by the vector $3\overrightarrow{OG}$, where G is the centroid of the triangle ABC.

**44** The three forces $\mathbf{F}_1$, $\mathbf{F}_2$ and $\mathbf{F}_3$ are in equilibrium when acting at the points whose position vectors are $\mathbf{r}_1$, $\mathbf{r}_2$ and $\mathbf{r}_3$ respectively. If

$$\begin{aligned} \mathbf{F}_1 &= 5\mathbf{i} + 6\mathbf{j}, & \mathbf{r}_1 &= c\mathbf{i} + \mathbf{j}, \\ \mathbf{F}_2 &= a\mathbf{i} - 4\mathbf{j}, & \mathbf{r}_2 &= 2\mathbf{i} - \mathbf{j}, \\ \mathbf{F}_3 &= -6\mathbf{i} + b\mathbf{j} \quad \text{and} \quad & \mathbf{r}_3 &= 3\mathbf{i} + 2\mathbf{j}, \end{aligned}$$

calculate the values of the constants $a$, $b$ and $c$ and the position vector of the point of concurrence of the lines of action of the three forces.

The force $\mathbf{F}_3$ is now reversed, $\mathbf{F}_1$, $\mathbf{F}_2$, $\mathbf{r}_1$, $\mathbf{r}_2$ and $\mathbf{r}_3$ remain unchanged, and a clockwise couple of magnitude 21 units is introduced into the system. Calculate the magnitude and direction of the resultant of this system and find a vector equation of its line of action. [AEB]

**45** Forces $\mathbf{F}_1 = -3\mathbf{i} + 2\mathbf{j} + \mathbf{k}$ and $\mathbf{F}_2 = \mathbf{j} - \mathbf{k}$ act at points $\mathbf{r}_1 = 2\mathbf{i} - 3\mathbf{j} + \mathbf{k}$ and $\mathbf{r}_2 = -\mathbf{i} + \mathbf{k}$ respectively, the units of force and distance being the newton and the metre respectively. A third force $\mathbf{F}_3$ is added to the system.
(i) If the system is in equilibrium, find the magnitude of $\mathbf{F}_3$ and the vector equation of its line of action.
(ii) If $\mathbf{F}_3$ acts through the origin and the system reduces to a couple, find $\mathbf{F}_3$ and the magnitude of the couple. [AEB]

# 8 Equilibrium of a rigid body

## 8.1 Two forces in equilibrium

When a rigid body is in equilibrium suspended by a string, it is acted on by two forces only. The tension in the string acts vertically upwards and the weight of the body acts vertically downwards. These two forces must be equal in magnitude and they must act along the same line, or else they would form a couple and rotate the body. This shows that the centre of gravity of the body must be in the same vertical line as the point of suspension.

Consider a uniform lamina in equilibrium suspended by a string attached to a point A of the lamina. A vertical line drawn through A must pass through G, the centre of gravity of the lamina. Now let the lamina be suspended from a point B of the lamina not on AG. In the new position of equilibrium a vertical line drawn through B must pass through G so that the point of intersection of the two lines will be the centre of gravity of the lamina.

When a book rests in equilibrium on a horizontal table, the upward force exerted by the table on the book must act in the vertical line through the centre of gravity of the book. Equilibrium will not be possible if the book projects over the edge so far that its centre of gravity is not over the table.

*Example 1*
The base radius of a solid cone is $a$ and its height is $4a$. The cone is suspended by a string attached to a point A on the rim of its base. Find the angle between the vertical and the axis of the cone in the position of equilibrium.

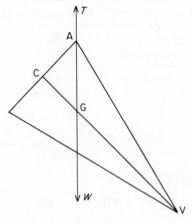

Fig. 8.1

In Fig. 8.1, C is the centre of the base and G is the centre of gravity. Since the height of the cone is $4a$, $CG = a$, i.e. $CG = CA$. It follows that the angle CGA equals $45°$, so that the axis of the cone makes an angle of $45°$ with the vertical.

*Example 2*

In the uniform rectangular lamina ABCD, $AB = 3a$ and $BC = 2a$. The triangular portion MCD, where M is the mid-point of BC, is removed from the lamina. The remainder is suspended by a string attached to the vertex A. Show that in the position of equilibrium the side AB makes an angle $\tan^{-1}(7/12)$ with the vertical.

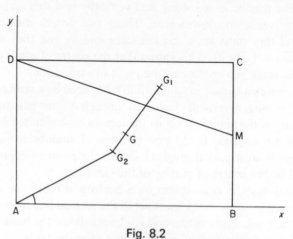

Fig. 8.2

Let $W$ be the weight of the lamina ABCD. Since the area of the triangle MCD is a quarter of the area of the rectangle ABCD, the weight of the portion removed is $W/4$ and the weight of the remainder is $3W/4$. Let AB be the $x$-axis and AD the $y$-axis. With the lamina horizontal, take moments about AB and AD.

| Shape | Weight | Centre of gravity | Moment about AD | Moment about AB |
|-------|--------|-------------------|-----------------|-----------------|
| ABMD | $3W/4$ | $G_2(\bar{x}, \bar{y})$ | $\frac{3}{4}W\bar{x}$ | $\frac{3}{4}W\bar{y}$ |
| MCD | $W/4$ | $G_1(2a, 5a/3)$ | $Wa/2$ | $5Wa/12$ |
| ABCD | $W$ | $G(3a/2, a)$ | $3Wa/2$ | $Wa$ |

Taking moments about AD

$$\tfrac{3}{4}W\bar{x} + Wa/2 = 3Wa/2$$
$$\bar{x} = 4a/3.$$

Taking moments about AB

$$\tfrac{3}{4}W\bar{y} + 5Wa/12 = Wa$$
$$\bar{y} = 7a/9.$$

When the remainder ABMD is suspended from A, the line $AG_2$ will be vertical and the angle between AB and the vertical will be the angle $BAG_2$. The tangent of this angle is $\bar{y}/\bar{x}$, i.e. 7/12. Therefore the required angle is $\tan^{-1}$ (7/12).

*Example 3*
The height of a uniform solid cone is $4a$ and the radius of its base is $3a$. The cone is cut by a plane through the mid-point of its axis parallel to its base. Find the position of the centre of gravity of the frustum cut off by the plane, and determine whether the frustum can stay in equilibrium with its curved surface resting on a horizontal table.

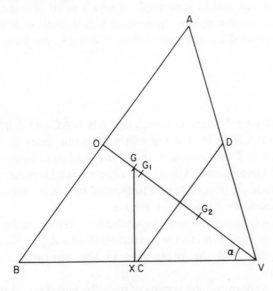

Fig. 8.3

In Fig. 8.3, O is the centre of the base of the cone. $OV = 4a$, $OA = OB = 3a$, $VA = VB = 5a$ and the angle $OVB = \alpha$. G is the centre of gravity of the frustum; $G_1$ and $G_2$ are the centres of gravity of the large cone and the small cone respectively.

Let $W$ be the weight of the large cone. The height of the small cone is half that of the large cone, and the area of its base is one-quarter of that of the large cone. Therefore the weight of the small cone is $W/8$, and the weight of the frustum is $7W/8$.

| Solid | Weight | Distance of centre of gravity from V | Moment about V |
|-------|--------|--------------------------------------|----------------|
| small cone | $W/8$ | $VG_2 = 3a/2$ | $(3/16)Wa \cos \alpha$ |
| frustum | $7W/8$ | $VG = x$ | $(7/8)Wx \cos \alpha$ |
| large cone | $W$ | $VG_1 = 3a$ | $3Wa \cos \alpha$ |

Taking moments about V

$$(3/16)Wa \cos \alpha + (7/8)Wx \cos \alpha = 3Wa \cos \alpha$$
$$7x + 3a/2 = 24a$$
$$x = 45a/14.$$

This is the distance from G to the vertex V. The vertical line through G will cut VB at a point X such that

$$VX = VG \cos \alpha = (45a/14)(4/5) = 2\tfrac{4}{7}a.$$

But $VC = 2.5a$ and so VX is greater than VC and the point X must lie between B and C. This shows that G is vertically above a point between B and C, and therefore the frustum can rest in equilibrium in this position. If the vertical line through G had crossed VB at a point between V and C, the frustum would have toppled over.

## Exercises 8.1

1 A uniform triangular lamina is such that $AB = AC$ and $\angle BAC = 90°$. The lamina is suspended by a string attached to the vertex B. Find the angle between AB and the vertical in the position of equilibrium.

2 Two equal uniform rods AB, AC are rigidly joined together at right angles at A. The rods are in equilibrium suspended by a string attached at B. Find the angle made by AB with the vertical.

3 A uniform solid hemisphere is suspended by a string attached to a point on the rim of its base. Find to the nearest tenth of a degree the angle between the horizontal and the plane face of the hemisphere when it is in equilibrium.

4 ABCD is a uniform square lamina of mass $2m$ and side $a$. A particle of mass $m$ is fastened to the lamina at the vertex B. If G is the centre of gravity of the lamina and particle, find the distance of G from AB and from AD. The lamina is suspended by a string attached to the vertex A. Find the angle between AB and the vertical in the position of equilibrium.

5 From the uniform square lamina ABCD with centre O, the triangle OCD is removed. The lamina is then suspended freely from the mid-point of AD. Find the angle between AB and the vertical in the position of equilibrium.

6 The height of a uniform solid cone is $12a$ and the radius of its base is $3a$. The cone is cut by a plane parallel to its base at a distance $4a$ from the base. Show that the centre of gravity of the frustum is at a distance $43a/19$ from the centre of the smaller plane face.

   Find whether the frustum can rest in equilibrium with its curved surface in contact with a horizontal table.

7 A uniform solid cube of edge $2a$ rests on a horizontal table. The cube is cut by a plane passing through one of its upper edges and making an angle $\tan^{-1}(5/8)$ with the vertical. The smaller part of the cube is removed. Find

the distance of the centre of gravity of the remainder from its vertical square face.

Deduce that this remainder will remain in equilibrium.

8   A uniform lamina ABCD is in the form of a trapezium with $AB = 5a$, $BC = DA = 2a$ and $CD = 3a$. Show that the distance of the centre of gravity of the lamina from AB is $11a/(8\sqrt{3})$.

Determine whether the lamina can rest in equilibrium in a vertical plane with the side AD in contact with a horizontal table.

9   Three particles P, Q, R are of equal mass. Show that the centre of gravity of the particles coincides with the centroid of the triangle PQR.

In the uniform triangular lamina ABC, $BC = 5a$ and $\tan ABC = 3/4$. The lamina is in equilibrium in a vertical plane with the side AB resting on a horizontal table. Find the least possible length of the side AB.

10  A uniform solid consists of a cylinder of radius $a$ and height $h$ with its base joined to the plane face of a hemisphere of radius $a$. Show that if $h$ is less than $a/\sqrt{2}$ the solid cannot rest in equilibrium on a horizontal table with its axis horizontal.

## 8.2   Parallel forces in equilibrium

When a rigid body is in equilibrium under the action of its weight $\mathbf{W}$ and two vertical forces $\mathbf{P}$ and $\mathbf{Q}$, the lines of action of the three forces must lie in one plane. For let $\mathbf{P}$ act through a point A and $\mathbf{Q}$ act through a point B, where AB is horizontal. If the line of action of $\mathbf{W}$ is not in the vertical plane through A and B, the moment of $\mathbf{W}$ about AB will not be zero and the body will not be in equilibrium. Therefore $\mathbf{W}$ must act in the vertical plane through AB. This shows that $\mathbf{W}$, $\mathbf{P}$ and $\mathbf{Q}$ must be coplanar forces.

Let a rigid body be in equilibrium under the action of a system of vertical forces (including its own weight), all acting in the same plane. Then the sum of the forces must be zero, upward forces being counted positive and downward forces being counted negative. Also the sum of the moments of the forces about any point in their plane must be zero.

*Example 1*
A uniform square lamina ABCD of side $2a$ and weight $W$ is suspended by two vertical strings attached at A and B respectively. The side AB is inclined at $30°$ to the horizontal, with A uppermost. Find the tension in each string.

In Fig. 8.4, AG and BG make angles of $75°$ and $15°$ respectively with the horizontal. Let $T_1$ and $T_2$ be the tensions in the strings.
Since the sum of the moments about B is zero,

$$T_1 \times AB \cos 30° - W \times BG \cos 15° = 0$$
$$T_1 a\sqrt{3} = Wa\sqrt{2} \cos 15°$$
$$\cos 15° = \cos 45° \cos 30° + \sin 45° \sin 30° = (\sqrt{3} + 1)/(2\sqrt{2}).$$

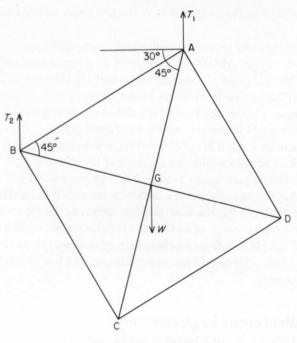

**Fig. 8.4**

Hence

$$T_1 = W(\sqrt{3} + 1)/(2\sqrt{3}).$$

Since the sum of the moments about A is zero,

$$T_2 \times AB \cos 30° - W \times AG \cos 75° = 0$$
$$\cos 75° = \cos 45° \cos 30° - \sin 45° \sin 30° = (\sqrt{3} - 1)/(2\sqrt{2}).$$

Hence

$$T_2 = W(\sqrt{3} - 1)/(2\sqrt{3}).$$

It can be checked that the sum of the tensions equals the weight of the lamina.

$$T_1 + T_2 = W[(\sqrt{3} + 1) + (\sqrt{3} - 1)]/(2\sqrt{3}) = W.$$

*Example 2*

A uniform triangular lamina ABC of mass $3m$ is such that AB = AC = 6.5 cm and BC = 5 cm. Two particles each of mass $m$ are attached to the lamina, one at D, the mid-point of AB, and the other at E, the mid-point of AC. The lamina is in equilibrium suspended in a horizontal position by three vertical strings, one attached at each vertex. Find the tension in each string.

The weights of the particles at D and E are equivalent to a force of magnitude $2mg$ acting vertically downward through O the mid-point of DE.

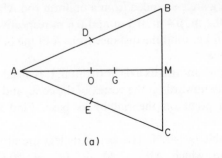

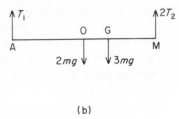

(a)                              (b)

Fig. 8.5

Let the tensions in the strings at A, B and C be $T_1$, $T_2$ and $T_3$ respectively. By symmetry $T_2$ and $T_3$ are equal, and so the tensions at B and C are equivalent to a force $2T_2$ acting vertically upward at M, the mid-point of BC.

The forces on the lamina have now been replaced by an equivalent set of forces all acting in the vertical plane through AM, as shown in Fig. 8.5(b). Taking moments about M

$$T_1 \times AM - 2mg \times OM - 3mg \times GM = 0.$$

Since AB = 6.5 cm and BM = 2.5 cm, AM = 6 cm and OM = 3 cm.

$$T_1 \times 6 = 2mg \times 3 + 3mg \times 2$$
$$T_1 = 2mg.$$

Taking moments about A

$$2T_2 \times AM - 2mg \times AO - 3mg \times AG = 0$$
$$2T_2 \times 6 = 2mg \times 3 + 3mg \times 4$$
$$T_2 = 3mg/2 = T_3.$$

As a check, $T_1 + T_2 + T_3 = 5mg$, which is the total weight of the lamina and the particles.

**Exercises 8.2**

1   A uniform rod AB of weight 20 N and length 1.2 m rests on a fulcrum at a distance 0.4 m from A. The rod is maintained in a horizontal position by a force **P** acting vertically downwards on the rod at A. Find the magnitude of **P** and the force exerted on the rod by the fulcrum.

2   A uniform plank of weight 400 N and length 2 m is supported in a horizontal position by two vertical ropes fastened to the plank at points 0.5 m from each end. A force of 100 N acts vertically downwards at one end of the plank. Find the tension in each rope.

3   A uniform plank 8 metres long lies on a flat roof with 3 metres of its length projecting over the edge. The mass of the plank is 100 kg. Find the vertical force which must be applied at the end of the plank on the roof to enable a boy of mass 50 kg to stand on the other end of the plank without it tipping.

4   Particles of mass 2, 3, 4 and 5 kg are suspended from a uniform rod AB of length 0.8 m at points distant 0.2 m, 0.4 m, 0.5 m and 0.6 m respectively from A. The mass of the rod is 6 kg. Find the distance from A of the point about which the rod will balance.

5   A uniform solid cone is in equilibrium with its axis horizontal. It is acted on by a force of 5 N acting vertically upwards at the vertex of the cone, and by a vertical force **P** acting at a point on the rim of its base. Find the magnitude of **P**.

6   The weight of a three-legged table is 240 N. The feet of its legs are at the corners of a triangle ABC in which AB = 40 cm, BC = 50 cm, CA = 30 cm. The vertical line through the centre of gravity of the table is at a distance 10 cm from AB and at a distance 20 cm from CA. Find the reaction between each leg and the floor.

7   A uniform rod AB of weight 50 N and length 2 m is hinged at A to a wall. The rod carries a load of weight 15 N at B, and the rod is supported in a horizontal position by a vertical force acting at a point of the rod 80 cm from A. Find the magnitude of this force. Find also the magnitude and direction of the force exerted on the rod by the hinge.

8   A uniform equilateral triangular lamina ABC of weight 6 N is suspended by two vertical strings attached at A and B respectively. Given that the side BC is vertical, find the tensions in the strings.

9   In the uniform rectangular lamina ABCD, AB = 3a and AD = 2a. The lamina is suspended by two vertical strings attached at A and B respectively. The lamina is in equilibrium with AB inclined to the vertical at the angle $\tan^{-1} 2$. Find the ratio of the tensions in the strings.

10   A uniform circular disc is supported in a horizontal position by three vertical strings attached at three points P, Q and R on the rim of the disc. Given that PQ = PR and that the angle QPR equals 30°, find the ratio of the tensions in the strings.

11   A uniform lamina is in the form of a triangle ABC with AB = BC = $a$ and the angle ABC = 120°. A straight cut is made parallel to BC meeting AB in X and AC in Y, where AX = $x$. Show that the trapezium XBCY can rest in equilibrium in a vertical plane with XB on a horizontal plane if $10x < a(1 + \sqrt{21})$.

12   In the uniform rectangular lamina ABCD, AB = 4a and BC = 2a. A square with sides parallel to the sides of the rectangle is cut out, the distances of the centre of the square from AB and BC being equal to the length of the side of the square. When the remainder is suspended from C, it hangs with CA vertical. Find the length of the side of the square.

## 8.3   Three forces in equilibrium

Consider a rigid body in equilibrium under the action of three forces **P**, **Q** and its weight **W**, all acting in the same vertical plane. If **P** and **Q** are not parallel, their lines of action will meet at a point O. The line of action of **W** must pass through O, otherwise **W** would have a turning moment about O. This shows that if three coplanar forces in equilibrium are not parallel, they must be concurrent.

It was shown in Section 4.4 that three concurrent forces in equilibrium can be represented in magnitude and direction by the sides of a triangle. Therefore a triangle of forces ABC can be drawn in which the forces **W**, **P** and **Q** are represented in magnitude and direction by the vectors $\overrightarrow{AB}$, $\overrightarrow{BC}$ and $\overrightarrow{CA}$ respectively.

*Example 1*

A uniform rod of weight 10 N is in equilibrium under the action of a force **P** of magnitude 8 N applied at one end of the rod and a force **Q** of magnitude 12 N applied at the other end. Find by drawing the directions of the forces **P** and **Q**, and find the inclination of the rod to the vertical.

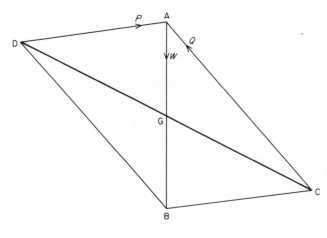

Fig. 8.6

The lines of action of **P** and **Q** must meet at a point A vertically above G, the mid-point of the rod. To construct the triangle of forces, a vertical line AB of length 10 units is drawn to represent the weight of the rod (Fig. 8.6).

Then the triangle ABC can be drawn with BC of length 8 units and CA of length 12 units. The vector $\overrightarrow{BC}$ represents **P** and the vector $\overrightarrow{CA}$ represents **Q**.

When the parallelogram BCAD is completed, its diagonals AB and CD will bisect one another. The line CD can represent the rod, with its mid-point vertically beneath A. By measurement from the diagram, the force **P** makes an angle of 83° with the vertical, the force **Q** makes an angle of 41° with the vertical (to the nearest degree). The angle between the rod and the vertical is 63°.

This case is equivalent to the construction of a triangle given the lengths of its sides.

*Example 2*

A uniform rod of weight 10 N is in equilibrium under the action of a force **P** applied at one end in a direction making 60° with the vertical and a force **Q** applied at the other end in a direction making 30° with the vertical. Find the magnitudes of the forces **P** and **Q**, and the inclination of the rod to the horizontal.

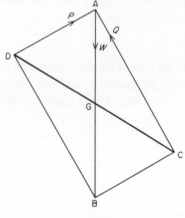

Fig. 8.7

A vertical line AB of length 10 units is drawn. The triangle ABC is constructed with angle ABC equal to 60° and angle BAC equal to 30° (Fig. 8.7).

The vectors $\overrightarrow{BC}$ and $\overrightarrow{CA}$ represent the forces **P** and **Q** respectively. By measurement from the diagram, $P = 5$ N and $Q = 8.7$ N.

As in the previous example, the diagonal CD of the parallelogram BCAD represents the rod. By measurement, the rod makes an angle of 30° with the horizontal.

This case is equivalent to the construction of a triangle given one side and two angles.

*Example 3*

A uniform rod of weight 10 N is in equilibrium under the action of a force **P** of magnitude 8 N applied at one end in a direction making 45° with the vertical and a force **Q** applied at the other end. Find the magnitude and direction of the force **Q**.

The weight of the rod is represented by a vertical line AB of length 10 units. Then the line BC is drawn of length 8 units with the angle ABC equal to 45° (Fig. 8.8).

The vector $\overrightarrow{CA}$ represents the force **Q** in magnitude and direction. By measurement from the diagram, **Q** is a force of 7.1 N inclined at approximately 52° to the vertical. The rod makes an angle of 6° approximately with the horizontal.

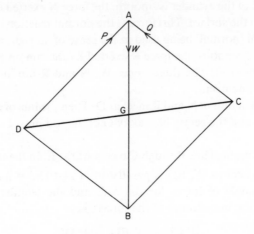

Fig. 8.8

This case is equivalent to the construction of a triangle given two sides and the included angle.

*Example 4*
A hollow cylinder is fixed with its axis vertical. A smooth rod AB of weight $W$ rests in equilibrium inclined at an angle $\theta$ to the horizontal with its end A in contact with the smooth inner surface of the cylinder and with a point C of the rod resting on the rim of the upper end of the cylinder. Find the forces exerted by the cylinder on the rod at A and C.

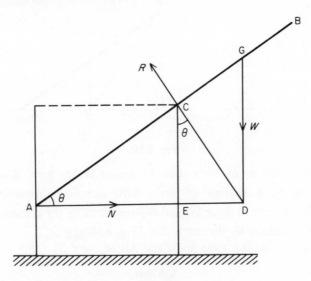

Fig. 8.9

Since the surface of the cylinder is smooth, the force **N** exerted on the rod at A is at right angles to the surface. This force is the normal reaction of the cylinder on the rod, the word 'normal' being used in the sense of 'at right angles' (Fig. 8.9).

Since the rod is smooth, the force **R** exerted by the rim on the rod at C is at right angles to the rod. The three forces **W**, **N** and **R** are in equilibrium and hence they are concurrent.

Let the lines of action of **N** and **R** meet at D. Then the line of action of **W** must pass through D, and the centre of gravity G of the rod must be vertically above D.

In Fig. 8.9 the vertical line through C meets AD at E. In the triangle CED, the vector $\overrightarrow{CE}$ is parallel to **W**, $\overrightarrow{ED}$ is parallel to **N** and $\overrightarrow{DC}$ is parallel to **R**. This triangle is a triangle of forces for the rod, and the lengths of its sides are proportional to the magnitudes of the forces, i.e.

$$W/CE = N/ED = R/DC.$$

In Fig. 8.9, $CE = DC \cos \theta$, $ED = DC \sin \theta$.

Hence
$$N = W(ED/CE) = W \tan \theta,$$
$$R = W(DC/CE) = W \sec \theta.$$

*Example 5*
A uniform rod PQ of weight $W$ can turn freely about a pivot at P. It is acted on at Q by a horizontal force **X** of magnitude $3W/4$, and is in equilibrium with Q lower than P. Find the magnitude of the reaction at the pivot and the angle between the rod and the horizontal.

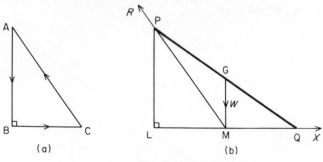

Fig. 8.10

The three forces acting on the rod are its weight **W**, the force **X** at Q and the reaction **R** at P. A triangle of forces ABC can be constructed with $\overrightarrow{AB}$ representing the vertical force **W**, and $\overrightarrow{BC}$ representing the horizontal force **X**. Then **R** is represented by the vector $\overrightarrow{CA}$ (Fig. 8.10(a)).
Since the angle ABC is a right angle and $AB/BC = 4/3$,

$$CA/AB = 5/4,$$
and
$$R = W(CA/AB) = 5W/4.$$

In Fig. 8.10(b), the weight $W$ acts vertically through G, the mid-poing of PQ, and through M, the mid-point of LQ.

The horizontal force $X$ acting at Q passes through M, and so the reaction $R$ must act through M.

It follows that

$$PL/LM = AB/BC = 4/3.$$

Since $LQ = 2LM$, this gives $PL/LQ = 2/3$. Therefore tan $PQL = 2/3$, and the angle between the rod and the horizontal is 34°, to the nearest degree.

*Example 6*
A rod AB of weight $W$ and length $8a$ has its centre of gravity G at a distance $5a$ from A. The rod is suspended by a light inextensible string of length $16a$ which is attached to the rod at A and B. The string passes over a small smooth peg O. Find the inclination of the rod to the vertical and the tension in the string.

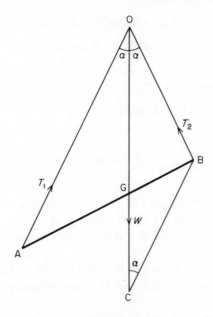

Fig. 8.11

Let $T_1$ be the tension in the string at A and $T_2$ be the tension in the string at B. Since the peg is smooth and the string is continuous, $T_1$ and $T_2$ are equal.

The horizontal component of $T_1$ at A acting on the rod at A must be counterbalanced by the horizontal component of $T_2$ acting on the rod at B. Since $T_1 = T_2$, the string at A must be inclined to the vertical at the same angle as the string at B.

In Fig. 8.11, Let $\angle AOG = \angle BOG = \alpha$.
By the sine rule,

$$\sin AGO/AO = \sin \alpha/AG,$$
$$\sin BGO/BO = \sin \alpha/GB.$$

Now $\sin AGO = \sin BGO$. Therefore $AO/BO = AG/GB$.
Since $AG = 5a$ and $GB = 3a$, this gives $AO/BO = 5/3$.
But $AO + BO = 16a$. Therefore $AO = 10a$ and $BO = 6a$. The third side AB of the triangle AOB equals $8a$. Hence the angle ABO is a right angle.

$$\tan OGB = OB/GB = 6a\,/3a = 2, \text{ and } \tan \alpha = 1/2.$$

Therefore the rod makes an angle of 63° 26' with the vertical. In Fig. 8.11, OG is produced to the point C such that CB is equal in length to BO. Then the triangle BOC is a triangle of forces for the rod.

The vector $\overrightarrow{OC}$ represents the weight of the rod, $\overrightarrow{CB}$ represents $T_1$ and $\overrightarrow{BO}$ represents $T_2$.
Since the magnitudes of the forces are proportional to the lengths of the corresponding sides of the triangle,

$$T_1/CB = T_2/BO = W/OC$$

i.e.
$$T_1 = T_2 = W(BO/OC).$$

Now
$$OC = 2BO \cos \alpha.$$
Since $\tan \alpha = 1/2$, $\cos \alpha = 2/\sqrt{5}$,

$$OC/BO = 4/\sqrt{5}.$$

This gives

$$T_1 = T_2 = W(\sqrt{5}/4).$$

### Exercises 8.3

1  A uniform rod AB of weight 100 N is in equilibrium under the action of a force **P** acting at A and a force **Q** acting at B.
   (a) Find $P$ and $Q$ given that **P** is at 30° to the vertical and **Q** is at 45° to the vertical.
   (b) Find **P** given that $Q = 100$ N with **Q** at 40° to the vertical.
   (c) Find the inclinations of **P** and **Q** to the vertical given that $P = 70$ N and $Q = 80$ N.
   Use a graphical method and give your answers to the nearest integer. In each case find the inclination of the rod to the horizontal.

2  Two smooth planes inclined at 30° and 60° to the horizontal intersect in a horizontal line. A uniform rod of weight 10 N rests at right angles to this line with an end on each plane. Show that the rod is inclined at 30° to the horizontal, and find the force exerted on the rod by each plane.

3  The lower end of a uniform rod of weight $W$ can turn freely about a pivot. The upper end rests against a smooth vertical wall, the angle between the rod and the wall being $\tan^{-1} 2$. Find the force exerted on the rod by the wall and the magnitude of the reaction at the pivot.

4   A uniform equilateral triangular lamina ABC is in equilibrium in a vertical plane with B vertically above C. It is acted on by a horizontal force of 4 N at A and by another force at B. Find the magnitude of this force and the weight of the lamina.

5   A uniform rod AB can turn freely in a vertical plane about a hinge at A. When the rod is inclined at 60° to the vertical it is in equilibrium under the action of a force exerted at B at right angles to the rod. By drawing an accurate diagram, show that the magnitude of the reaction at the hinge is approximately equal to two-thirds of the weight of the rod.

6   A uniform rod AB of length one metre hangs in equilibrium with its upper end A against a smooth vertical wall, the lower end B being attached to one end of a light inextensible string. The other end of the string is fastened to a point of the wall vertically above A. Given that the angle between the rod and the wall is 60°, show by an accurate drawing that the length of the string is approximately 1.32 m.

7   A uniform triangular lamina ABC is equilateral with sides of length 40 cm. It rests in a horizontal position on a smooth table. A force of 6 N acts on the lamina at B towards C, and a force of 2 N acts at A towards C. The lamina is kept in equilibrium by a third force acting at a point X on AB. Find the magnitude of this force and the distance of X from A.

8   A uniform rod AB of weight 10 N and length $4a$ can turn freely in a vertical plane about a hinge at A. One end of a light inextensible string is fastened to a point C of the rod, where $AC = 3a$, and the other end of the string is fastened to a point D vertically above A, where $AD = a$. The rod is in equilibrium with the string horizontal. Show by drawing that the tension in the string is approximately 19 N and that the force exerted on the rod at A makes an angle with the vertical approximately equal to 62°.

9   A rod AB of length one metre and weight 20 N is suspended in equilibrium by a light inextensible string attached to the rod at A and B. The string passes over a smooth peg O. Given that $AO = 1.5$ m and $BO = 1$ m, show that the centre of gravity of the rod is at a distance 0.6 m from A and that the tension in the string is 10.7 N approximately.

10  A uniform rod AB is in equilibrium with the end B in contact with a smooth vertical wall. The rod makes an angle $\alpha$ with the wall and is in a vertical plane which is at right angles to the wall. It is supported by a light inextensible string of length $a$ which is attached to a point of the wall vertically above B and to a point C of the rod such that $BC = a$. Show that $AB = 4a$ and that equilibrium is possible for any value of $\alpha$ between 0° and 90°.

## 8.4   General conditions for equilibrium

Let the force $\mathbf{F}$ acting in the $x$–$y$ plane have a moment $C$ about the origin O. Then as shown in Section 7.5 this force is equivalent to a force $\mathbf{F}$ acting at O together with a couple of moment $C$.

A system of $n$ forces $\mathbf{F}_1, \mathbf{F}_2, \ldots, \mathbf{F}_n$ acting in the $x$–$y$ plane will be equivalent to $n$ forces $\mathbf{F}_1, \mathbf{F}_2, \ldots, \mathbf{F}_n$ all acting at the origin O together with $n$ couples.

The forces at O have a resultant $\mathbf{R}$ and the $n$ couples can be combined to form a couple $G$. The system will be in equilibrium if and only if both $\mathbf{R}$ and $G$ are zero.

Let $X$ and $Y$ be the sums of the components of the $n$ forces parallel to the $x$-axis and the $y$-axis respectively. For the resultant $\mathbf{R}$ to be zero, both $X$ and $Y$ must be zero. The conditions for equilibrium are therefore

$$X = 0, \ Y = 0, \ G = 0.$$

In tackling a problem on the equilibrium of a rigid body acted upon by a system of coplanar forces, three independent equations can always be obtained.
(a) The sum of the components of the forces parallel to the $x$-axis is zero.
(b) The sum of the components of the forces parallel to the $y$-axis is zero.
(c) The sum of the moments of the forces about the origin is zero.

Note that any point in the plane can be chosen to be the origin O. It is an advantage to take moments about a point through which two forces act, for this will give a simpler equation not involving these two forces.

Note also that the sum of the components of the forces in any given direction will be zero when the resultant $\mathbf{R}$ is zero. By resolving in any two directions two independent equations can be obtained.

The equation obtained by resolving in a third direction will not be independent of the two equations already obtained.

*Example 1*
The sides of a uniform square lamina ABCD of weight $W$ are of length $2a$. The lamina is in equilibrium in a vertical plane with the vertex A resting on a smooth horizontal table and with AB making an angle $\tan^{-1} (1/2)$ with the table. A force $\mathbf{P}$ acts at B in the direction BC and a force $\mathbf{Q}$ acts at M, the mid-point of DA, towards the centre of the lamina. Express $P$ and $Q$ in terms of $W$ and find the force exerted on the lamina at A.

The four forces acting on the lamina are shown in Fig. 8.12. Since the table is smooth the reaction $\mathbf{R}$ at A is vertical.
Let $\alpha$ be the angle between AB and the table.
Then $\tan \alpha = 1/2$, $\sin \alpha = 1/\sqrt{5}$, $\cos \alpha = 2/\sqrt{5}$.
By taking moments about N, the mid-point of BC, an equation is obtained not involving $P$ and $Q$.

$$Wa \cos \alpha - R(2a \cos \alpha - a \sin \alpha) = 0$$
$$W - R(2 - \tan \alpha) = 0$$
$$R = 2W/3.$$

By resolving the forces parallel to AD, an equation is obtained which does not involve $Q$.

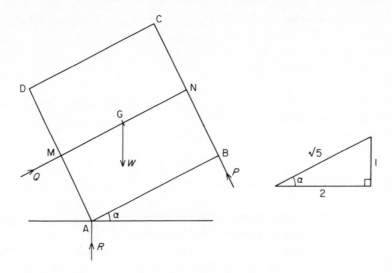

Fig. 8.12

$$P + R \cos \alpha - W \cos \alpha = 0$$
$$P = (W - R) \cos \alpha = 2W/(3\sqrt{5}).$$

By resolving the forces parallel to AB, an equation is obtained which does not involve $P$.

$$Q + R \sin \alpha - W \sin \alpha = 0$$
$$Q = (W - R) \sin \alpha = W/(3\sqrt{5}).$$

*Example 2*
A uniform bar AB of weight $W$ and length $4a$ can turn freely in a vertical plane about a hinge at A. The bar is supported by a light chain of length $5a$, one end of which is attached to the bar at C, where $AC = 3a$, the other end being fastened to a point D vertically above A, where $AD = 4a$. The bar carries a load of weight $10W$ suspended from B. Find the tension in the chain and the magnitude and direction of the force exerted on the bar at the hinge.

The sides of the triangle CDA are $5a$, $4a$ and $3a$. Hence the angle DAC is a right angle and the bar is horizontal.
   Let $X$ and $Y$ be the components of the force on the bar at A, and let **T** be the tension in the chain at C (Fig. 8.13).
By taking moments about C, an equation is obtained which does not involve the forces **X** and **T**.

$$Y \times 3a - W \times a + 10W \times a = 0$$
$$Y = -3W.$$

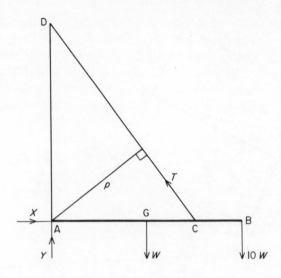

Fig. 8.13

Resolving vertically

$$Y + T \cos \text{CDA} - W - 10W = 0$$
$$4T/5 = 14W$$
$$T = 35W/2.$$

Resolving horizontally

$$X - T \cos \text{ACD} = 0$$
$$X = 3T/5 = 21W/2.$$

The force **R** exerted on the bar at A is the resultant of **X** and **Y**.

$$R^2 = X^2 + Y^2 = (441/4 + 9)W^2$$
$$R = (\sqrt{477})W/2.$$

The angle $\theta$ which **R** makes with AB is given by

$$\tan \theta = Y/X = -2/7.$$

Therefore **R** acts in the direction shown in Fig. 8.14, where the angle MON is $\tan^{-1} (2/7)$.

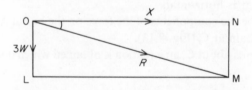

Fig. 8.14

In this example, $T$ can be found directly by taking moments about A. There are two ways of finding the moment of **T** about A.

(a) Multiply $T$ by $p$, the perpendicular distance from A to CD. Now $p$ equals AC sin ACD, i.e. $12a/5$. Therefore the moment of **T** about A is $12aT/5$.

(b) Resolve **T** into its vertical and horizontal components at C. The horizontal component will have zero moment about A. Hence the moment of **T** about A is $(4T/5) \times 3a$, i.e. $12aT/5$.

Taking moments about A

$$12aT/5 - (W \times 2a) - (10W \times 4a) = 0$$
$$12T/5 = 42W$$
$$T = 35W/2.$$

A system of coplanar forces will be in equilibrium if the sum of the moments of the forces is zero about each of three points A, B and C not in a straight line.

For if the sum of the moments about A is zero, the system does not reduce to a couple. If also the sum of the moments about B is zero, the system may reduce to a force along AB, but this cannot be true if the sum of the moments about C is also zero.

*Example 3*
A uniform square lamina ABCD of weight $W$ and side $2a$ is in equilibrium in a vertical plane with its lowest side AB horizontal. A force **P** acts at C towards D, a force **Q** acts vertically upwards at D and a force **R** acts at A towards C. Find $P$, $Q$ and $R$ in terms of $W$.

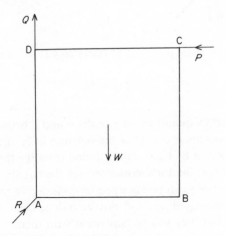

Fig. 8.15

Taking moments about A

$$P \times 2a - W \times a = 0, \quad \text{i.e.} \quad P = W/2.$$

Taking moments about C

$$W \times a - Q \times 2a = 0, \quad \text{i.e.} \quad Q = W/2.$$

Taking moments about D

$$R \times a\sqrt{2} - W \times a = 0, \quad \text{i.e.} \quad R = W/\sqrt{2}.$$

*Example 4*

Two uniform rods AB and BC, each of length $2a$ and weight $W$, are smoothly hinged together at B. The rods are suspended from a pivot at the end A of the rod AB, and are in equilibrium with a horizontal force of magnitude $W/2$ acting at C. Find the angle which each rod makes with the vertical and the magnitude of the reactions at A and B.

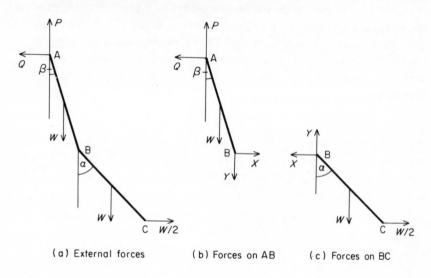

(a) External forces     (b) Forces on AB     (c) Forces on BC

Fig. 8.16

There are six unknown quantities, the angles $\alpha$ and $\beta$ between the rods and the vertical, the components $P$ and $Q$ of the reaction at A and the components $X$ and $Y$ of the reaction at B. These can be found from the three equations for the equilibrium of AB and the three equations for the equilibrium of BC.

Three more equations can be obtained by considering the equilibrium of the whole system. These equations will not be independent of the six equations mentioned above, but they will be consistent with them.

Consider the equilibrium of the rod BC (Fig. 8.16(c)).

Resolving horizontally, $X = W/2$.
Resolving vertically, $Y = W$.
Taking moments about B

$$\tfrac{1}{2}W(2a \cos \alpha) - W(a \sin \alpha) = 0.$$

This gives $\tan \alpha = 1, \alpha = 45°$.

Consider the equilibrium of the rod AB (Fig. 8.16(b)).

Resolving horizontally,   $Q = X = W/2$.

Resolving vertically,      $P = W + Y = 2W$.

Taking moments about A

$$X(2a \cos \beta) - W(a \sin \beta) - Y(2a \sin \beta) = 0.$$

This gives $\tan \beta = 1/3$.

Magnitude of the reaction at B $= \sqrt{(X^2 + Y^2)} = \sqrt{5}W/2$.

Magnitude of the reaction at A $= \sqrt{(P^2 + Q^2)} = (\sqrt{17})W/2$.

As a check, consider the equilibrium of the whole system (Fig. 8.16(a)).

Resolving horizontally,   $Q = W/2$.

Resolving vertically,      $P = 2W$.


**Exercises 8.4**

1   A uniform ladder AB of weight $W$ rests with the end A against a smooth vertical wall and the end B on a smooth horizontal floor. The ladder is kept from sliding by a force **P** acting at its mid-point. Find the magnitude of **P** when the angle between the ladder and the floor is $\theta$
 (a) given that **P** is horizontal and $\tan \theta = 4$
 (b) given that **P** acts along BA and $\theta = 30°$
 (c) given that **P** is at right angles to AB and $\tan \theta = 2$.

2   A uniform square lamina ABCD of weight $W$ and side $2a$ is in equilibrium in a vertical plane with the vertex A resting on a smooth horizontal table. The angle between AB and the table is $30°$. A force **P** acts at B towards C and a force **Q** acts at D towards C. Find the values of $P$ and $Q$ in terms of $W$, and find also the reaction between the lamina and the table.

3   A uniform rectangular lamina ABCD of weight 80 N can turn about hinges at A and D, where D is vertically above A. Given that AB = 60 cm and AD = 40 cm and that the force exerted by the hinge at D is horizontal, find the vertical and horizontal components of the force exerted by the hinge at A. Hence find the magnitude of this force and the angle it makes with AB.

4   A uniform square lamina ABCD of weight $W$ is in equilibrium in a vertical plane with AB horizontal and CD uppermost. It is acted on by a vertical force **P** at B, a force **Q** along DC and a force **R** along BA. Express $P$, $Q$ and $R$ in terms of $W$.

5   A uniform ladder AB of weight $W$ rests with the end A on a smooth horizontal floor and the end B against a smooth vertical wall. It is maintained in equilibrium at an angle $\tan^{-1} 2$ with the floor by a horizontal force **P** applied at A. Find the magnitude of **P**. Find also the horizontal force required at A to allow a man of weight $W$ to stand at the top of the ladder.

6   Two uniform rods AB and BC, each of weight $W$, are smoothly hinged together at B. They are suspended from a pivot at the end A of the rod AB

and are in equilibrium with a horizontal force of magnitude $W$ acting at C. Find, to the nearest degree, the angle between each rod and the vertical and the magnitude of the reactions at A and B.

7   A rod AB rests in equilibrium on a smooth horizontal table. A horizontal force of magnitude 30 N acts at right angles to AB at its mid-point, and a horizontal force **P** acts at right angles to AB at a point C of AB such that $3AC = CB$. A horizontal force **Q** acts at A towards B and a string attached to the rod at B makes an angle of $30°$ with BA. Find the magnitudes of **P** and **Q** and the tension in the string.

8   A uniform rectangular lamina ABCD of weight 8 N with AB = 40 cm and BC = 30 cm can turn freely in a vertical plane about a pivot at A. It is in equilibrium with D vertically above A under the action of a force of 10 N acting at B towards D, a force of 4 N acting at C towards B and a couple. Show that the pivot exerts on the lamina a force of magnitude 10 N and find the moment of the couple.

9   A uniform solid cone of weight $W$ rests with its base on a smooth horizontal table. Its base radius is $3a$ and its height is $4a$. The cone is tilted until the vertex is vertically above O, the point of contact between the cone and the table. The cone is kept in equilibrium in this position by a force **P** acting at the vertex in the direction of the axis of the cone and a force **Q** acting directly away from O at the highest point of the base. Find in terms of $W$ the magnitudes of **P** and **Q** and of the reaction at O.

10  A uniform solid hemisphere of weight $W$ rests in equilibrium on a smooth horizontal table with its plane face inclined at an angle $\tan^{-1}(4/3)$ to the horizontal. A force **P** acts at right angles to the plane face at its highest point A, and a force **Q** acts at the lowest point B of the plane face along the diameter BA. Find $P$ and $Q$ in terms of $W$.

## 8.5   Friction

When a rigid body is in contact with a rough plane, the force exerted by the plane on the body can be resolved into two component forces, the normal reaction **N** at right angles to the plane, and the force of friction **F** parallel to the plane.

When the body is on the point of sliding, equilibrium is said to be limiting, and **F** is then the force of limiting friction (see Section 4.7).

The coefficient of friction $\mu$ is defined by

$$\mu = \frac{\text{force of limiting friction}}{\text{normal reaction}}.$$

In limiting equilibrium, $F = \mu N$.
When equilibrium is not limiting, $F < \mu N$.

The angle between the resultant reaction **R** and the normal reaction **N** when equilibrium is limiting is known as the angle of friction, denoted by $\lambda$. It is clear from Fig. 8.17 that

$$\tan \lambda = (\mu N)/N = \mu.$$

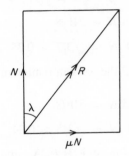

**Fig. 8.17**

*Example 1*
A uniform rod AB inclined at an angle $\alpha$ to the horizontal is in equilibrium with one end A on a horizontal table. A string attached to the other end B is at right angles to the rod. Show that if the coefficient of friction between the rod and the table is 0.3, it is possible for $\alpha$ to equal $60°$ but not for $\alpha$ to equal $30°$.

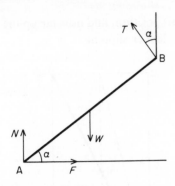

**Fig. 8.18**

Let **W** be the weight of the rod and $2a$ its length. Let **T** denote the tension in the string, **N** the normal reaction at A and **F** the force of friction. These four forces form a system in equilibrium (Fig. 8.18).
Taking moments about A

$$2aT - Wa \cos \alpha = 0,$$
$$T = \tfrac{1}{2}W \cos \alpha.$$

Resolving horizontally

$$F - T \sin \alpha = 0,$$
$$F = \tfrac{1}{2}W \sin \alpha \cos \alpha.$$

Resolving vertically

$$N + T \cos \alpha - W = 0$$
$$N = W - T \cos \alpha = \tfrac{1}{2}W(2 - \cos^2 \alpha).$$

If $\alpha = 60°$, $F = \sqrt{3}W/8$ and $N = 7W/8$.

$$F/N = \sqrt{3/7} \approx 0.25.$$

Since this value is less than 0.3 the rod can remain in equilibrium inclined at 60°
to the table.
If $\alpha = 30°$, $F = \sqrt{3}W/8$ and $N = 5W/8$.

$$F/N = \sqrt{3/5} \approx 0.34.$$

But the greatest possible value of $F$ is $0.3N$, and so the rod cannot be in
equilibrium inclined at 30° to the table.

*Example 2*
The length of a uniform ladder AB of weight $W$ is $2a$. The end A rests against a
smooth vertical wall and the end B rests on a rough horizontal floor. The
coefficient of friction between the ladder and the floor is 0.3.
(a) Show that the ladder can remain in equilibrium inclined at an angle $\tan^{-1} 2$
to the floor.
(b) With the ladder in this position, find how far up the ladder a boy of weight
$W$ can stand without the ladder slipping.

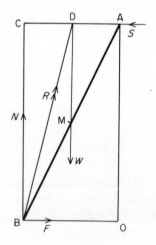

Fig. 8.19

(a) Let $\mathbf{N}$ and $\mathbf{S}$ be the normal reactions at B and A respectively, and let $\mathbf{F}$ be the
force of friction at B when $\tan ABO = \tan \theta = 2$ (Fig. 8.19).
Resolving horizontally, $\quad F = S$.
Resolving vertically, $\quad\quad N = W$.

Taking moments about B

$$S(2a \sin \theta) = Wa \cos \theta,$$
$$F/N = S/W = \tfrac{1}{2} \cot \theta = 0.25.$$

Since $\mu = 0.3$ the force of limiting friction would be $0.3N$.

F is less than its greatest possible value and the ladder is in equilibrium.

*Alternative method*

In equilibrium the resultant reaction **R** at B must pass through D, the mid-point of CA, since **S** and **W** meet at D.

$$\tan \text{DBC} = \text{CD/BC} = \tfrac{1}{2}\text{CA/BC} = \tfrac{1}{2} \cot \theta,$$

i.e. $$\tan \text{DBC} = 1/4.$$

Since tan DBC is less than 0.3, the angle between the resultant reaction **R** and the normal reaction **N** is less than the angle of friction and the ladder is in equilibrium.

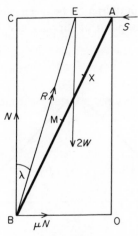

Fig. 8.20

(b) Let the ladder be in limiting equilibrium when the boy of weight $W$ stands on the ladder at a point X where $BX = x$ (Fig. 8.20). The symbols $N$, $S$ and $F$ will be used as before, but their values will be changed.

Resolving horizontally, $F = S$.

Resolving vertically, $N = W + W = 2W$.

Since the ladder is on the point of slipping

$$F = 0.3N = 0.6W,$$
$$S = F = 0.6W.$$

Taking moments about B

$$S(2a \sin \theta) = Wa \cos \theta + Wx \cos \theta,$$
$$(0.6W)(2a \tan \theta) = W(a + x).$$

Since $\tan \theta = 2$, this gives $x = 1.4a$.

Therefore the boy can stand at X, where $BX = 1.4a$.

*Alternative method*

In limiting equilibrium the tangent of the angle between the resultant reaction **R** at B and BC will be 0.3. Then **R** and **S** will meet at E, where CE equals 0.3CB, i.e. $0.3(2a \sin \theta)$.

The line of action of the combined weight $2W$ must pass through E.

Therefore
$$\tfrac{1}{2}(a \cos \theta + x \cos \theta) = CE = 0.6a \sin \theta$$
$$a + x = 1.2a \tan \theta = 2.4a$$
$$x = 1.4a$$

*Example 3*

Two uniform rods AB and BC, of weights $3W$ and $W$ respectively and each of length $2a$, are freely jointed at B. They are in equilibrium in a vertical plane with the ends A and C resting on a rough horizontal plane and with $\angle ABC = \angle ACB = 45°$. The coefficient of friction between each rod and the plane is 2/3. Show that one rod is on the point of slipping and find the magnitude of the reaction between the rods.

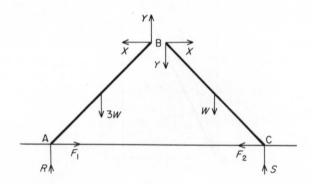

Fig. 8.21

Figure 8.21 shows the forces on the rods. **R** and **S** are the normal reactions at A and C; $\mathbf{F_1}$ and $\mathbf{F_2}$ are the forces of friction. $X$ and $Y$ are the components of the reaction at B.

Consider first the equilibrium of the whole system.

Taking moments about C

$$R(4a/\sqrt{2}) - 3W(3a/\sqrt{2}) - W(a/\sqrt{2}) = 0.$$

This gives
$$R = 5W/2.$$

Taking moments about A

$$S(4a/\sqrt{2}) - W(3a/\sqrt{2}) - 3W(a/\sqrt{2}) = 0.$$

This gives
$$S = 3W/2.$$

As a check, $R + S = 4W = $ total weight of the rods.
Resolving horizontally

$$F_1 = F_2.$$

Next consider the equilibrium of the rod BC.
Taking moments about B for the forces on the rod BC

$$F_2(2a/\sqrt{2}) + W(a/\sqrt{2}) - S(2a/\sqrt{2}) = 0.$$
$$F_2 = S - \tfrac{1}{2}W = W.$$

At the point C, the force of friction $F_2$ equals $W$ and the normal reaction $S$ equals $3W/2$. Therefore $F_2/S = 2/3$. Since the coefficient of friction is 2/3, the rod BC is in limiting equilibrium.

At the point A, the force of friction $F_1$ equals $W$ and the normal reaction $R$ equals $5W/2$. Therefore $F_1/R = 2/5$. This is less than the coefficient of friction, and so the rod AB is not on the point of slipping at A.

Consider the equilibrium of the rod AB.
Resolving vertically,     $R + Y = 3W, \ Y = W/2.$
Resolving horizontally,     $X = F_1 = W.$

Magnitude of the reaction at B $= \sqrt{(X^2 + Y^2)}$
$\qquad\qquad\qquad\qquad\qquad\qquad = \sqrt{5}W/2.$

*Example 4*
A uniform solid cone of weight $W$, height $2a$ and base radius $a$, stands on a horizontal table. A horizontal force **P** is applied to the cone at its vertex V, the magnitude of the force increasing from zero until equilibrium is broken. Show that if the coefficient of friction $\mu$ between the cone and the table is less than 1/2, the cone will slide, but if $\mu$ is greater than 1/2 the cone will tilt.

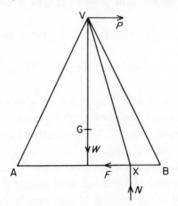

Fig. 8.22

The resultant of the forces **P** and **W** passes through V. While the cone is in equilibrium, the line of action of this resultant will meet the base of the cone at a point X, and the resultant of the normal reaction **N** and the force of friction **F** must act along XV. As $P$ increases, the point X moves towards **B**.

Resolving horizontally, $P = F$.

Resolving vertically, $W = N$.

The cone will be on the point of sliding when **F** is the force of limiting friction. This occurs when $F = \mu N$, i.e. when $P = \mu W$.

The cone will be on the point of tilting about **B** when the moment of **P** about **B** balances the moment of **W** about **B**. This occurs when $2aP = aW$, i.e. when $P = W/2$.

If $\mu < 1/2$, $P$ reaches the value $\mu W$ before it reaches the value $W/2$. Therefore the cone will slide as soon as $P$ is greater than $\mu W$.

If $\mu > 1/2$, $P$ reaches the value $W/2$ before it reaches the value $\mu W$. Therefore the cone will tilt as soon as $P$ is greater than $W/2$.

## Exercises 8.5

1   A uniform ladder AB of weight $W$ rests with the end A against a smooth vertical wall and its foot B on rough level ground. The ladder is inclined at an angle $\theta$ to the horizontal and is in limiting equilibrium.

   (a) Given that $\theta = 45°$, find the coefficient of friction.

   (b) Given that the coefficient of friction is 0.2, find the angle $\theta$ to the nearest degree.

   (c) Given that the magnitude of the reaction at A is $W/4$, find the angle of friction to the nearest degree.

2   The end A of a uniform rod AB of weight $W$ and length $2a$ rests on a rough horizontal table. The rod rests against a smooth peg C, where $AC = a + x$, and is in limiting equilibrium inclined at an angle $\alpha$ to the table. The coefficient of friction between the rod and the table is $\mu$.

   (a) Given that $\alpha = 45°$ and $\mu = 1/2$, find $x$ and the magnitude of the reaction at C.

   (b) Given that $\alpha = 60°$ and $x = a/4$, find $\mu$ and the magnitude of the reaction at C.

   (c) Given that $x = 3a/5$ and the magnitude of the reaction at C is $W/2$, find $\alpha$ and $\mu$.

3   A ladder of length 4 m and weight 200 N rests against a smooth vertical wall. Its foot stands on rough horizontal ground and the angle between the ladder and the ground is $\tan^{-1} 3$. The centre of gravity of the ladder is at a distance 1 m from the foot. When a boy of weight 400 N stands at the top of the ladder, the ladder does not slip. Show that the coefficient of friction between the ladder and the ground is at least $1/4$.

4   A uniform rod AB of weight 10 N and length $2a$ is in equilibrium in a horizontal position with its end B against a rough vertical wall. Its end A is supported by a string which is fastened to a point C of the wall at a distance $b$ vertically above B. Show that the force of friction at B is 5 N, and that the coefficient of friction between the rod and the wall must be at least $b/(2a)$.

5   Two equal uniform rods AB and AC are freely jointed at A. The rods are in a vertical plane with the ends B and C on a horizontal table. The coefficient of friction between each rod and the table is 0.3. Show that the rods can rest in equilibrium with $\angle$ ABC and $\angle$ ACB equal to 60°, but not with these angles equal to 59°.

6   A uniform rod AB of length $2a$ and weight $W$ rests against a smooth horizontal rail. The rod is inclined at 30° to the horizontal with its end B against a rough vertical wall. The distance from B to the rail is $a/2$. Show that the wall exerts on the rod a force of magnitude $W$ at an angle of 30° to the horizontal. Show also that the coefficient of friction between the rod and the wall is at least $1/\sqrt{3}$.

7   Two uniform rods AB and AC of equal length are smoothly jointed at A. The rods are in equilibrium in a vertical plane with the ends B and C resting on a rough horizontal plane, each rod making an angle $\tan^{-1} 3$ with the horizontal. Find the magnitude of the reaction between the rods at A and the least possible value of the coefficient of friction $\mu$ between the rods and the plane
    (a) given that each rod weighs 60 N
    (b) given that one rod weighs 60 N and the other weighs 120 N.

8   A uniform solid cylinder of height $5a$ and radius $a$ stands on a rough horizontal plane, the coefficient of friction between the cylinder and the plane being $1/2$. By considering the components of the weight of the cylinder parallel to the plane and at right angles to the plane, show that, if the plane is slowly tilted, the cylinder will topple before it slides.

9   A uniform solid sphere of weight $W$ rests on a plane which is inclined to the horizontal at an angle $\alpha$. The sphere is kept in equilibrium by a horizontal force applied at its highest point. Show that the magnitude of the normal reaction between the sphere and the plane equals $W$, and that the coefficient of friction between the sphere and the plane is at least $\tan(\alpha/2)$.

10   A uniform cube of weight 10 N stands on a horizontal table. A gradually increasing horizontal force **P** is applied to the mid-point of an upper edge of the cube at right angles to this edge. Find the value of $P$ when the cube is on the point of moving if the coefficient of friction between the cube and the table is (a) 0.4 (b) 0.6.

## Miscellaneous exercises 8

1   In a uniform rectangular lamina ABCD, AB $= 2a$ and BC $= 4a$. The triangular lamina BOC is removed, where O is the point of intersection of the diagonals. The remainder hangs in equilibrium freely suspended from A. Find the angle between AD and the vertical.

2   A square lamina ABCD of side $2a$ is at rest on a smooth horizontal table. Forces of magnitude $2P$ act along AB and CD, and forces of magnitude $P$ act along BC and AD, the direction of the forces being given by the order of the letters. A fifth force **Q** keeps the lamina in equilibrium. Show that

$Q = 2P$, and find the distance of the line of action of $\mathbf{Q}$ from the centre of the lamina.

3   A cylindrical tube of length $4a$, radius $a$ and weight $W$, stands with its axis vertical on a horizontal table. A smooth uniform rod inclined at 45° to the horizontal is in equilibrium in contact with the upper rim and the inner surface of the tube. Given that the tube is on the point of toppling, show that the weight of the rod is $W/2$.

4   A uniform rod AB of weight $W$ and length $2a$ is free to turn in a vertical plane about a pivot at A. One end of a light inextensible string is attached to a point C on the same level as A, where $AC = a$. The other end of the string carries a light smooth ring which can slide on the rod. A load of weight $2W$ is suspended from the rod at B. Given that the angle BAC is 60°, find the tension in the string and the magnitude of the force exerted on the rod at A.

5   A square lamina ABCD can turn freely in a horizontal plane about a pivot at A. A force $8P$ acts along BA, a force $12P$ acts along CB, a force $10P$ acts along CD and a force $6P$ acts along DA. Find the magnitude and direction of the least force acting at C that will keep the lamina at rest.

6   A uniform rod AB of weight 20 N can turn freely in a vertical plane about a pivot at A. A load of weight 10 N is suspended from B, and the rod is in equilibrium with AB making an angle of 30° with the upward vertical, under the action of a force acting at B at right angles to AB. Find the magnitude of this force. Find also, to the nearest degree, the angle between the reaction at the pivot and the vertical.

7   A uniform rod is of weight $W$ and length $6a$. The end A of the rod rests on a smooth horizontal floor, and the rod leans against a smooth peg C, where $AC = 2a$. The rod is maintained in equilibrium inclined at 45° to the floor by a horizontal force $\mathbf{P}$ applied to the rod at A. Find the magnitude of $\mathbf{P}$ and of the force exerted on the rod by the floor.

8   Three equal uniform rods AB, BC, CD each of weight $W$ are freely jointed at B and C, and are freely pivoted at A and D to two fixed points at the same level, where $AD = 2BC$. Find the magnitude of the forces exerted at A and at B on the rod AB when BC is horizontal and below AD.

9   A uniform solid hemisphere is placed with its curved surface in contact with a plane inclined at an angle $\tan^{-1}(0.4)$ to the horizontal. Show that, if the coefficient of friction between the hemisphere and the plane is not less than 0.4, there are two positions in which the hemisphere can rest in equilibrium. Show that in each case the angle between its plane face and the vertical is approximately 8°.

10   A uniform solid hemisphere rests with its curved surface in contact with a horizontal floor and a vertical wall. The coefficient of friction between the hemisphere and both wall and floor is $1/7$. Show that, when equilibrium is limiting, the plane face of the hemisphere is inclined to the horizontal at an angle between 25° and 26°.

11  The axis of a smooth uniform cylinder is horizontal, and its curved surface is in contact with a smooth vertical wall and with a wedge of angle $\alpha$. The wedge rests on a rough horizontal floor, and the weight of the wedge is twice the weight of the cylinder. Given that the wedge is about to slip, show that the coefficient of friction between the wedge and the floor is $\frac{1}{3} \tan \alpha$.

12  A uniform bar of length 1.5 m and weight 12 N is suspended horizontally by three vertical springs of equal length and modulus. The springs are attached one at each end and one at the mid-point of the bar. A load of weight 18 N is then suspended from the bar at a point 0.5 m from one end. Find the tension in each spring.

13  Two uniform rods AB and AC, of equal weight and each of length $2a$, are freely jointed at A. They rest in equilibrium with A above the level of B and of C, and with each rod in contact with a smooth peg. The two pegs are at a distance $b$ apart and are at the same level. Given that each rod makes an angle of 60° with the horizontal, show that $a = 4b$.

14  A uniform rectangular lamina ABCD rests in a vertical plane supported by two smooth pegs X and Y. Given that AC is parallel to XY, show that BD is vertical.

15  A uniform circular disc of weight 64 N rests in a horizontal position supported at points A, B and C on its rim. Given that AB $=$ AC $=$ 0.5 m and BC $=$ 0.6 m, find the force exerted on the disc at A.

16  Two uniform rods AB, BC of equal length are freely jointed at B. The rods stand on a smooth horizontal table at right angles to one another with B vertically above AC. They are kept in equilibrium by horizontal forces **P** and **Q** applied at A and C respectively. Given that the weights of AB and BC are 4 N and 8 N respectively, find the magnitudes of **P** and **Q** and of the reaction between the rods at B.

17  A smooth sphere of weight $W$ and radius $a$ rests against a smooth plane inclined at 30° to the vertical. The sphere is supported by a string of length $a$, one end of the string being fastened to a point on the plane and the other end to a point on the surface of the sphere. Find the tension in the string and the magnitude of the force exerted on the sphere by the plane.

18  A uniform rod AB of length $2a$ and weight $W$ is hinged at A to a vertical wall. The rod is supported in a horizontal position by a strut MN hinged at M to the mid-point of the rod, while the end N rests on the ground at the foot of the wall, the distance AN being $3a/4$. The rod carries a load of weight $4W$ suspended from B. Show that the magnitude of the force exerted on the rod by the strut is $15W$, and find the magnitude of the force exerted on the rod at A.

19  A light rod carrying a particle of weight 49 N at each end is balanced horizontally across a rough cylinder, the axis of the cylinder being horizontal and at right angles to the rod. The rod is 2 metres long and the radius of the cylinder is 10 cm. A particle of weight 2 N is now suspended from one end of the rod. Given that the rod does not slip, find to the nearest

degree the inclination of the rod to the horizontal in the new position of equilibrium. Show that in this position the magnitude of the force of friction at the point of contact is approximately 20 N.

20  A uniform square lamina ABCD of side 40 cm and weight 6 N can turn freely in a vertical plane about a pivot at A. The lamina is in equilibrium with B vertically below A under the action of a force 6 N acting along BC and a force acting along DE, where E is the point on AB such that AE = 30 cm. Find the horizontal and vertical components of the force exerted on the lamina by the pivot.

21  Four equal rods AB, BC, CD, DA are freely hinged together to form a square ABCD which lies on a smooth horizontal table. The mid-points of AB and AD are connected by a string of length $\frac{1}{2}$BD. A force of 10 N is applied to the hinge A in the direction $\overrightarrow{AC}$, and a force of 10 N is applied to the hinge C in the direction $\overrightarrow{CA}$. Find the tension in the string, and show in a diagram the forces acting on the rod AB.

22  Two equal uniform rods AB and BC, each of weight 5 N, freely jointed at B, are suspended by a string attached to A and to a fixed point. A horizontal force of 15 N is applied at C. Show that in the position of equilibrium the angle ABC is approximately $163°$.

23  A uniform cube of weight $2W$, whose vertical section is the square ABCD of side $a$, has the base AD fixed horizontally. It is cut along the diagonal AC, and the coefficient of friction between the two faces thus made is 1/4. A horizontal force **P** is applied at B so that the portion ABC is on the point of slipping upwards. Find $P$ in terms of $W$, and find the distance from C of the point on AC through which the resultant reaction between the two faces acts.

24  A uniform ladder of weight $W$ and length $2a$ rests with its upper end against a smooth vertical wall. Its lower end stands on rough horizontal ground, and the coefficient of friction between the ladder and the ground is 1/2. Given that the ladder is in limiting equilibrium, find the angle which the ladder makes with the horizontal. Find also the magnitude of the reaction at the wall.

   A man of weight $W$ climbs the ladder. Find how far up it he can go before the ladder will slip. Find also how far up the ladder he can go when a load of weight $W$ is placed on the foot of the ladder.  [L]

25  A uniform square lamina ABCD of weight $W$ and side $2a$ rests in a vertical plane with the vertex A in contact with a rough horizontal plane. The lamina is kept in equilibrium by a force $P$ acting at the vertex C in the direction of DC produced. Given that the height of the vertex D above the horizontal plane is $6a/5$, show that $P = W/10$.

   Find the coefficient of friction between the lamina and the plane if the lamina is in limiting equilibrium.  [L]

26  The foot of a uniform ladder of weight $W$ rests on rough horizontal ground and the top of the ladder rests against a smooth vertical wall. The ladder

makes an angle of 60° with the horizontal. If a man of weight $3W$ can stand at the top of the ladder, show that the coefficient of friction between the ladder and the ground is not less than $(7\sqrt{3})/24$.

Find the frictional force between the ladder and the ground when this man is halfway up the ladder. [L]

27 A uniform rod AB of weight $W$ and length $4a$ rests with its end A on a rough horizontal plane, and with a point C of the rod in contact with a smooth peg. If AC $= 3a$ and if the rod is inclined at 30° to the horizontal, find the magnitude of the force exerted on the rod by the peg. Show also that the coefficient of friction between the rod and the plane cannot be less than $1/\sqrt{3}$.

28 A uniform rod AB of length $2l$ and weight $W$ is smoothly hinged to a vertical wall at A. It is held at rest in a horizontal position by a light inextensible string attached at B and to a point C which is vertically above and at a distance $l$ from A. Calculate the tension in the string and show that the reaction at A is equal in magnitude to the tension.

When a load of weight $4W$ is attached to the rod at a point D, the horizontal and vertical components of the reaction at A are equal in magnitude. Find AD. [L]

29 A smooth horizontal rail is fixed at a height $a/2$ above rough horizontal ground. A solid uniform right circular cone of base radius $a$ and height $2a$ rests in contact with the ground and the rail, its axis being horizontal and perpendicular to the rail. Given that the cone is about to slip, find the coefficient of friction between the cone and the ground.

30 A uniform rod of weight $4W$ and length $2a$ is maintained in a horizontal position by two light inextensible strings each of length $a$ attached to the ends of the rod. The other ends of the strings are attached to small rings each of weight $W$ which can slide on a fixed horizontal bar. The coefficient of friction between each ring and the bar is $1/2$. Show that in equilibrium the greatest possible distance between the rings is $16a/5$.

31 Two uniform rods OA and AB each of length $2a$ and weight $W$ are freely jointed together at A. The rod OA can turn freely in a vertical plane about a pivot at its lower end O, and the end B of the rod AB can move in the same vertical plane along a horizontal plane which passes through O. The coefficient of friction between the rod AB and the plane is $1/2$. The rods are in equilibrium with tan AOB $= 2$. Find the smallest horizontal force applied to rod AB at B such that B is about to slip (a) away from O (b) towards O.

32 Two uniform rods AB and BC, each of weight $W$, are freely jointed at B and hang from a smooth fixed hinge at A. The rods are in a straight line making an acute angle $\alpha$ with the downward vertical, and rest in equilibrium with a point K on BC in contact with a smooth fixed peg. Show that BK $= \frac{1}{3}$BC. Show also that the magnitude of the reaction between the rods at B is $\frac{1}{2}W\sqrt{(1 + 3\cos^2 \alpha)}$. [L]

33  Show that the centre of mass of a uniform solid hemisphere of radius $a$ is at a distance $3a/8$ from the centre of its plane face.

The hemisphere rests in equilibrium with its plane face vertical and its curved surface on a rough inclined plane. Find, correct to the nearest degree, the angle between the plane and the horizontal. [JMB]

34  A uniform wire of length $\pi a$ is bent to form a semicircle of radius $a$. Prove that the distance of the centre of mass of the wire from the centre of the circle is $2a/\pi$.

Another uniform wire of length $(2 + \pi)a$ is bent into the form of a semicircular arc together with the diameter AB joining the ends of the arc, and hangs in equilibrium from the point A. Show that AB makes an angle $\theta$ with the vertical, where $\tan \theta = 2/(2 + \pi)$. [JMB]

35  A uniform wire of weight $W$, bent into a semicircular arc, has a particle of weight $W$ attached to one end and is suspended freely from the other end. Show that when the system is in equilibrium the straight line through the ends of the wire makes an angle $\tan^{-1} (3\pi/2)$ with the horizontal.

(The centre of mass of a uniform semicircular arc of wire is at a distance $2a/\pi$ from the centre of the circle of radius $a$ of which it forms part.)

[JMB]

36  A rigid body is in equilibrium under the action of three non-parallel coplanar forces. Show that the lines of action of the forces are concurrent.

Two smooth inclined planes are fixed perpendicular to each other with their line of intersection horizontal. A uniform rod is in equilibrium perpendicular to the line of intersection and with one end resting on each plane. Show that the mid-point of the rod is vertically above the line of intersection of the planes. [JMB]

37  A uniform rod AB is placed inside a rectangular box with the ends A and B in contact with adjacent faces X and Y respectively of the box and with the vertical plane through AB perpendicular to both faces X and Y. The coefficients of friction between the rod and the faces X and Y are $\mu$ and $2\mu$ respectively. When the face X is horizontal the rod is on the point of slipping when at an angle $\theta$ to the horizontal, but when the face Y is horizontal the rod is about to slip when at an angle $\phi$ to the horizontal. Find $\tan \theta$ in terms of $\mu$ and show that $\tan \theta = 2 \tan \phi$.

The rod is replaced by a non-uniform rod CD whose centre of gravity is at G, and the rod CD is placed inside the box as before with C in contact with the face X which is horizontal. If the coefficients of friction at the faces X and Y remain unchanged and CD is on the point of slipping when at an angle $\phi$ to the horizontal, find the ratio CG:CD in terms of $\mu$. [AEB]

38  Show that the centre of gravity of a uniform semicircular lamina of radius $r$ is at a distance $4r/(3\pi)$ from the centre.

A uniform solid right circular cone of height $h$ has vertex O, and ACB is a diameter of the base whose radius is $r$ and whose centre is C. The cone is cut into two equal parts by a plane perpendicular to ACB and containing the axis of the cone. Find the distance of the centre of gravity of the half AOC of the cone from (i) OC (ii) AC.

If this portion is freely suspended from C, the line OA is horizontal. Find the ratio $r:h$.                              [AEB]

39  The two sides AB, BC of a pair of step-ladders are of the same length and are smoothly hinged together at B. The side BC is of negligible weight and the side AB is uniform and of weight $W$. The mid-points of AB and BC are connected by a light inextensible string and a weight $W$ is attached to the mid-point of the string. The ladders are at rest forming an equilateral triangle ABC with A and C in contact with rough horizontal ground. The attached weight hangs freely with each portion of the string perpendicular to the side of the ladder to which it is connected. If the coefficient of friction $\mu$ between the ladders and the ground is the same at A and C, calculate the least possible value of $\mu$.

If $\mu = 1/3$ and if the weight attached to the string is gradually increased, calculate the least value of this weight necessary to disturb equilibrium and state how equilibrium will be broken.                              [AEB]

40  Two uniform rods, AB and BC, are of the same length and weight $3W$ and $W$ respectively. They are smoothly jointed at B and stand in a vertical plane with A and C on a rough horizontal plane. The coefficient of friction between each rod and the plane is $2/3$. Equilibrium is about to be broken by one of the rods slipping on the plane. Find which rod will slip and calculate the angle each rod makes with the plane. Calculate also the reaction at the hinge B in magnitude and direction.                              [AEB]

41  A uniform sphere of radius $a$ and weight $W$ is supported by three equal fixed rough rods joined at their ends to form a horizontal equilateral triangle of side $b$. Show that the centre of the sphere is at a height of $\sqrt{(a^2 - b^2/12)}$ above the level of the rods.

A couple is applied to the sphere about a vertical axis through the centre of the sphere. If the sphere is on the point of slipping, show that the moment of the couple about this vertical axis is $\mu Wab/\sqrt{(12a^2 - b^2)}$, where $\mu$ is the coefficient of friction between each rod and the sphere.                              [AEB]

42  Three uniform rods are rigidly jointed to form an obtuse-angled triangle ABC in which AB $= 5a$, BC $= 3a$ and AC $= 7a$. Show that $\cos CAB = 13/14$.

The triangle is simply supported at A and B and rests in a vertical plane with AB horizontal and the vertex C uppermost. If the weight per unit length of each rod is $w$, calculate the horizontal distance of the centre of gravity of the triangle from A and the magnitude of the reaction at each support. When a particle of weight $kwa$ is hung from C, the reaction at the support A is zero. If equilibrium is maintained, calculate the value of $k$.                              [AEB]

43  A uniform straight scaffold pole of weight $W$ has one end A on rough horizontal ground and the other end B against a rough vertical wall. The

vertical plane containing AB is perpendicular to the wall. The pole is inclined at an angle $\alpha$ to the horizontal where $\tan \alpha = 3/4$. The ground and the wall are equally rough and the pole is resting in limiting equilibrium. Calculate the coefficient of friction at either contact point.

The pole is kept in the same vertical plane, leaning against the wall, but its angle of inclination to the horizontal is increased to $\beta$, where $\tan \beta$ is $4/3$. A body of weight $kW$ is now hung from B and equilibrium is again limiting. Calculate the value of $k$. [AEB]

**44** Using a diagram, explain the meaning of the term 'angle of friction'.

A uniform straight pole AB of length 2.5 metres stands in limiting equilibrium with the end A on horizontal ground and the end B against a vertical wall. The vertical plane containing AB is perpendicular to the wall and A is at a distance 0.7 metres from the wall. The angles of friction between the pole and the ground and between the pole and the wall are equal. Show that the line of action of the resultant force exerted on the pole by the ground at A is perpendicular to the line of action of the resultant force exerted on the pole by the wall at B. Hence, or otherwise, show that the coefficient of friction at A and B is $1/7$.

Given that the pole weighs 50 N, find the magnitude of the resultant reactions
(i) between the pole and the ground
(ii) between the pole and the wall. [AEB]

**45** A uniform rod AB, of weight 120 N and length 4 m, is in equilibrium with the end A on a rough horizontal floor at an angle $\cos^{-1}(3/4)$ to the horizontal. The rod is resting against a smooth fixed peg at a point C in the rod such that AC $= 3$ m. Show that $\mu$, the coefficient of friction between the rod and the floor, is greater than $\sqrt{7}/5$.

When a particle of weight 60 N is attached to the rod at a distance $x$ metres measured along the rod from A, equilibrium is limiting. If $\mu = \sqrt{7}/4$, find the value of $x$. [AEB]

**46** Three non-parallel coplanar forces act on a rigid body which is in equilibrium. Show that the lines of action of these forces are concurrent.

The end A of a uniform rod AB, of weight $W$ and length $2l$, is smoothly hinged to a vertical wall. One end of a light string is tied to the rod at B and the other end is tied to the wall at C, where C is at a distance $2l$ vertically above A. The rod is in equilibrium when AB makes an angle $2\theta$ with the downward vertical through A, where $\theta$ is less than $45°$. Calculate, in terms of $W$ and $\theta$,
(i) the tension in the string
(ii) the magnitude and direction of the reaction at the hinge.

Given that the string BC is elastic, with natural length $2l$ and modulus of elasticity $3W/2$, find the value of $\theta$. [AEB]

**47** The uniform rectangular plate ABCD has AB $= 3a$ and BC $= 5a$. The point X is on AD such that DX $= 3a$. The triangular lamina XDC is cut away leaving the trapezium-shaped plate ABCX which is of weight $W$. Calculate the distances of the centre of gravity of the plate ABCX from BC and AB.

(i) If the plate ABCX is freely suspended from X, calculate the angle, in degrees and minutes, which CX makes with the vertical.

(ii) A particle of weight $W$ is attached at A and the plate ABCX is freely suspended from X. When a second particle of weight $\lambda W$ is attached at C, the edge BC is horizontal. Calculate the numerical value of $\lambda$.     [AEB]

48   A uniform lamina ABCD is in the form of a trapezium in which the angles A and B are right angles, BC $= 2$AB $= 2a$ and AD $= x$. Find the distance of the centre of gravity of the lamina from AB.

The lamina is placed in a vertical plane with AD on a horizontal plane. If $x = \frac{3}{4}a$, show that the lamina will not topple about D.     [L]

49   The edges of a uniform square lamina ABCD are each of length $3a$. A portion of the lamina in the form of a square APQR, where P lies on AB, R lies on AD and AP is of length $a$, is removed. Find the distance from AB and the distance from BC of the centroid of the remainder of the lamina.

This remainder is suspended freely from B. Show that, if $\theta$ is the inclination of BC to the vertical, then $\tan \theta = 11/13$.     [L]

50   Show that the centre of mass of a uniform right circular solid cone of height $h$ is at a distance $3h/4$ from the vertex.

A uniform solid spinning top has the shape of a right circular cone of radius $3r$ and height $4r$ surmounted by a cylinder of base radius $3r$ and height $6r$. Find the distance of the centre of mass of the top from the vertex of the cone. Hence show that if the top is placed with the curved surface of the cone on a horizontal plane, it will topple.     [L]

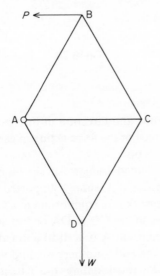

Fig. 8.23

**51** A light framework ABCD consists of 5 smoothly-jointed rods of equal lengths (Fig. 8.23). The framework carries a load $W$ at D and is smoothly hinged and fixed at A. The framework is kept in equilibrium in a vertical plane with AC horizontal by a force $P$ applied at B in a direction parallel to CA. Find the magnitude of $P$ and the magnitude and direction of the reaction at A. Find, graphically or otherwise, the forces in the five rods and state which rods are in compression. [AEB]

**52** Figure 8.24 shows a smoothly-jointed framework consisting of 5 light rigid rods with A smoothly hinged at a fixed point; $BD = 5a$, $AD = BC = 3a$ and $AB = DC = 4a$. When a load of weight $W$ is hung from B, the framework is kept in equilibrium in a vertical plane with AB horizontal by a force of magnitude $P$ acting at C parallel to BD.

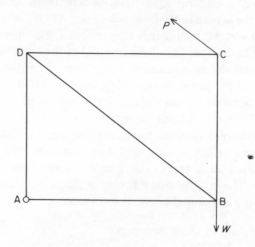

Fig. 8.24

(i) Show that $P = 5W/6$.

(ii) Find the magnitude and direction of the reaction on the hinge at A.

(iii) Find the magnitude of the force acting in each rod and name the rods which are in compression. [AEB]

**53** The light smoothly-jointed framework shown in Fig. 8.25 is smoothly hinged to a fixed support at A, simply supported at B and carries a load of 50 N at C. The framework is in equilibrium in a vertical plane with AB and DC horizontal; $AB = BC = CD = DA = 5$ m and $BD = 6$ m. Show that the reaction on the hinge at A is vertical and calculate its magnitude.

Using a graphical method, or otherwise, find the forces acting in each of the members of the framework stating which are in tension and which are in compression. [AEB]

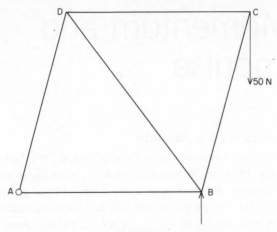

Fig. 8.25

**54** Figure 8.26 shows a framework consisting of five light rigid rods which are smoothly jointed at their ends; $AD = 3a$, the angles DAC and ADC are each $30°$ and the angles ACB and BAD are each $90°$. Show that the length of AB is $2a\sqrt{3}$. A load of 100 N is hung from D and the framework is kept in equilibrium in a vertical plane with AD horizontal by a horizontal force at A and a force at B. Using a graphical method, or otherwise, find
(i) the magnitude of the force at A
(ii) the magnitude and the direction of the force at B
(iii) the magnitudes of the forces acting in the five rods, stating which rods are in tension. [AEB]

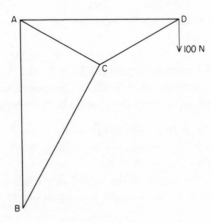

Fig. 8.26

# 9 Momentum and impulse

## 9.1 Momentum of a particle

When a particle of mass $m$ is moving with velocity $\mathbf{u}$, the product $m\mathbf{u}$ is known as the momentum of the particle, or more precisely, as its linear momentum. This is a vector in the same direction as $\mathbf{u}$, along a straight line passing through the particle. Momentum is not a free vector, but is a localised vector. Since mass is measured in kg and velocity in $m\,s^{-1}$, we measure momentum in $kg\,m\,s^{-1}$.

When a rigid body of mass $M$ is moving without rotating, all the particles composing the body have the same velocity. If this velocity is $\mathbf{v}$, the momentum of the body is given by the vector $M\mathbf{v}$ passing through the centre of mass of the body.

*Example 1*

A particle of mass $m$ is moving in a straight line with velocity $\mathbf{v}$. The particle is given a blow which trebles its speed. Find the change in momentum (a) if the particle continues to move in the same direction as before (b) if the particle now moves in the opposite direction.

(a) Momentum before the blow $= m\mathbf{v}$.
      Momentum after the blow $= 3m\mathbf{v}$.
          Change in momentum $= 3m\mathbf{v} - m\mathbf{v} = 2m\mathbf{v}$.
(b) Momentum before the blow $= m\mathbf{v}$.
      Momentum after the blow $= -3m\mathbf{v}$.
          Change in momentum $= -3m\mathbf{v} - m\mathbf{v} = -4m\mathbf{v}$.

*Example 2*

The initial velocity of a particle of mass 0.1 kg is $40\ m\,s^{-1}$ in the direction of the vector $\mathbf{j}$. A blow is given to the particle which changes its velocity to $30\ m\,s^{-1}$ in the direction of the vector $\mathbf{i}$. Find the change in momentum.

Momentum before the blow $= (0.1)(40\mathbf{j}) = 4\mathbf{j}$.
Momentum after the blow $\ = (0.1)(30\mathbf{i}) = 3\mathbf{i}$.
Change in momentum $\qquad = 3\mathbf{i} - 4\mathbf{j}$.
Since $3^2 + 4^2 = 5^2$, the length of the vector $3\mathbf{i} - 4\mathbf{j}$ is 5.

Therefore the change in momentum is $5\ kg\,m\,s^{-1}$ in the direction of the vector $3\mathbf{i} - 4\mathbf{j}$.

**Exercises 9.1**

1  A particle of mass $m$ is moving in a horizontal line when its velocity is

suddenly changed from **u** to **v**. Find the change in momentum of the particle.

(a) $m = 0.4$ kg, $\mathbf{u} = 3$ m s$^{-1}$ due east, $\mathbf{v} = 8$ m s$^{-1}$ due east.

(b) $m = 0.2$ kg, $\mathbf{u} = 5$ m s$^{-1}$ due east, $\mathbf{v} = 15$ m s$^{-1}$ due west.

(c) $m = 0.3$ kg, $\mathbf{u} = 12$ m s$^{-1}$ due west, $\mathbf{v} = 18$ m s$^{-1}$ due east.

(d) $m = 0.2$ kg, $\mathbf{u} = 5$ m s$^{-1}$ due north, $\mathbf{v} = 5$ m s$^{-1}$ due east.

2   The velocity of a particle of mass $m$ moving in the plane of the vectors **i** and **j** is changed from **u** to **v**. Find the magnitude of the change in momentum in kg m s$^{-1}$. In each case state the unit vector in the direction of the change in momentum.

(a) $m = 0.2$ kg, $\mathbf{u} = 2\mathbf{i} - \mathbf{j}$, $\mathbf{v} = 12\mathbf{i} + 9\mathbf{j}$.

(b) $m = 0.5$ kg, $\mathbf{u} = 4\mathbf{i} + 3\mathbf{j}$, $\mathbf{v} = 4\mathbf{i} - 3\mathbf{j}$.

(c) $m = 0.5$ kg, $\mathbf{u} = -2\mathbf{i} + 6\mathbf{j}$, $\mathbf{v} = 4\mathbf{j}$.

(d) $m = 0.4$ kg, $\mathbf{u} = 5\mathbf{i} - 2\mathbf{j}$, $\mathbf{v} = 2\mathbf{i} + 2\mathbf{j}$.

3   A ball of mass 1.5 kg falling vertically strikes a horizontal floor with speed 5 m s$^{-1}$. It rebounds with speed 3 m s$^{-1}$. Find the change in momentum.

4   The initial speed of a particle of mass 5 kg is 12 m s$^{-1}$ and its final speed is 13 m s$^{-1}$. Find the magnitude of the change in momentum given that this change is at right angles to the initial momentum.

## 9.2   Impulse

Let a constant force **P** act on a particle of mass $m$ for a time $t$. In this time let the velocity of the particle change from **u** to **v**. The constant force will give the particle a constant acceleration $(\mathbf{v} - \mathbf{u})/t$ in the direction of the force. Since the force equals the product of the mass and the acceleration,

$$\mathbf{P} = m(\mathbf{v} - \mathbf{u})/t,$$

i.e.
$$\mathbf{P}t = m\mathbf{v} - m\mathbf{u}.$$

The product **P**$t$ is known as the impulse of the force **P** acting for time $t$. Since force is measured in newtons and time in seconds, the unit of impulse is the newton-second (N s). The equation

$$\mathbf{P}t = m\mathbf{v} - m\mathbf{u}$$

shows that the impulse of the force equals the change in momentum of the particle. In brief,

$$\text{impulse} = \text{change in momentum}.$$

When a cricket ball is struck by a bat, a large force acts for a very short time. It is not easy to measure the force or the time, but by measuring the change in momentum of the ball, the impulse of the force can be found.

Let a force **P** which is not constant act on a particle of mass $m$ from the instant when $t = 0$ until the instant when $t = T$. In this time let the velocity **v** of the particle change from **U** to **V**. The impulse of the force **P** is defined to be

$$\int_0^T \mathbf{P} \, dt.$$

Since $\mathbf{P} = m\dfrac{d\mathbf{v}}{dt}$, the impulse is given by

$$\int_0^T m\frac{d\mathbf{v}}{dt}dt = \left[mv\right]_U^V = m\mathbf{V} - m\mathbf{U},$$

i.e. impulse = change in momentum.

*Example 1*

The speed of a car travelling in a straight line is reduced from $25\ \mathrm{m\,s^{-1}}$ to $5\ \mathrm{m\,s^{-1}}$ by a constant force acting for 4 s. The mass of the car is 1200 kg. Find the force.

Initial momentum = $1200 \times 25\ \mathrm{kg\ m\,s^{-1}} = 30\,000\ \mathrm{kg\ m\,s^{-1}}$.
Final momentum = $1200 \times 5\ \mathrm{kg\ m\,s^{-1}} = 6000\ \mathrm{kg\ m\,s^{-1}}$.
Change in momentum = $24\,000\ \mathrm{kg\ m\,s^{-1}}$.

Since the force is constant,

force × time = impulse = change in momentum
force × 4 = 24 000
force = 6000 N.

*Example 2*

A blow given to a particle of mass 0.2 kg changes its velocity from $30\ \mathrm{m\,s^{-1}}$ in the direction of the vector $\mathbf{j}$ to $50\ \mathrm{m\,s^{-1}}$ in the direction of the vector $\mathbf{i}$. The blow lasts for 0.1 s. Find the magnitude of the constant force that would give the particle the same impulse in this time.

Initial momentum = $0.2 \times 30\mathbf{j} = 6\mathbf{j}$.
Final momentum = $0.2 \times 50\mathbf{i} = 10\mathbf{i}$.
Change in momentum = final momentum − initial momentum
= $10\mathbf{i} - 6\mathbf{j}$.

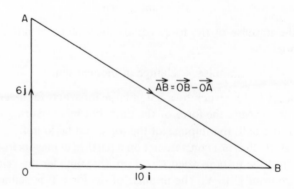

Fig. 9.1

In Fig. 9.1, the vector $\overrightarrow{OA}$ represents the initial momentum, $\overrightarrow{OB}$ represents the final momentum and $\overrightarrow{AB}$ the change in momentum.

Let **P** be the constant force required.

$$\text{force} \times \text{time} = \text{impulse} = \text{change in momentum}$$
$$\mathbf{P} \times 0.1 = 10\mathbf{i} - 6\mathbf{j}$$
$$\mathbf{P} = 100\mathbf{i} - 60\mathbf{j} = 20(5\mathbf{i} - 3\mathbf{j})$$
$$P = 20\sqrt{34} \text{ N}.$$

*Example 3*

A bat strikes a ball of mass 0.2 kg at an instant when the ball is falling vertically with speed 10 m s$^{-1}$. When the ball leaves the bat, it is travelling horizontally with speed 25 m s$^{-1}$. Find the impulse given to the ball.

Initial momentum $= 0.2 \times 10 = 2$ kg m s$^{-1}$ vertically downwards.
Final momentum $= 0.2 \times 25 = 5$ kg m s$^{-1}$ horizontally.

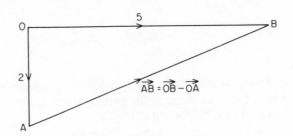

Fig. 9.2

In Fig. 9.2, the vector $\overrightarrow{OA}$ represents the initial momentum and the vector $\overrightarrow{OB}$ represents the final momentum.

Change in momentum = final momentum − initial momentum
$$= \overrightarrow{OB} - \overrightarrow{OA} = \overrightarrow{AB}.$$

Now OA = 2, OB = 5. Hence AB $= \sqrt{29}$ and tan OAB $= 2.5$. By the principle

$$\text{impulse} = \text{change in momentum},$$

the impulse given to the ball is $\sqrt{29}$ N s at approximately 68° to the upward vertical.

*Example 4*

A horizontal force **P** acts in a constant direction on a particle of mass 0.4 kg. At time $t$ the magnitude of the force is $(3 + 2t)$ N. When $t = 0$ the particle is at rest. Find the speed of the particle after four seconds.

$$\text{Impulse} = \int_0^4 (3 + 2t)\,dt = \left[ 3t + t^2 \right]_0^4 = 28 \text{ N s}.$$

Therefore the change of momentum is 28 kg m s⁻¹. Since the particle starts from rest, its final momentum is 28 kg m s⁻¹. Its mass is 0.4 kg and hence its final speed is 70 m s⁻¹.

Figure 9.3 shows the graph of $P$ against $t$. The area of the trapezium OABC represents the impulse of the force in the given time.

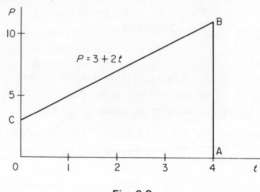

Fig. 9.3

**Exercises 9.2**

1  A particle of mass $m$ initially at rest is acted on by a constant horizontal force **P**. After time $t$ the speed of the particle is $v$.
   (a) $m = 4$ kg, $P = 6$ N, $t = 10$ s. Find $v$.
   (b) $m = 2$ kg, $P = 5$ N, $v = 10$ m s⁻¹. Find $t$.
   (c) $m = 0.5$ kg, $P = 100$ N, $v = 20$ m s⁻¹. Find $t$.
   (d) $m = 0.2$ kg, $v = 8$ m s⁻¹, $t = 0.01$ s. Find $P$.

2  A horizontal force **P** acts in a constant direction on a particle of mass $m$ for time $T$. The initial speed of the particle is $u$ in the same direction as the force and its final speed is $v$.
   (a) $P = 3t^2$ N, $T = 4$ s, $m = 2$ kg, $u = 8$ m s⁻¹. Find $v$.
   (b) $P = (1 + 2t)$ N, $T = 3$ s, $m = 0.4$ kg, $u = 10$ m s⁻¹. Find $v$.
   (c) $P = (2 + 4t)$ N, $T = 5$ s, $m = 0.3$ kg, $v = 250$ m s⁻¹. Find $u$.
   (d) $P = (1 + 6t)$ N, $m = 0.1$ kg, $u = 20$ m s⁻¹, $v = 320$ m s⁻¹. Find $T$.

3  When a car of mass 800 kg is pushed along a level road its speed is increased from 2 m s⁻¹ to 5 m s⁻¹ in 6 seconds. Find the constant force that would achieve this.

4  A railway truck of mass 6000 kg is brought to rest in 3 minutes 20 seconds from a speed of 30 km h⁻¹ by a constant force. Find its magnitude.

5  A particle moving due east is given a blow such that it then moves with the same speed in the direction N 30° E. Find graphically the direction of the blow.

6  A constant force acting for 0.1 s changes the velocity of a particle of mass 2.5 kg from 10 m s⁻¹ in the direction of the vector $3\mathbf{i} + 4\mathbf{j}$ to 20 m s⁻¹ in the direction of the vector $4\mathbf{i} - 3\mathbf{j}$. Find the magnitude of the force.

7  A tennis ball of mass 0.03 kg travelling horizontally at 25 m s$^{-1}$ is struck by a racquet. The ball is returned with speed 40 m s$^{-1}$ in the opposite direction. If the impact lasts for 0.13 s find the magnitude of the average force exerted on the ball.

8  The mass of the head of a hammer is 4 kg. When the hammer is moving vertically downwards at 6 m s$^{-1}$ it strikes a nail and rebounds vertically at 1 m s$^{-1}$. The average force exerted on the nail by the hammer during impact is 140 N. Find the time for which the impact lasts.

9  A particle of mass 0.3 kg is moving horizontally due north at 10 m s$^{-1}$ when it is given a blow such that its velocity becomes 10 m s$^{-1}$ due east. Find the magnitude and direction of the blow.

10  A stone of mass 0.5 kg strikes a smooth floor and then skids along the floor at 10 m s$^{-1}$. Its velocity before impact is 20 m s$^{-1}$ at 30° to the vertical. Find the magnitude and direction of the impulse given to the stone.

## 9.3  Continuous change of momentum

Consider a horizontal jet of water striking a vertical wall at right angles. If the water runs down the wall after impact, the horizontal momentum of the water is lost. The water will exert a horizontal force on the wall and by Newton's third law the wall exerts an equal force on the water. To find the magnitude of this force, apply the principle

$$\text{impulse} = \text{change in momentum.}$$

Let $P$ be the magnitude of the horizontal force exerted by the wall on the water. In time $t$ the impulse of this force is $Pt$. Let $m$ be the mass of the water striking the wall in time $t$ and let $v$ be its horizontal speed. The momentum lost by the water in time $t$ is $mv$ and therefore

$$Pt = mv.$$

*Example 1*
A jet of water from a hose-pipe impinges on a vertical wall at right angles. The speed of the water on impact is 4 m s$^{-1}$ and the area of cross-section of the pipe is 15 cm$^2$. Assuming that the water does not rebound from the wall, find the force exerted by the jet on the wall.

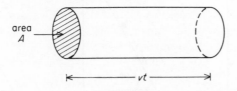

area
$A$

$vt$

Fig. 9.4

When the speed of the water is $v$ and the area of cross-section is $A$, the volume of water leaving the pipe in $t$ seconds equals the volume of a cylinder of length $vt$ and cross-section $A$, i.e. it equals $vtA$ (Fig. 9.4).
Now $v = 4 \text{ m s}^{-1}$ and $A = 15 \text{ cm}^2 = 15 \times 10^{-4} \text{ m}^2$.
Volume of water in time $t = 4 \times 15 \times 10^{-4} \times t = 6t \times 10^{-3} \text{ m}^3$.
The mass of a cubic metre of water is $10^3$ kg.
Mass of water in time $t = 6t$ kg.
Momentum lost in time $t = 6t \times 4 \text{ kg m s}^{-1}$.

$$\text{Impulse} = \text{change in momentum}$$
$$\text{force} \times t = 24t$$
$$\text{force} = 24 \text{ N}.$$

*Example 2*
Every second a pump delivers 20 kg of water through a nozzle of diameter 5 cm. The jet of water strikes a vertical wall at right angles and does not rebound. Find the force exerted on the wall.

The volume of 1 kg of water is $10^{-3} \text{ m}^3$.
Volume of water per second $= 20 \times 10^{-3} = 0.02 \text{ m}^3$.
The area of cross-section of the nozzle is $\pi(2.5)^2 \text{ cm}^2$, i.e. $\pi(6.25) \times 10^{-4} \text{ m}^2$.
   Let the speed of the water be $v \text{ m s}^{-1}$. The water delivered each second will fill a cylinder of length $v$ m and of cross-section $\pi(6.25) \times 10^{-4} \text{ m}^2$. Therefore

$$v \times \pi(6.25) \times 10^{-4} = 0.02$$
$$v = 32/\pi \text{ m s}^{-1}.$$

Momentum lost in $t$ seconds $= (20t)(32/\pi) \text{ kg m s}^{-1}$.

$$\text{Impulse} = \text{change in momentum}$$
$$\text{force} \times t = (20t)(32/\pi)$$
$$\text{force} = 640/\pi \approx 204 \text{ N}.$$

*Example 3*
Sand falls vertically from rest through a distance 5 m on to a fixed smooth plane inclined at 60° to the horizontal. The quantity of sand falling per second is

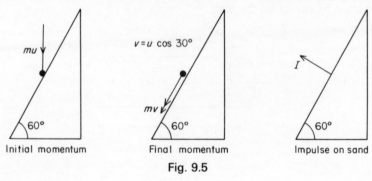

| Initial momentum | Final momentum | Impulse on sand |

Fig. 9.5

0.1 kg, and after impact it runs down the plane. Find the force exerted by the sand on the plane. (Take $g$ as $10 \text{ m s}^{-2}$.)

Let $u$ be the speed of the sand on impact.
Since the sand falls through 5 m,

$$u^2 = 2g \times 5$$
$$u = 10 \text{ m s}^{-1}.$$

The vertical velocity of the sand can be resolved into two components. The component $u \cos 30°$ down the plane is unchanged. The component at right angles to the plane is $u \cos 60°$, i.e. $5 \text{ m s}^{-1}$. Each second the momentum of 0.1 kg of sand moving at $5 \text{ m s}^{-1}$ is destroyed.

Momentum lost in $t$ seconds $= (0.1)(5)t \text{ kg m s}^{-1}$
$$= 0.5t \text{ kg m s}^{-1}.$$

Impulse $=$ change in momentum
force $\times t = 0.5t$
force $= 0.5$ N.

This force is exerted on the sand by the plane in the direction of the impulse shown in Fig. 9.5. By Newton's third law the force exerted on the plane by the sand will be 0.5 N in the opposite direction.

## Exercises 9.3

1 A jet of water delivering 4 litres per second strikes a vertical wall at right angles at a speed of $15 \text{ m s}^{-1}$ and does not rebound. Find the magnitude of the force exerted on the wall by the jet.

2 Each minute 400 kg of sand flows vertically on to a conveyor belt moving horizontally at $0.6 \text{ m s}^{-1}$. Find the horizontal force exerted by the belt on the sand.

3 A fire-pump forces a horizontal jet of water through a nozzle of cross-section $25 \text{ cm}^2$ at the rate of $1.5 \text{ m}^3$ per minute. Find the force which the jet would exert on a vertical wall at right angles to the jet. Find also the magnitude of the force needed to prevent the pump from moving backward.

4 The upper ends of two vertical pipes are connected by a short horizontal pipe so that water flowing up one pipe flows down the other. Find the vertical force exerted on the pipes by the water given that the radius of each pipe is 4 cm and the speed of flow is $2.5 \text{ m s}^{-1}$.

5 A fire-float throws four jets of water in the same direction at 60° to the horizontal. Each jet delivers 600 litres of water per minute through a nozzle of $4 \text{ cm}^2$ cross-section. Find the horizontal thrust on the fire-float.

6 Hailstones falling with velocity $8 \text{ m s}^{-1}$ at 15° to the vertical strike a horizontal roof of area $12 \text{ m}^2$ and do not rebound. If 50 hailstones fall on a square metre each second and if the average mass of the hailstones is 2.2 g,

show that the vertical force on the roof due to the impact is approximately 10 N.

## 9.4 Conservation of linear momentum

When two particles moving in the same straight line collide, the collision is said to be direct. In this case there is no reason to use vector notation. We will adopt the convention of representing velocities before impact by a single arrow and velocities after impact by a double arrow. The velocities before and after a collision between a particle P of mass $m_1$ and a particle Q of mass $m_2$ are set out below.

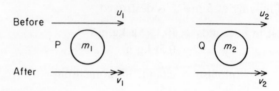

Velocities towards the right are counted positive, and velocities towards the left are counted negative.

Change in momentum of P $= m_1v_1 - m_1u_1$.
Change in momentum of Q $= m_2v_2 - m_2u_2$.

These changes are equal in magnitude but they are in opposite directions, because the impulse given to P by Q is equal in magnitude to the impulse given to Q by P but is in the opposite direction. This gives

$$m_1v_1 - m_1u_1 = -(m_2v_2 - m_2u_2),$$

i.e.
$$m_1v_1 + m_2v_2 = m_1u_1 + m_2u_2.$$

This shows that there is no change in the linear momentum of the system. The momentum after the collision equals the momentum before the collision. This illustrates the *Principle of Conservation of Linear Momentum* which states that if no external force acts in a given direction the momentum of a system of particles in that direction is constant.

### Example 1

Two particles P and Q are moving in the same straight line towards one another. The mass of P is $3m$ and its speed is $2u$, while the mass of Q is $2m$ and its speed is $u$. Given that the particles coalesce on impact, find their speed after the collision.

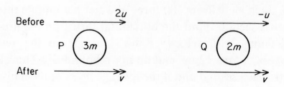

Let the velocity of P be $2u$ in the positive direction and the velocity of Q be $-u$ in the same direction. Let $v$ be their common velocity after the collision.

Momentum before collision $= (3m)(2u) + (2m)(-u) = 4mu$.
Momentum after collision $\;= (3m)(v) + (2m)(v) = 5mv$.

Momentum after collision $=$ momentum before collision
$$5mv = 4mu$$
$$v = 4u/5.$$

*Example 2*
A particle P of mass $2m$ is attached to one end of a light inextensible string and a particle Q of mass $3m$ is attached to the other end. The particles are at rest on a smooth horizontal table when the heavier particle is given a speed $u$ directly away from the lighter particle. Find the impulse given to each particle by the string when it becomes taut.

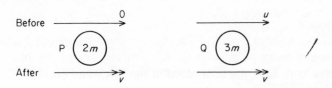

Let $v$ be the speed of Q after the string becomes taut. Since the distance between P and Q will then remain constant, the speed of P will also be $v$.

Initial momentum $= 3mu$.
 Final momentum $= 2mv + 3mv = 5mv$.

Since no external forces act in the direction of the string, the final momentum equals the initial momentum.

$$5mv = 3mu$$
$$v = 3u/5.$$

Change in momentum of P $= 2mv = 6mu/5$.
 Impulse given to P $= 6mu/5$.

This impulse is in the same direction as $u$.
Change in momentum of Q $= 3mv - 3mu = -6mu/5$.
Therefore the impulse given to Q by the string is $6mu/5$ in the opposite direction to $u$.

These impulses are given to the particles by a very large tension in the string acting for a very short time.

*Example 3*

A particle P of mass $2m$ moving with velocity $3\mathbf{i} + 4\mathbf{j}$ collides with a particle Q of mass $3m$ moving with velocity $-2\mathbf{i} - \mathbf{j}$. After the collision P and Q move together with velocity $\mathbf{v}$. Find $\mathbf{v}$ and the change in momentum of each particle.

No external force acts on the system, and therefore the momentum before the collision equals the momentum after the collision.

Initial momentum $= 2m(3\mathbf{i} + 4\mathbf{j}) + 3m(-2\mathbf{i} - \mathbf{j}) = 5m\mathbf{j}$.
 Final momentum $= (2m + 3m)\mathbf{v} = 5m\mathbf{v}$.

Since the final momentum equals the initial momentum, the common velocity $\mathbf{v}$ is equal to $\mathbf{j}$.

$$\begin{aligned}
\text{Change in momentum of P} &= 2m\mathbf{j} - 2m(3\mathbf{i} + 4\mathbf{j}) \\
&= -6m(\mathbf{i} + \mathbf{j}).
\end{aligned}$$

$$\begin{aligned}
\text{Change in momentum of Q} &= 3m\mathbf{j} - 3m(-2\mathbf{i} - \mathbf{j}) \\
&= 6m(\mathbf{i} + \mathbf{j}).
\end{aligned}$$

*Example 4*

A block of wood of mass $M$ is sliding on a smooth horizontal table with speed $u$. A lump of putty of mass $m$ falling vertically strikes the block of wood and adheres to it. Find the final speed of the block and putty.

Vertically, the table will give the block an impulse equal to the vertical component of the impulse given to the block by the putty.

Horizontally no external forces act, and therefore there will be no change in momentum in this direction.

The final speed $v$ will be in the same direction as the initial speed. Therefore

$$(M + m)v = Mu$$
$$v = Mu/(M + m).$$

## Exercises 9.4

1  A particle of mass $m$ moving with speed $6u$ collides with a particle of mass $2m$ moving with speed $u$ in the same direction. The lighter particle is brought to rest by the impact. Find the speed of the heavier particle after the collision.

2  A bullet of mass $0.01$ kg is fired into a stationary block of wood of mass $0.8$ kg and remains embedded in the wood. The speed of the bullet was $540$ m s$^{-1}$. Find the final speed of the block and bullet.

3  A particle of mass $0.4$ kg moving with speed $15$ m s$^{-1}$ collides directly with a particle moving in the opposite direction at $12$ m s$^{-1}$. Both particles remain at rest after the impact. Find the mass of the second particle.

4  A railway truck of mass $1000$ kg travelling at $12$ m s$^{-1}$ overtakes a truck of

mass 600 kg travelling at 8 m s$^{-1}$ in the same direction. After the impact the trucks are coupled together. Find their common speed.

5   A lorry of mass 1000 kg is travelling at 3 m s$^{-1}$ when four sacks of rubbish each of mass 50 kg are dropped vertically into the lorry. Find the final speed of the lorry.

6   Two particles, one of mass $m$ and the other of mass $4m$, are connected by a light inextensible string. The particles are moving apart in the same line each with speed $u$. Find the change in momentum of each particle when the string becomes taut.

7   A boy of mass 50 kg standing at rest on smooth ice catches a ball of mass 0.5 kg, the horizontal component of the velocity of the ball being 10 m s$^{-1}$. The boy returns the ball to the thrower with the same horizontal component of its velocity. Find the momentum acquired by the boy and his final speed.

8   A particle of mass $3m$ moving with velocity $(3\mathbf{i} - 2\mathbf{j})$ collides with and coalesces with a particle of mass $km$ moving with velocity $(2\mathbf{i} + 5\mathbf{j})$. The new particle moves in the direction of the vector $(\mathbf{i} + \mathbf{j})$. Find the value of $k$.

9   A boy of mass 40 kg sliding on ice with speed 15 m s$^{-1}$ overtakes and collides with a boy of mass 60 kg moving in the same direction at 5 m s$^{-1}$. If they move together after the collision, find their common speed.

10   A car of mass 1000 kg moving at 20 m s$^{-1}$ collides with a car of mass 500 kg moving at 30 m s$^{-1}$. Before impact the cars were travelling at right angles to one another. After impact they are locked together. Find their speed immediately after the collision.

## 9.5   Impact with a fixed plane

When a ball falls from rest through a height $h$ on to a level floor, it rebounds to a height less than $h$. Its speed $v$ when it leaves the floor must be less than its speed $u$ when it strikes the floor. While the ball is in contact with the floor it is first compressed and then regains its shape.

During compression the ball is brought to rest by an impulse of magnitude $mu$, where $m$ is its mass. During expansion the ball is given an impulse of magnitude $mv$. The ratio of $mv$ to $mu$ depends on the elasticity of the ball and the floor. If the ball is dropped from various heights and the height of the rebound is measured, it can be verified that for a particular ball and floor the ratio $v/u$ is constant. This experimental result is expressed by the equation

$$v = eu,$$

where the constant $e$ is called the coefficient of restitution.

Whenever a ball strikes a fixed plane at right angles this equation applies.

If no energy were lost on impact, the equation would become $v = u$ and $e = 1$. In practice, energy is lost in the form of heat and sound and in consequence $e$ is less than 1.

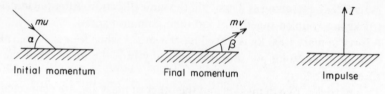

Fig. 9.6

Consider next the oblique impact of a ball, moving on a horizontal plane, with a smooth vertical wall. Let the initial velocity **u** of the ball make an angle $\alpha$ with the wall.

Let $m$ be the mass of the ball and let its velocity **v** after impact make an angle $\beta$ with the wall. Since the wall is smooth, the impulse **I** on the ball acts at right angles to the wall. The principle

$$\text{impulse} = \text{change in momentum}$$

at right angles to the wall can be applied.

$$\text{Change in momentum} = mv \sin \beta - (-mu \sin \alpha),$$
$$\text{impulse} = m(v \sin \beta + u \sin \alpha).$$

Parallel to the wall there is no impulse, and the component of the velocity of the ball parallel to the wall is unchanged. This gives

$$v \cos \beta = u \cos \alpha.$$

Let $e$ be the coefficient of restitution between the ball and the wall. Then

$$v \sin \beta = eu \sin \alpha.$$

From the last two equations it follows that

$$\tan \beta = e \tan \alpha.$$

The situation is summarised in Fig. 9.7, which shows the components of the velocity of the ball before and after impact.

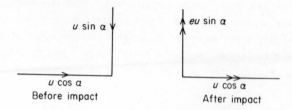

Fig. 9.7

*Example 1*
The coefficient of restitution between a golf ball and a concrete floor is found by experiment to be 0.92. The ball is dropped from rest at a point above the floor

and takes one second to reach the floor. Show that the ball comes to rest 24 seconds after being released.

The ball strikes the floor with speed $g$ m s$^{-1}$ and leaves the floor with speed $eg$ m s$^{-1}$. In the first bounce it reaches its highest point after $e$ seconds and falls to the floor in another $e$ seconds. If $t_1$ denotes the time for the first bounce,

$$t_1 = 2e \text{ s.}$$

In the second bounce, the ball leaves the floor with speed $e^2 g$ m s$^{-1}$ and reaches its highest point in $e^2$ seconds. If $t_2$ denotes the time for the second bounce,

$$t_2 = 2e^2 \text{ s.}$$

The time for each bounce is $e$ times that for the previous bounce. The total time before the ball comes to rest is the sum of the series

$$1 + 2e + 2e^2 + 2e^3 + \ldots + 2e^n + \ldots$$

Apart from the first term, this is a geometric series with common ratio $e$. Since $e = 0.92$, the series is convergent.

$$
\begin{aligned}
\text{Total time} &= 1 + 2e/(1 - e) \\
&= (1 + e)/(1 - e) \\
&= 1.92/0.08 \\
&= 24 \text{ s.}
\end{aligned}
$$

*Example 2*
A ball strikes a smooth horizontal floor when travelling at an angle $\alpha$ to the vertical, and rebounds in a direction making an angle $\beta$ to the vertical. After its second bounce it leaves the floor in a direction making an angle $\theta$ with the vertical. Express $\tan \theta$ in terms of $\tan \alpha$ and $\tan \beta$.
Given that $\alpha = 35°$ and $\beta = 50°$, find $\theta$ to the nearest degree.

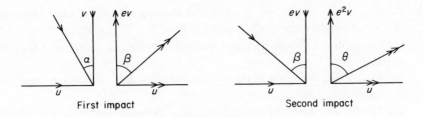

First impact  Second impact

Fig. 9.8

When the ball first strikes the floor let $v$ be the vertical component of its velocity and let $u$ be the horizontal component. This gives

$$u = v \tan \alpha.$$

Immediately after the first impact, the vertical component of the velocity will be $ev$, where $e$ is the coefficient of restitution. The horizontal component $u$ is unchanged. It follows that

$$u = ev \tan \beta.$$

Immediately after the second impact, the vertical component of the velocity will be $e^2v$, and hence

$$u = e^2v \tan \theta.$$

From the first and second equations

$$e \tan \beta = \tan \alpha.$$

From the second and third equations

$$e \tan \theta = \tan \beta.$$

Therefore

$$\tan \theta = \tan^2 \beta / \tan \alpha.$$

When $\alpha = 35°$ and $\beta = 50°$,

$$\tan \theta = \tan^2 50° / \tan 35° \approx 2.03$$
$$\theta \approx 64°.$$

## Exercises 9.5

1   A ball falls from rest at a height of 10 m above level ground. On its first bounce it rises to a height $h_1$ and on its second bounce it rises to a height $h_2$.
(a) Given that $h_1 = 2.5$ m, find the coefficient of restitution.
(b) Given that $h_2 = 1.6$ cm, find the coefficient of restitution.
(c) Given that the coefficient of restitution is 0.6, find $h_1$.
(d) Given that $h_2 = 8.1$ cm, find $h_1$.

2   A ball is dropped from rest at a height $h$ above a level floor. The coefficient of restitution between the ball and the floor is 0.4. (Take $g$ as $10$ m s$^{-2}$.)
(a) Given that $h = 5$ m, find the speed of the ball immediately after the first impact.
(b) Given that the speed of the ball immediately after the second impact is $0.32$ m s$^{-1}$, find $h$.
(c) Given that $h = 20$ m, find the time between the first and second impacts.
(d) Given that the time between the first and third impacts is $1.12$ s, find $h$.

3   A ball is released from rest and strikes a level floor after falling through 1.8 m. On its first bounce it rises to a height of 45 cm above the floor. Show that the total time taken before the ball comes to rest is 1.8 s approximately.

4 A ball travelling at $10 \text{ m s}^{-1}$ strikes a smooth wall at an angle of $45°$. The coefficient of restitution between the ball and the wall is 0.8. Show that the velocity of the ball immediately after impact makes an angle of $39°$ approximately with the wall and that the speed of the ball is then $\sqrt{82} \text{ m s}^{-1}$.

5 A ball falls from rest at a height of 6 m above a horizontal plane. On its first bounce it rises to a height of 1.5 m above the plane. Show that the ball travels a total distance of 10 m before coming to rest.

6 A ball strikes a smooth level floor when its velocity makes an angle of $60°$ with the floor. When it rebounds, its velocity makes an angle of $30°$ with the floor. Find the coefficient of restitution between the ball and the floor. Find also to the nearest degree the angle between the velocity of the ball and the floor immediately after the second bounce.

7 A ball strikes a smooth vertical wall when moving at an angle of $60°$ to the wall. The ball rebounds at an angle of $45°$ to the wall. Show that one half of the kinetic energy of the ball is lost on impact.

8 A small sphere of mass 0.6 kg strikes a smooth vertical wall when it is moving horizontally with velocity $10 \text{ m s}^{-1}$ at $30°$ to the wall. The coefficient of restitution between the sphere and the wall is $1/3$. Find the magnitude of the impulse given to the sphere by the wall.

9 A ball is thrown on to a smooth horizontal floor and bounces at the points A, B, C, D, .... Show that the lengths AB, BC, CD, ... are in geometric progression.

10 When a ball strikes a smooth vertical wall at right angles, its kinetic energy after impact is one-half of its initial kinetic energy. Find the coefficient of restitution between the ball and the wall. Find also the fraction of its kinetic energy that would be lost if the initial velocity of the ball were to make an angle of $45°$ with the wall.

## 9.6 Direct impact of two spheres

Let $m_1$ and $m_2$ be the masses of two spheres A and B, whose centres are moving in the same straight line with velocities $u_1$ and $u_2$ respectively. If the spheres collide, let the velocities of their centres after the collision be $v_1$ and $v_2$ respectively.

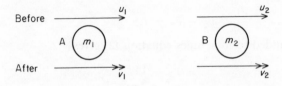

By the principle of conservation of linear momentum applied along the line of centres,

$$m_1 v_1 + m_2 v_2 = m_1 u_1 + m_2 u_2.$$

When $m_1$, $m_2$, $u_1$ and $u_2$ are known, a second equation is needed in order to find $v_1$ and $v_2$. This is provided by Newton's experimental law, which can be stated as follows:

The relative velocity of A to B after impact is $(-e)$ times the relative velocity of A to B before impact, where $e$ is a positive constant less than 1.

The value of $e$, the coefficient of restitution, depends on the elasticity of the two spheres. Newton's experimental law is expressed by the equation

$$v_2 - v_1 = -e(u_2 - u_1).$$

A diagram is essential in solving a problem on this topic. The symbols $u_1$, $u_2$, $v_1$, $v_2$ are used in the diagram on p. 315 to denote velocities in the positive direction, i.e. from left to right. If a sphere is moving to the left, the symbol representing its velocity will have a negative numerical value.

*Example 1*
A sphere A of mass 4 kg travelling at 5 m s$^{-1}$ collides directly with a sphere B of mass 3 kg travelling in the opposite direction at 4 m s$^{-1}$. The coefficient of restitution between the spheres is 1/6. Find their velocities after impact.

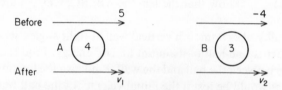

Since the momentum along the line of centres is unchanged,

$$4v_1 + 3v_2 = 4 \times 5 + 3 \times (-4) = 8. \qquad \ldots(1)$$

By Newton's experimental law,

$$v_2 - v_1 = -e(-4 - 5).$$

Since $e = 1/6$, this gives

$$v_2 - v_1 = 1.5. \qquad \ldots(2)$$

Equation (1) added to four times equation (2) gives

$$7v_2 = 14$$

i.e.
$$v_2 = 2 \text{ m s}^{-1}.$$

By substitution in equation (2), $v_1 = 0.5$ m s$^{-1}$.

After the collision, the spheres move in the same direction with velocities 0.5 m s$^{-1}$ and 2 m s$^{-1}$.

*Example 2*

A sphere A of mass $m$ collides directly with a sphere B of mass $2m$ at rest. Show that (a) if $e$, the coefficient of restitution, is 0.5, the sphere A is brought to rest (b) if $e > 0.5$, the velocity of A is reversed in direction by the impact.

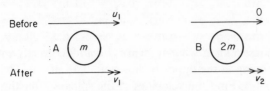

Since the momentum along the line of centres is unchanged,

$$mv_1 + 2mv_2 = mu_1,$$

i.e.
$$v_1 + 2v_2 = u_1. \quad \ldots(1)$$

By Newton's experimental law,

$$v_2 - v_1 = -e(0 - u_1) = eu_1. \quad \ldots(2)$$

By subtracting twice equation (2) from equation (1), $v_2$ can be eliminated and the equation

$$3v_1 = (1 - 2e)u_1 \quad \ldots(3)$$

obtained.

(a) If $e = 0.5$, $(1 - 2e) = 0$ and so $v_1 = 0$. This means that when $e = 0.5$, the sphere A is brought to rest.

(b) If $e > 0.5$, $(1 - 2e)$ is negative.
From equation (3), the sign of $v_1$ will be opposite to the sign of $u_1$. Therefore the direction of the velocity of the sphere A is reversed by the impact.

**Exercises 9.6**

1   A sphere of mass $m_1$ moving with velocity $u_1$ collides directly with a sphere of mass $m_2$ moving with velocity $u_2$ in the same direction. The coefficient of restitution is $e$. Find the velocities of the spheres after the collision.
(a) $m_1 = 2$ kg, $m_2 = 3$ kg, $u_1 = 12$ m s$^{-1}$, $u_2 = 6$ m s$^{-1}$, $e = 2/3$.
(b) $m_1 = 4$ kg, $m_2 = 5$ kg, $u_1 = 25$ m s$^{-1}$, $u_2 = 10$ m s$^{-1}$, $e = 0.2$.
(c) $m_1 = 4$ kg, $m_2 = 1$ kg, $u_1 = 8$ m s$^{-1}$, $u_2 = 4$ m s$^{-1}$, $e = 0.25$.
(d) $m_1 = 3$ kg, $m_2 = 3$ kg, $u_1 = 6$ m s$^{-1}$, $u_2 = 2$ m s$^{-1}$, $e = 0.5$.

2   A sphere of mass 2 kg moving with velocity $u_1$ collides directly with a sphere of mass 3 kg moving with velocity $u_2$ in the same line. After the impact the velocities of the spheres are $v_1$ and $v_2$ respectively. The coefficient of restitution is $e$.
(a) $u_1 = 6$ m s$^{-1}$, $u_2 = 2$ m s$^{-1}$, $v_1 = 3$ m s$^{-1}$. Find $v_2$ and $e$.
(b) $u_2 = 6$ m s$^{-1}$, $v_1 = 9$ m s$^{-1}$, $e = 1/3$. Find $u_1$ and $v_2$.
(c) $u_1 = 12$ m s$^{-1}$, $u_2 = 3$ m s$^{-1}$, $v_2 = 7$ m s$^{-1}$. Find $v_1$ and $e$.
(d) $u_1 = 9$ m s$^{-1}$, $v_2 = 8$ m s$^{-1}$, $e = 2/3$. Find $u_2$ and $v_1$.

3   Two spheres A and B collide directly. Their velocities are $u_1$, $u_2$ before impact and $v_1$, $v_2$ after impact respectively. Find the coefficient of restitution and the ratio of the mass of sphere A to the mass of sphere B.

(a) $u_1 = \phantom{0}4 \text{ m s}^{-1}, u_2 = -4 \text{ m s}^{-1}, v_1 = \phantom{0}3 \text{ m s}^{-1}, v_2 = 4 \text{ m s}^{-1}$.

(b) $u_1 = 10 \text{ m s}^{-1}, u_2 = \phantom{-}2 \text{ m s}^{-1}, v_1 = \phantom{0}1 \text{ m s}^{-1}, v_2 = 5 \text{ m s}^{-1}$.

(c) $u_1 = \phantom{0}6 \text{ m s}^{-1}, u_2 = -2 \text{ m s}^{-1}, v_1 = \phantom{0}1 \text{ m s}^{-1}, v_2 = 3 \text{ m s}^{-1}$.

(d) $u_1 = \phantom{0}3 \text{ m s}^{-1}, u_2 = -6 \text{ m s}^{-1}, v_1 = -1 \text{ m s}^{-1}, v_2 = 2 \text{ m s}^{-1}$.

4   A sphere of mass 5 kg moving at 36 m s$^{-1}$ collides directly with a sphere of mass 4 kg moving in the opposite direction at 18 m s$^{-1}$. The coefficient of restitution is 1/3. Find the velocities of the spheres after the collision.

5   A sphere of mass 6 kg collides directly with a sphere of mass 4 kg travelling with the same speed in the opposite direction. One sphere is brought to rest by the impact. Find the coefficient of restitution.

6   Two spheres of equal mass collide directly when travelling in the same direction, one with velocity **u** and one with velocity 2**u**. After the impact, the velocity of one sphere is 4**u**/3. Find the coefficient of restitution.

7   A sphere of mass 2 kg and a sphere of mass 4 kg are travelling in the same direction in the same line. The velocity of the lighter sphere is 6 m s$^{-1}$ and the velocity of the heavier sphere is 3 m s$^{-1}$. The spheres collide directly. Given that the coefficient of restitution is 1/2, find the velocities of the spheres after impact. Show that one-twelfth of the initial kinetic energy of the spheres is lost in the collision.

8   A sphere of mass $m$ collides directly with a sphere of mass $4m$, their speeds being $3u$ and $u$ respectively in the same direction. One of the spheres is brought to rest by the collision. Find the coefficient of restitution.

9   A sphere of mass 2 kg travelling at 60 m s$^{-1}$ collides directly with a sphere of mass 10 kg which is at rest. Show that if the coefficient of restitution is greater that 0.2 the spheres will move in opposite directions after the collision.

Given that the coefficient of restitution is 0.4, find the ratio of the final kinetic energy to the initial kinetic energy.

10  A sphere of mass 4 kg moving at 8 m s$^{-1}$ collides directly with a sphere of mass 2 kg which is at rest. Their relative velocity after impact is 2 m s$^{-1}$. Find the loss of kinetic energy due to the impact.

## Miscellaneous exercises 9

1   A fire engine picks up water from a lake and delivers it at the same level through a circular nozzle of diameter 8 cm at a speed of 30 m s$^{-1}$. The jet strikes a vertical wall at right angles and the water does not rebound from the wall. Show that the force exerted on the wall is approximately 4524 N and that the effective rate of working of the engine is nearly 68 kW.

2 Two smooth vertical walls stand on a smooth horizontal floor and intersect at an acute angle $\theta$. A particle on the floor is projected horizontally at right angles to one wall and away from it. After one impact with each wall the particle is moving parallel to the wall it struck first. The coefficient of restitution between the particle and each wall is 1/2. Show that $\tan \theta = 1/(2\sqrt{2})$.

3 A straight groove PQ on a smooth horizontal table is 1.2 m long, and a small sphere A rests in the groove 0.4 m from P. A second sphere B is projected from Q with speed $6 \text{ m s}^{-1}$, and collides with sphere A. Given that the coefficient of restitution is 3/4 and that the spheres reach P and Q simultaneously, find the ratio of the mass of sphere A to the mass of sphere B.

4 Two smooth spheres, of equal size but of masses $2m$ and $m$, move directly towards one another on a smooth horizontal plane with equal speeds. The impact brings the heavier sphere to rest. Show that two-thirds of the original kinetic energy is destroyed.

5 Two spheres of masses 2 kg and 3 kg are moving directly towards one another with speeds $50 \text{ m s}^{-1}$ and $100 \text{ m s}^{-1}$ respectively. The impact gives the lighter sphere an impulse of magnitude 216 N s. Find the coefficient of restitution.

6 Three particles P, Q, R each of mass 0.4 kg lie in a straight line in that order on a smooth horizontal table. P and Q are connected by a light inextensible string; Q and R are connected by a light inextensible string. Initially both strings are slack. P is given a velocity of $30 \text{ m s}^{-1}$ so that it moves directly away from Q. Find the magnitude of the impulse given to P at the instant when R is jerked into motion.

7 A ball of mass 0.4 kg falls vertically from rest through a distance of 5 m and strikes a horizontal floor. It then rises through a distance of 1.25 m. Find the magnitude of the impulse given to the ball when it strikes the floor for the second time. (Take $g = 10 \text{ m s}^{-2}$.)

8 A ball falling vertically strikes a smooth plane inclined at $30°$ to the horizontal. It then rebounds horizontally. Find the coefficient of restitution between the ball and the plane.

9 A particle is projected horizontally and after time $T$ it strikes a horizontal plane. It rebounds and strikes the plane again after a further time $\frac{1}{2}T$. Find the coefficient of restitution between the particle and the plane.

10 A particle of mass $2m$, moving with speed $3u$, overtakes a particle of mass $m$ moving in the same direction with speed $u$. Given that the coefficient of restitution is 1/2, show that the momentum of each particle is changed by an amount $2mu$, and that the loss of kinetic energy at impact is $mu^2$.

11 A sphere moving on a smooth horizontal table strikes in turn two smooth vertical planes which are at right angles. The coefficient of restitution is the same for each impact. Show that its final velocity is in the opposite direction to its initial velocity.

12 Two particles of equal mass, connected by a light inextensible string of length $a$, rest side by side on a rough horizontal table. The coefficient of friction between each particle and the table is 1/4. One particle is given a horizontal velocity of magnitude $\sqrt{(ga)}$. Show that it will travel a distance $5a/4$ before coming to rest. Show also that the kinetic energy lost by friction is three times that lost owing to the jerk in the string.

13 A particle of mass $m$ is moving horizontally with speed $u$. It is given a blow such that the direction of its velocity is turned through a right angle and its kinetic energy is multiplied by 3. Find the magnitude of the blow.

14 A particle which is moving in the direction of the vector $3\mathbf{i} + 4\mathbf{j}$ is given a blow. The speed of the particle is doubled and it now moves in the direction of the vector $4\mathbf{i} + 3\mathbf{j}$. Show that the blow is in the direction of the vector $5\mathbf{i} + 2\mathbf{j}$.

15 A small target of mass $M$ is suspended by a light string of length $a$. A bullet of mass $m$ moving horizontally strikes and becomes embedded in the target, which first comes to rest when the string makes $60°$ with the downward vertical. Find the initial speed of the bullet.

16 A pump raises water from a depth of 10 m and discharges it horizontally through a pipe of 0.1 m diameter at a speed of $8\ \mathrm{m\,s^{-1}}$. Show that the rate of working of the pump is 8.2 kW approximately. The jet of water impinges normally on a vertical wall and does not rebound. Show that the force exerted on the wall is approximately 500 N.

17 A horizontal jet of water of cross-section $5\ \mathrm{cm^2}$ impinges normally on a vertical wall. The jet exerts a force of 50 N on the wall. Assuming that the water does not rebound from the wall, find the speed of the jet.

18 Every minute a machine-gun fires 900 bullets each of mass 0.015 kg. The initial velocity of the bullets is $800\ \mathrm{m\,s^{-1}}$ at $20°$ to the horizontal. Show that the horizontal force exerted on the gun is nearly 170 N.

19 Two small spheres of masses $m$ and $2m$ are suspended from a fixed point by two light inextensible strings of equal length. The spheres are raised until they are at their greatest distance apart with the strings horizontal, and are then released from rest. The coefficient of restitution between the spheres is 1/2. Show that the first impact brings the heavier sphere to rest, and that the second impact brings the lighter sphere to rest.

20 Two particles A and B each of mass $m$ are at rest on a smooth horizontal table at a distance $a$ apart. They are connected by a light inextensible string of length $2a$. The particle A is given a velocity $u$ along the table in a direction perpendicular to AB. Find the magnitude of the impulse given to A when the string becomes taut (a) if B is fixed (b) if B is free to move.

21 A pump working at the effective rate of 2.5 kW raises 10 kg of water per second from a depth of 20 m and delivers it through a horizontal pipe. Find the force which the jet will exert on an obstacle placed at right angles to its path. (Take $g = 10\ \mathrm{m\,s^{-2}}$.)

22 A particle is moving with uniform velocity on a smooth horizontal table

when a light inelastic string joining it to a particle of equal mass at rest on the table becomes taut. Immediately afterwards the two particles are moving in directions inclined at 45° to one another. Show that two-fifths of the original kinetic energy has been lost.

23  A sphere of mass $m$ moving with speed $2u$ on a smooth horizontal plane collides directly with a second sphere of the same radius but of mass $3m$ moving in the same direction with speed $u$.
(a) Find the speeds of the two spheres after the impact if they are perfectly elastic.
(b) Find the coefficient of restitution between the spheres if after the impact the speed of the second sphere is twice that of the first.  [L]

24  Two spheres moving in opposite directions along the same path collide directly. Their masses are $2m$ and $3m$, and their speeds immediately before collision are $7u$ and $3u$ respectively. Show that after impact the speed of the heavier sphere lies between $u$ and $5u$.
   Find the ratio of their kinetic energies after impact if nine-tenths of the total kinetic energy is lost in the collision.  [L]

25  A solid wooden cube of mass $M$ and edge $a$ at rest on a smooth horizontal table is struck by a bullet of mass $M/5$ moving with velocity $u$. The bullet enters the cube at the centre of one vertical face and leaves it with velocity $5u/6$ at the centre of the opposite face. Find the final velocity of the cube.
   If it can be assumed that the force exerted by the bullet on the cube is constant, draw a velocity–time graph for the cube and for the bullet. Find the time the bullet takes to pass through the cube and the distance the cube moves in this time.  [L]

26  A sphere of mass $2m$ moving with speed $u$ on a smooth horizontal plane strikes directly a stationary sphere of the same radius but of mass $3m$. The second sphere then strikes a vertical wall at right angles and rebounds to meet the first sphere directly again. If the coefficient of restitution between the two spheres is $1/2$ and that between the second sphere and the wall is $2/3$, find the final speeds of the spheres.  [L]

27  A wooden cube of mass $M$ is free to slide on a smooth horizontal table. A bullet of mass $m$ travelling horizontally strikes the cube and becomes embedded in it. Assuming that the cube does not rotate, express the loss of energy as a fraction of the original kinetic energy of the bullet.

28  A water pump raises 50 kg of water a second through a height of 20 m. The water emerges as a jet with speed 50 m s$^{-1}$. Find the kinetic energy and the potential energy given to the water each second and hence find the effective power developed by the pump.
   Given that the jet is directed at 30° above the horizontal, find the further height attained by the water.
   The jet of water impinges at its highest point directly against a vertical wall. Show that the force exerted on the wall is at least $1250\sqrt{3}$ N.
(Take $g = 10$ m s$^{-2}$.)  [L]

**29** A particle A of mass $m$, moving with speed $2u$, collides directly with a particle B of mass $M$ moving in the same direction with speed $u$. After the collision, A continues to move in the same direction but with speed $u$. Show that the coefficient of restitution between the particles is $m/M$ and deduce that after the collision the kinetic energy of A is less than that of B. If the ratio of these energies is 1:6, show further that $m/M = 2 - \sqrt{3}$. [JMB]

**30** A particle of mass $m$ strikes a fixed smooth plane while moving at speed $u$ in a direction making an angle $\theta$ ($\neq 0$) with the normal to the plane. It rebounds at an angle $\phi$ to the normal. Find the coefficient of restitution between the particle and the plane.

Show that the impulse on the plane is $mu \sin(\theta + \phi)/\sin \phi$. [JMB]

**31** A particle of mass $m$ moving in a horizontal line with speed $v$ is subjected to a horizontal impulse. The impulse turns the direction of motion of the particle through $30°$ and reduces its speed to $v/\sqrt{3}$. Find the magnitude of the impulse.

**32** A particle moving along a smooth horizontal floor hits a smooth vertical wall and rebounds in a direction at right angles to its initial direction of motion. The coefficient of restitution is $e$. Find, in terms of $e$, the tangent of the angle between the initial direction of motion and the wall.

Prove that the kinetic energy after the rebound is $e$ times the initial kinetic energy. [JMB]

**33** A smooth circular horizontal table has a smooth vertical rim around its edge. A particle is projected horizontally with velocity $u$ across the table from a point A on the edge. The direction of projection makes an angle $\theta$ with the diameter AB and the coefficient of restitution between the particle and the rim is $e$. After $n$ impacts with the rim the particle then strikes the rim at B.

(i) If $n = 1$, show that $\tan^2 \theta = e$, and find in terms of $u$ and $e$ the speed with which the particle reaches B.

(ii) If $n = 2$, show that $\tan^2 \theta = e + e^2 + e^3$. [AEB]

**34** Two particles A and B, each of mass $m$, moving on a smooth horizontal table in opposite directions with speeds of $6u$ and $3u$ respectively, collide directly. As a result of the collision the velocity of B is reversed in direction but remains unchanged in magnitude. Find

(i) the coefficient of restitution between A and B

(ii) the magnitude and direction of the impulse received by A.

The particle B continues to move at speed $3u$ until it overtakes, strikes and coalesces with a particle C, of mass $2m$, which is moving along the same line. Given that the total kinetic energy of B and C before they coalesce exceeds the kinetic energy of the combined particle by $\frac{3}{4}mu^2$, find

(i) the initial speed of C

(ii) the speed of the combined particle. [AEB]

**35** Three particles A, B, C with masses $10m$, $5m$, $3m$ respectively lie at rest in a straight line on a smooth horizontal table. The particle A is projected horizontally with speed $u$ and undergoes a direct perfectly elastic collision with B. Show that B acquires speed $4u/3$.

Subsequently B is in direct, perfectly-elastic collision with C. Show that immediately after this collision the velocities of A and B are equal.

[JMB]

36  Two small beads, A and B, of masses $2m$ and $3m$ respectively are threaded on a fixed smooth horizontal straight wire, which has a metal stop S firmly fixed at one end. The bead A is made to move along the wire and it strikes B which is initially at rest. As a result of this collision B moves with speed $u$ and hits S from which it rebounds with speed $2u/3$. If the coefficient of restitution between A and B is $1/5$, show that the speed of A is reduced by $3u/2$ but that it continues to move in the same direction after the first impact with B. Find
(i) the magnitude of the impulse absorbed by the stop S
(ii) the final speeds of A and B, in terms of $u$, after their second collision.

[AEB]

37  Three small spheres A, B and C, of equal radii and masses $2m$, $3m$ and $4m$ respectively, lie at rest on a frictionless horizontal plane with their centres in a straight line. The sphere A is made to move towards B with speed $2u$ and after the collision B begins to move towards C with speed $u$. The sphere B is brought to rest by its collision with C. Calculate the coefficients of restitution between A and B and between B and C.

Show that, in all, three collisions only take place and find the final speeds of each sphere in terms of $u$.

[AEB]

38  A particle A of mass $m$ moving in a straight line on a horizontal plane has its initial velocity of $7u$ reduced to $3u$ in time $t$ while moving through a distance $2d$. Find the retarding force, assumed constant, acting on the particle and also the value of $t$.

At the instant when A is moving with speed $3u$ it collides directly with a particle B of mass $2m$ which is moving with speed $u$ in the same line and in the same direction. The speed of A immediately after the collision is $7u/5$ in the same direction as before. Find
(i) the coefficient of restitution between A and B
(ii) the magnitude of the impulse received by A
(iii) the kinetic energy lost in this collision.

[AEB]

39  Three small spheres A, B and C, of equal radii and masses $m$, $2m$ and $3m$ respectively, are placed with their centres in a straight line on a smooth horizontal table with B between A and C. The sphere A is given a velocity $5u$ in the direction $\overrightarrow{AB}$ and, as a result of the ensuing collision between A and B, B moves towards C with speed $3u$. Calculate
(i) the magnitude and the sense of the velocity of A after the collision
(ii) the coefficient of restitution between A and B
(iii) the loss of kinetic energy due to the collision between A and B.

The sphere B strikes C and as a result C receives an impulse of $4mu$. Calculate

(iv) the velocities of B and C after their collision

(v) the coefficient of restitution between B and C. [AEB]

40   A body of mass 100 kg falls freely from rest a distance of 5 metres on to a pile of mass 1000 kg. The body does not rebound and the pile is driven into the ground a distance of 0.2 metres. Calculate (i) the common speed of the body and the pile immediately after the blow (ii) the resistance of the ground, supposed uniform.

On another occasion the body of mass 100 kg falls freely from rest through a distance of 5 metres on to fixed horizontal rock and rebounds to a height of 0.8 metres. Find the coefficient of restitution between the body and the rock and the loss in kinetic energy.

(Take the acceleration due to gravity to be $10 \text{ m s}^{-2}$.) [AEB]

# 10 Motion in a circle and simple harmonic motion

## 10.1 Uniform motion in a circle

Let the point P be moving in a circle of radius $r$ with centre at the origin O. At time $t$ let the radius OP make an angle $\theta$ with the $x$-axis. Then the angular velocity $\omega$ of P about O is given by

$$\omega = \frac{d\theta}{dt}.$$

When $\omega$ is constant, the angular velocity of P about O is said to be uniform.

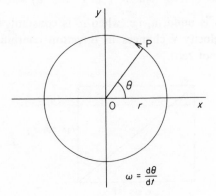

Fig. 10.1

Let the vector $\overrightarrow{OP}$ be denoted by $\mathbf{r}$, and let $\mathbf{i}$ and $\mathbf{j}$ be unit vectors in the directions of the positive $x$-axis and $y$-axis respectively. The vector $\mathbf{r}$ can be resolved into a component vector $r \cos \theta \, \mathbf{i}$ along Ox and a component vector $r \sin \theta \, \mathbf{j}$ along Oy, i.e.

$$\mathbf{r} = (r \cos \theta)\mathbf{i} + (r \sin \theta)\mathbf{j} = r\begin{pmatrix} \cos \theta \\ \sin \theta \end{pmatrix}.$$

### Velocity of P

The velocity $\mathbf{v}$ of P is found by differentiating $\mathbf{r}$ with respect to $t$.

$$\mathbf{v} = \frac{d\mathbf{r}}{dt} = \dot{\mathbf{r}},$$

where the dot signifies differentiation with respect to $t$.
Since $r$ is constant,

$$\frac{d}{dt}(r \cos \theta) = -r \sin \theta \frac{d\theta}{dt} = -\omega r \sin \theta,$$

$$\frac{d}{dt}(r \sin \theta) = r \cos \theta \frac{d\theta}{dt} = \omega r \cos \theta.$$

Therefore the velocity $\mathbf{v}$ is given by

$$\mathbf{v} = \frac{d}{dt}(r \cos \theta \, \mathbf{i} + r \sin \theta \, \mathbf{j})$$

$$= -(\omega r \sin \theta)\mathbf{i} + (\omega r \cos \theta)\mathbf{j} = \omega r \begin{pmatrix} -\sin \theta \\ \cos \theta \end{pmatrix}.$$

The speed $v$ of P equals the modulus of the velocity $\mathbf{v}$.

$$v = \sqrt{(\omega^2 r^2 \sin^2 \theta + \omega^2 r^2 \cos^2 \theta)} = \omega r.$$

When the motion is uniform, i.e. when $\omega$ is constant, the speed $v$ of P is constant, but its velocity $\mathbf{v}$ changes in direction continually. Therefore the acceleration of P is not zero.

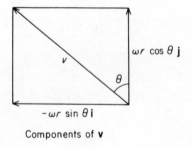

$\omega r \cos \theta \, \mathbf{j}$

$v$

$\theta$

$-\omega r \sin \theta \, \mathbf{i}$

Components of $\mathbf{v}$

Fig. 10.2

## Acceleration of P

The acceleration of P is found by differentiating the velocity $\mathbf{v}$ with respect to the time $t$. When $\omega$ is constant, this gives

$$\dot{\mathbf{v}} = \ddot{\mathbf{r}} = -\omega^2 r \cos \theta \, \mathbf{i} - \omega^2 r \sin \theta \, \mathbf{j}$$

$$\dot{\mathbf{v}} = -\omega^2 r \begin{pmatrix} \cos \theta \\ \sin \theta \end{pmatrix} = -\omega^2 \mathbf{r}.$$

Now the vector $-\mathbf{r}$ is in the opposite direction to the vector $\mathbf{r}$. Therefore the acceleration of P is directed along the radius towards the centre O. The magnitude of this acceleration is $\omega^2 r$, which equals $v^2/r$ since $v = \omega r$.

Consider a particle of mass $m$ which is moving with constant speed $v$ in a circular path of radius $r$. The particle has an acceleration of magnitude $v^2/r$ towards the centre of the circle. By the principle

$$\text{force} = \text{mass} \times \text{acceleration},$$

the force on the particle must act along the radius towards the centre of the circle and it must be of magnitude $mv^2/r$.

*Example 1*
A particle P of mass 2 kg is attached to a fixed point O on a smooth horizontal table by a string of length 5 m. The particle is moving in a circle on the table with constant speed $4\ \text{m s}^{-1}$. Find the tension in the string.

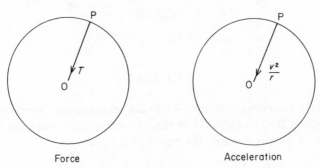

Force          Acceleration

Fig. 10.3

Let $T$ be the magnitude of the force acting towards O giving the particle its acceleration, i.e. $T$ is the tension in the string.

$$T = mv^2/r = 2 \times 16/5 = 6.4\ \text{N}.$$

*Example 2*
A circular disc, which is horizontal, is rotating about its centre O with constant angular velocity. A particle P of mass 0.8 kg is attached to the disc at a distance 0.2 m from the centre. If the disc turns through 15 radians each minute, find the magnitude $F$ of the horizontal force exerted by the disc on the particle.

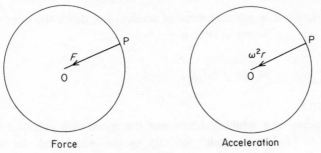

Force          Acceleration

Fig. 10.4

Angular velocity $= 15/60 = 0.25$ rad s$^{-1}$.

Acceleration towards O $= \omega^2 r = (0.25)^2 \times (0.2)$ m s$^{-2}$.

Force required $= m\omega^2 r = (0.8)(0.25)^2(0.2) = 0.01$ N.

*Example 3*
A car of mass $m$ is travelling with constant speed $v$ in a circle of radius $r$ on a smooth banked racing track. Assuming that the car can be treated as a particle, find $\alpha$, the angle of inclination of the track.

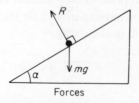
Forces

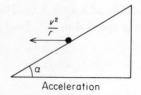
Acceleration

Fig. 10.5

Since the track is smooth, only two forces act on the car, its weight $m\mathbf{g}$ and the normal reaction $\mathbf{R}$. The resultant of these two forces produces the horizontal acceleration of the car towards the centre of the circle.

*Method 1*
Apply Newton's second law (a) vertically (b) horizontally.
Resolving vertically

$$R \cos \alpha - mg = 0,$$

since there is no vertical acceleration.
Resolving horizontally

$$R \sin \alpha = mv^2/r.$$

From these two equations, $\tan \alpha = v^2/(gr)$.
*Method 2*
Apply Newton's second law down the slope.
Resolving down the slope

$$mg \sin \alpha = m(\text{component of acceleration down the slope})$$
$$= m(v^2/r) \cos \alpha.$$

Hence         $\tan \alpha = v^2/(gr)$
$$\alpha = \tan^{-1} (v^2/gr).$$

**Exercises 10.1**

1   The radius of a wheel is 1.5 m and the speed of a point on the rim is 6 m s$^{-1}$. Find the angular velocity of the wheel and the number of revolutions it makes per second.

**2** An aircraft travelling at $200 \text{ m s}^{-1}$ is moving in a horizontal circle of radius 2 km. Show that the magnitude of its acceleration is approximately $2g$.

**3** Show that to one significant figure the angular velocity of the earth about its axis is $7 \times 10^{-5} \text{ rad s}^{-1}$.

Assuming that the earth is a sphere of radius $6.4 \times 10^6$ m, show that the percentage fall in the weight of a body when it is moved from the north pole to the equator is between 0.3 and 0.4.

**4** A circular disc revolving steadily in a horizontal plane about its centre O makes one complete revolution each second. A particle is placed on the disc at a distance of 20 cm from O. Given that the coefficient of friction between the particle and the disc is 0.5, show that the particle will slip.

**5** A car is travelling at $12 \text{ m s}^{-1}$ on horizontal ground in a circle of radius 30 m. The coefficient of friction between the tyres and the ground is 0.5. Assuming that the car may be treated as a particle, show that it will not slip.

**6** A smooth metal cone is fixed with its axis vertical and vertex downwards. A particle moves on the inner surface of the cone in a horizontal plane at a height $h$ above the vertex. Find the speed of the particle in terms of $h$ and $g$.

**7** A particle is moving with constant speed $v$ in a horizontal circle on the smooth inner surface of a sphere of radius $a$. The force exerted by the sphere on the particle equals twice the weight of the particle. Show that $v^2 = 3ag/2$.

**8** A car is moving round a circular track which is banked at an angle $\tan^{-1} 0.2$ to the horizontal. The car is travelling in a circle of radius 800 m at a speed greater than $40 \text{ m s}^{-1}$. Assuming that the car may be treated as a particle, show that it will tend to slip outwards.

**9** A smooth wire bent in the shape of a circle of radius $a$ is rotating with angular velocity $\sqrt{(2g/a)}$ about a vertical diameter. Show that a bead threaded on the wire can remain at rest relative to the wire at a point which is at a vertical distance $a/2$ above the lowest point of the wire.

**10** Two particles of mass $m$ and $2m$ are connected by a light inelastic string of length $6a$. The particles are moving on a smooth horizontal plane with the string taut. The path of each particle is a circle and the speed of the heavier particle is $v$. Find the tension in the string.

## 10.2 Non-uniform motion in a circle

Let the point P move in a circle of radius $r$ and centre O with variable speed. Then the angular velocity of P about O is not constant. Let the position vector $\mathbf{r}$ of P relative to O be given by

$$\mathbf{r} = r \cos \theta \, \mathbf{i} + r \sin \theta \, \mathbf{j}.$$

As $\dfrac{\mathrm{d}}{\mathrm{d}t} \cos \theta = (-\sin \theta)\dot{\theta}$ and $\dfrac{\mathrm{d}}{\mathrm{d}t} \sin \theta = (\cos \theta)\dot{\theta}$, the velocity $\mathbf{v}$ of P is given by

$$\mathbf{v} = \dot{\mathbf{r}} = (-r \sin \theta)\dot{\theta} \, \mathbf{i} + (r \cos \theta)\dot{\theta} \, \mathbf{j}.$$

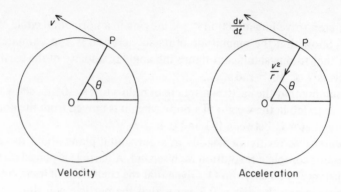

Velocity               Acceleration

Fig. 10.6

The acceleration of P is found by differentiating $\mathbf{v}$ with respect to $t$.

$$\mathbf{v} = \ddot{\mathbf{r}} = (-r\dot\theta^2 \cos\theta - r\ddot\theta \sin\theta)\mathbf{i} + (-r\dot\theta^2 \sin\theta + r\ddot\theta \cos\theta)\mathbf{j}$$
$$= -r\dot\theta^2(\cos\theta\, \mathbf{i} + \sin\theta\, \mathbf{j}) + r\ddot\theta(-\sin\theta\, \mathbf{i} + \cos\theta\, \mathbf{j}),$$

where $\ddot\theta$ denotes $\dfrac{d\dot\theta}{dt}$.

Thus the acceleration of P is the sum of two vectors which are at right angles to one another.

The vector $-r\dot\theta^2(\cos\theta\, \mathbf{i} + \sin\theta\, \mathbf{j})$ equals $-\omega^2\mathbf{r}$. Its magnitude is $\omega^2 r$ and it is directed along the radius towards O. The vector $r\ddot\theta\,(-\sin\theta\, \mathbf{i} + \cos\theta\, \mathbf{j})$ is directed along the tangent to the circle at P. Its magnitude $r\ddot\theta$ equals $\dfrac{d}{dt}(r\dot\theta)$, i.e. $\dfrac{dv}{dt}$. These vectors are shown in Fig. 10.6.

*Example*
A point moves in a circle of radius 24 m. At time $t$ seconds its speed $v$ equals $(t^2 + 4t)\ \mathrm{m\,s^{-1}}$. Find its acceleration when $t = 2$.

When $t = 2$, $v = 12$ and $v^2/r = 144/24 = 6$.
Therefore the component of the acceleration towards the centre is $6\ \mathrm{m\,s^{-2}}$.

Also when $t = 2$, $\dfrac{dv}{dt} = \dfrac{d}{dt}(t^2 + 4t) = 2t + 4 = 8$.

Therefore the component of the acceleration along the tangent is $8\ \mathrm{m\,s^{-2}}$.

The magnitude of the resultant acceleration is $\sqrt{(8^2 + 6^2)}$, i.e. $10\ \mathrm{m\,s^{-2}}$. Its direction makes an angle $\alpha$ with the radius where

$$\tan\alpha = 8/6 = 4/3.$$

**330** Advanced Mathematics 2

**Exercises 10.2**

1 A point is moving in a circle of radius 4 m, its speed at time $t$ seconds being $t^2$ $\text{ms}^{-1}$. Find the magnitude of its acceleration at the instant when the acceleration is equally inclined to the radius and the tangent.

2 A point moves from rest round a circle in such a way that its speed increases at the constant rate 5 $\text{ms}^{-2}$. Find the component of its acceleration along the radius at the instant when it has travelled half-way round the circle.

3 A point P moves on a semicircular arc AB with centre O in such a way that at time $t$ the radius OP makes an angle $\pi(1 - \cos t)$ with the radius OA. Find the times at which the resultant acceleration of the point P is
   (a) directed along the radius OP towards O
   (b) directed along the tangent to the circle at P.

4 A point P describes a circle of radius $a$, its speed at time $t$ being $2\pi a/(T + t)$, where $T$ is a constant. Show that at the instant when the speed is $\pi a/T$ the distance travelled is $2\pi a \ln 2$.
   Find the speed of P when it has moved once round the circle.

## 10.3 The conical pendulum

A particle P of mass $m$ is attached to a fixed point O by a light inextensible string of length $l$. When the particle describes a horizontal circle with the string taut, the string sweeps out a cone with vertex O. The particle and the string are then said to form a conical pendulum.

Let C be the centre of the circle, so that OC is vertical and CP is horizontal. Let $v$ be the speed of the particle and let the angle COP equal $\theta$. (Fig. 10.7)

There are two forces acting on the particle, its own weight and the tension **T** in the string. Apply the principle

$$\text{force} = \text{mass} \times \text{acceleration}$$

(a) horizontally, (b) vertically.

(a) Radius of circle $= \text{CP} = l \sin \theta$.
Acceleration towards C $= v^2/(l \sin \theta)$.
Horizontal component of **T** $= T \sin \theta$.
Therefore

$$T \sin \theta = mv^2/(l \sin \theta).$$

(b) Since there is no acceleration vertically,

$$T \cos \theta = mg.$$

By the elimination of $T$,

$$\tan \theta = v^2/(gl \sin \theta)$$
$$v^2 = gl \sin \theta \tan \theta$$
$$= gl \sin^2 \theta/\cos \theta.$$

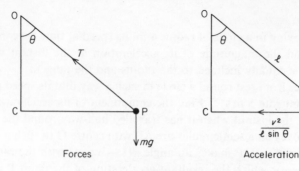

Forces                    Acceleration

Fig. 10.7

Since the particle is moving with speed $v$ in a circle of radius $l \sin \theta$, its angular velocity $\omega$ about C, the centre of the circle, is given by

$$\omega = v/(l \sin \theta).$$

The time taken to travel once round the circle is $2\pi/\omega$. This is the periodic time of the conical pendulum. If $n$ complete revolutions are made in unit time, $\omega = 2\pi n$.

Since the radius of the circle is $l \sin \theta$, the acceleration of the particle towards C is $\omega^2 l \sin \theta$. By $F = ma$,

$$T \sin \theta = m\omega^2 l \sin \theta.$$

This gives

$$T = m\omega^2 l.$$

Now $T \cos \theta = mg$. Therefore

$$m\omega^2 l \cos \theta = mg$$
$$\omega^2 = g/(l \cos \theta) = g/\text{OC}.$$

If the particle is not hanging at rest, OC is less than $l$. Therefore

$$\omega^2 > g/l.$$

This shows that for a conical pendulum of length $l$ the angular velocity must be greater than $\sqrt{(g/l)}$.

*Example 1*
In a conical pendulum, the tension in the string is equal to twice the weight of the bob. Given that the bob makes one revolution per second, find the length of the string.
With the notation of Fig. 10.7,

$$T \cos \theta = mg.$$

But $T = 2mg$, so that $\cos \theta = 1/2$ and $\theta = 60°$.

Resolving horizontally

$$T \sin \theta = mv^2/(l \sin \theta)$$
$$mv^2 = 2mgl \sin^2 \theta$$
$$v^2 = 2gl \sin^2 \theta = 3gl/2.$$

Each second the bob travels a distance $2\pi l \sin \theta$, i.e. $v = \pi l\sqrt{3}$.

$$v^2 = 3\pi^2 l^2.$$

But $v^2$ equals $3gl/2$.

Hence
$$l = g/(2\pi^2).$$

*Example 2*
A particle P of mass 0.8 kg is attached to the points Q and R by two strings of lengths 3 m and 5 m respectively. Q is vertically beneath R and QR equals 4 m. The particle describes a horizontal circle with speed 6 m s$^{-1}$ with both strings taut. Find the tension in each string. (Take $g = 10$ m s$^{-2}$.)

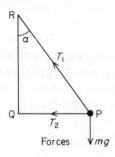
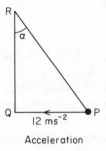

Fig. 10.8

Since the sides of the triangle PQR are in the ratio 3:4:5, the triangle is right-angled and PQ is horizontal. Let the angle QRP $= \alpha$ and let $T_1$ and $T_2$ be the tensions in PR and PQ respectively (Fig. 10.8).
Since there is no vertical acceleration,

$$T_1 \cos \alpha - mg = 0$$
$$T_1 = mg \sec \alpha = (0.8 \times 10)(5/4) = 10 \text{ N}.$$

Resolving in the direction PQ

$$T_1 \sin \alpha + T_2 = mv^2/r,$$

where $v$ is the speed of P and $r$ is the radius of the circle.
Since $v = 6$ and $r = 3$, $v^2/r = 12$.

$$10 \sin \alpha + T_2 = 0.8 \times 12 = 9.6$$
$$T_2 = 9.6 - 6 = 3.6 \text{ N}.$$

Motion in a circle and simple harmonic motion **333**

*Example 3*

A particle of mass $m$ is attached to the mid-point P of a light inextensible string of length $2a$. The upper end of the string is fastened at a point O while the lower end Q carries a ring of mass $m$. The ring can slide freely on a vertical rod passing through O. The particle is moving in a horizontal circle with $OQ = 3a/2$. Find the acceleration of the particle.

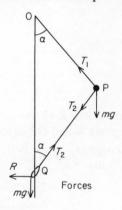
Forces

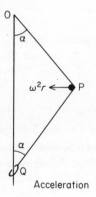
Acceleration

Fig. 10.9

In the triangle OPQ, OP and QP make the same angle $\alpha$ with OQ.
Since $OQ = 3a/2$, $2a \cos \alpha = 3a/2$ and so $\cos \alpha = 3/4$ (Fig. 10.9).

The ring at Q is in equilibrium under the action of three forces, its weight, the tension $\mathbf{T}_2$ in the string PQ and the horizontal reaction $\mathbf{R}$ between the ring and the rod.
Resolving vertically for the forces on the ring

$$T_2 \cos \alpha - mg = 0$$
$$T_2 = mg \sec \alpha = 4mg/3.$$

The particle is acted on by three forces, its weight and the tensions $\mathbf{T}_1$ and $\mathbf{T}_2$. The resultant of these forces gives the particle its acceleration.
Since the particle has no vertical acceleration,

$$T_1 \cos \alpha - T_2 \cos \alpha - mg = 0$$
$$T_1 = T_2 + mg \sec \alpha$$
$$= 8mg/3.$$

By applying Newton's second law horizontally

$$T_1 \sin \alpha + T_2 \sin \alpha = m \times \text{acceleration}$$
$$\text{acceleration} = (T_1 + T_2)(\sin \alpha)/m$$
$$= 4g \sin \alpha$$
$$= g\sqrt{7}.$$

This acceleration is towards the centre of the circle in which the particle moves, i.e. towards the mid-point of OQ.

## Exercises 10.3

1  A particle of mass 0.5 kg is attached by a light inextensible string of length 2 m to a fixed point, and moves as a conical pendulum with the string inclined at 60° to the vertical. Find the speed of the particle in terms of $g$.

2  The mass of the bob of a conical pendulum is 2 kg and the length of the string is 12.5 cm. Show that when the bob makes three revolutions per second the tension in the string is $9\pi^2$ newtons.

3  Two particles are suspended from the same point by separate strings of different lengths, and they describe horizontal circles with the same angular velocity. Prove that the two circles are in the same plane.

4  A particle attached by a light inextensible string to a fixed point O describes a horizontal circle with constant speed. The particle takes two seconds to move once round the circle. Show that the centre of the circle is approximately one metre below O.

5  A particle of mass 0.1 kg is attached to a string 40 cm long, and revolves at constant speed in a horizontal circle making 75 revolutions per minute. Show that the tension in the string is $\pi^2/4$ newtons.

6  A particle of mass 2 kg is suspended from a fixed point by an elastic string of unstretched length one metre. When the particle revolves as a conical pendulum making 40 revolutions per minute, the length of the string is 1.8 m. Show that when the particle hangs at rest the extension of the string is $(g/4\pi^2)$ m.

7  A particle of mass $m$ is fastened to the mid-point of a light inextensible string of length $4a$. The ends of the string are attached to two points A and B, where $AB = 2a$ and A is vertically above B. The particle describes a horizontal circle with constant speed $\sqrt{(6ga)}$. Find the tension in each half of the string.

8  One end of a light inextensible string of length $2a$ is attached to a fixed point A, and the other end is attached to a fixed point B at a distance $a$ vertically below A. A smooth ring threaded on the string is rotating in a horizontal circle of centre B with constant angular velocity. Show that the magnitude of the acceleration of the ring is $2g$.

9  A particle P of mass $m$ is attached to one end of a light inextensible string of length $3a$, and a particle Q of mass $2m$ is attached to the other end. The string passes through a smooth fixed ring O and the particle Q hangs at rest vertically beneath O. The particle P describes a horizontal circle with angular velocity $\sqrt{(g/a)}$. Find the distance of the particle Q from the ring.

10  In a conical pendulum a particle of mass $m$ is attached to one end of a light elastic string and the other end of the string is fastened to a fixed point. The natural length of the string is $a$ and its modulus of elasticity is $mg$. The particle describes a horizontal circle with constant speed with the string inclined at 60° to the vertical. Show that the time taken for one revolution is $\pi\sqrt{(6a/g)}$.

## 10.4 Simple harmonic motion

A particle P moves with simple harmonic motion in a straight line when
(a) its acceleration is always directed towards a fixed point O in this line, and
(b) the magnitude of its acceleration is proportional to the distance OP.

(a) x positive          (b) x negative

Fig. 10.10

If the line is the x-axis and O is the origin, this definition leads to the differential equation

$$\ddot{x} = \frac{d^2x}{dt^2} = -\omega^2 x.$$

This equation states that
(a) the acceleration is towards the origin, both when $x$ is positive and when $x$ is negative
(b) the acceleration is proportional to $x$.
The constant $\omega^2$ is chosen because a positive constant is required and because of the connection between simple harmonic motion and uniform motion in a circle.

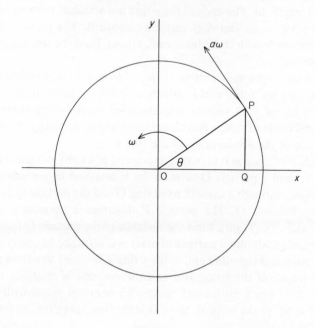

Fig. 10.11

Let P be the point $(x, y)$ on the circle $x^2 + y^2 = a^2$, and let the radius OP rotate with constant angular velocity $\omega$. Let OP make an angle $\theta$ with the $x$-axis and let $Q(x, 0)$ be the foot of the perpendicular from P to the $x$-axis.

Acceleration of $P = a\omega^2$ towards O.

Acceleration of Q = horizontal component of acceleration of P
$$= -a\omega^2 \cos \theta$$
$$= -\omega^2 x.$$

Therefore the acceleration of Q is proportional to $x$, i.e. proportional to the distance of Q from O, and it is always directed towards O.

This shows that the point Q moves with simple harmonic motion on the diameter of the circle.

Since P moves once round the circle in time $2\pi/\omega$, this is the time taken by Q to make one complete oscillation. This time $2\pi/\omega$ is known as the periodic time or the period of the motion.

Velocity of $P = a\omega$ along the tangent at P.

Velocity of Q = horizontal component of velocity of P
$$= -a\omega \sin \theta.$$

If $v$ denotes the velocity of Q,
$$v^2 = a^2\omega^2 \sin^2 \theta$$
$$= \omega^2(a^2 - a^2 \cos^2 \theta)$$

Since $x = a \cos \theta$, this gives $v^2 = \omega^2(a^2 - x^2)$.

*Solution of the equation* $\ddot{x} = -\omega^2 x$
If $v$ denotes the velocity of the particle, its acceleration $\ddot{x}$ can be expressed as $v\dfrac{dv}{dx}$, and the equation becomes

$$v\frac{dv}{dx} = -\omega^2 x.$$

This is a first-order differential equation in which the variables can be separated. Integration gives

$$\int v\,dv = -\int \omega^2 x\,dx$$

$$\tfrac{1}{2}v^2 = -\tfrac{1}{2}\omega^2 x^2 + c$$

where $c$ is a constant.
Let the particle come to rest when $x = a$, i.e. let $v = 0$ when $x = a$. Then the constant $c$ equals $\tfrac{1}{2}\omega^2 a^2$ and

$$v^2 = \omega^2(a^2 - x^2).$$

The particle will oscillate between two points A and B given by $x = a$ and $x = -a$. The distance $a$ is called the amplitude of the motion.

From the equation $v^2 = \omega^2(a^2 - x^2)$ it is clear that the particle has its maximum speed $a\omega$ when $x = 0$, i.e. when it is passing through O, the mid-point of its path. From the differential equation $\ddot{x} = -\omega^2 x$ it is clear that the acceleration has its greatest magnitude $a\omega^2$ when $x$ equals $a$ or $-a$, i.e. when the particle is at A or B.

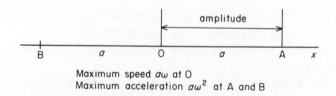

Maximum speed $a\omega$ at O
Maximum acceleration $a\omega^2$ at A and B

The next step is to express $x$ in terms of $t$. Since $v = \dot{x}$,

$$\dot{x}^2 = \omega^2(a^2 - x^2).$$

Let $x = a \cos \theta$. Then $\dot{x} = -a \sin \theta \, \dot{\theta}$ and $(a^2 - x^2) = a^2 \sin^2 \theta$. Therefore

$$a^2 \sin^2 \theta \, \dot{\theta}^2 = \omega^2 a^2 \sin^2 \theta$$
$$\dot{\theta} = \pm\omega$$
$$\theta = \pm\omega t + \text{a constant.}$$

Since $x = a \cos \theta$, this gives

$$x = a \cos (\omega t + \alpha)$$

where $\alpha$ is a constant.

Let $x = 0$ when $t = 0$. Then $\cos \alpha = 0$ and $\alpha = \pm\pi/2$.
This gives $x = \pm a \sin (\omega t)$.
The solution

$$x = a \sin (\omega t)$$

satisfies the conditions $x = 0$ and $\dot{x} = a\omega$ when $t = 0$.
The solution

$$x = -a \sin (\omega t)$$

satisfies the conditions $x = 0$ and $\dot{x} = -a\omega$ when $t = 0$.

In many examples it is convenient to take the particle to be at the origin when $t = 0$ and to take its velocity to be positive at this instant. These conditions are satisfied by the solution

$$x = a \sin (\omega t).$$

This equation gives the displacement of the particle at time $t$, measured from the instant when it passes through the mid-point of its path.

## Periodic time

Since the sine function is periodic with period $2\pi$, the value of $a \sin(\omega t)$ is unchanged when $t$ is replaced by $t + 2\pi/\omega$. This is also true of the general solution

$$x = a \cos(\omega t + \alpha)$$

obtained above.

At intervals of time equal to $2\pi/\omega$, the particle will pass through the same point with the same speed in the same direction.

The period or periodic time $T$ is given by

$$T = 2\pi/\omega.$$

The value of $T$ depends only on $\omega$, and is independent of the initial position and the initial velocity of the particle.

## Frequency

The number of complete oscillations made in unit time is called the frequency and is denoted by $n$. This is the reciprocal of $T$, i.e. $n = 1/T$.

## Summary

| | | |
|---|---|---|
| Acceleration | $\ddot{x} = -\omega^2 x$ | ...(1) |
| Velocity | $v^2 = \omega^2(a^2 - x^2)$ | ...(2) |
| Displacement | $x = a \sin(\omega t)$ | |
| if | $x = 0$ and $\dot{x} = a\omega$ when $t = 0$ | ...(3) |
| Periodic time | $T = 2\pi/\omega$ | ...(4) |
| Frequency | $n = 1/T = \omega/2\pi$ | ...(5) |

*Example 1*
A particle in simple harmonic motion takes 4 seconds to travel from one end A of its path to the other end B. Its greatest speed is $3\pi$ m s$^{-1}$. Find the amplitude of the motion and the distance of the particle from the mid-point of AB when its speed is $\pi$ m s$^{-1}$.

Since the particle takes 4 seconds to travel from A to B, the time for a complete oscillation is 8 seconds, i.e. $T = 8$.

Equation (4)   $T = 2\pi/\omega$
$$\omega = 2\pi/8 = \pi/4.$$

Let AB be the $x$-axis, with the origin at the mid-point of AB. Let the distance of the particle from the origin at time $t$ be given by

$$x = a \sin(\omega t).$$

Then $x = 0$ and $\dot{x} = a\omega$ when $t = 0$.

$$\text{Maximum speed} = a\omega$$
$$a\omega = 3\pi$$
$$a = (3\pi)/(\pi/4) = 12.$$

Equation (2) $\quad v^2 = \omega^2(a^2 - x^2).$

When $v = \pi$,

$$\pi^2 = (\pi/4)^2(12^2 - x^2)$$
$$16 = 144 - x^2$$
$$x = \pm 8\sqrt{2}.$$

Hence the amplitude is 12 m and the required distance is $8\sqrt{2}$ m.

*Example 2*

A point in simple harmonic motion makes five complete oscillations per second and its path is 4 cm long. Find the greatest magnitude of its acceleration.

Equation (5) $\quad n = 1/T = \omega/2\pi.$

Since $n = 5$, $\omega = 10\pi$.

Equation (1) $\quad \ddot{x} = -\omega^2 x.$

The acceleration is greatest when $x$ is greatest. Since twice the amplitude is 0.04 m, the amplitude $a$ equals 0.02.

$$\text{Greatest acceleration} = a\omega^2$$
$$= (0.02)(10\pi)^2$$
$$= 2\pi^2 \text{ m s}^{-2}.$$

*Example 3*

A point executes simple harmonic motion in a straight line, its greatest speed being $10 \text{ m s}^{-1}$. When it is 4 m from the mid-point of its path its speed is $6 \text{ m s}^{-1}$. Find its speed when it is 3 m from the mid-point.

Equation (3) $\qquad\qquad\qquad\qquad x = a \sin (\omega t).$

$$\text{Greatest speed} = a\omega = 10.$$

Equation (2) $\qquad\qquad\qquad\qquad v^2 = \omega^2(a^2 - x^2)$
$$= 100 - \omega^2 x^2.$$

When $x = 4$, $v = 6$

$$36 = 100 - 16\omega^2$$

Hence $\omega = 2$, $a = 5$.

When $x = 3$, $\qquad\qquad\qquad\qquad v^2 = 4(25 - 9) = 64$
$$v = 8 \text{ m s}^{-1}.$$

*Example 4*

At time $t$ the displacement $x$ metres of a point moving on the $x$-axis is given by $x = 10 \sin (4\pi t)$. Find the time taken by the point to travel directly from the point A at which $x = 5\sqrt{3}$ to the point B at which $x = -5\sqrt{2}$.

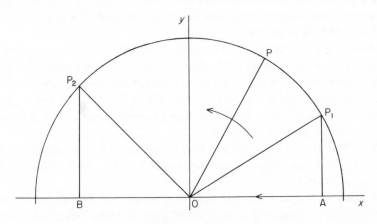

Fig. 10.12

By comparison with the equation $x = a \sin (\omega t)$, it can be seen that the amplitude $a$ is 10 m and that $\omega$ equals $4\pi$.

In Fig. 10.12 let the radius OP of the circle with centre O and radius 10 m rotate with angular velocity $4\pi$ rad s$^{-1}$. Let A and B be the projections on the $x$-axis of the points $P_1$ and $P_2$ respectively. The time taken by the point on the $x$-axis to move from A to B equals the time taken by the radius OP to rotate from $OP_1$ to $OP_2$.

$$\cos P_1Ox = OA/OP_1 = \sqrt{3}/2$$
$$\text{angle } P_1Ox = \pi/6.$$

$$\cos P_2Ox = OB/OP_2 = -1/\sqrt{2}$$
$$\text{angle } P_2Ox = 3\pi/4.$$

$$\text{angle } P_1OP_2 = 3\pi/4 - \pi/6 = 7\pi/12.$$

Angular velocity of OP $= 4\pi$.
Time taken by OP to rotate through $7\pi/12$ radians $= (7/48)$ s.

**Exercises 10.4**

1  A point P is moving in a straight line with simple harmonic motion about a point O. The speed of P is 4 m s$^{-1}$ when OP $= 3$ m, and it is 3 m s$^{-1}$ when OP $= 4$ m. Find the amplitude and the period.

**2** The greatest speed of a point in simple harmonic motion is $8 \text{ m s}^{-1}$ and its period is $\pi$ seconds. Find the distance of the point from the mid-point of its path when its speed is $4 \text{ m s}^{-1}$.

**3** A point in simple harmonic motion makes 120 complete oscillations per minute. Its greatest acceleration is $20 \text{ m s}^{-2}$. Find its greatest speed.

**4** A point moving in a straight line with simple harmonic motion makes 300 complete oscillations per minute. Its maximum speed is $\pi \text{ m s}^{-1}$. Show that the amplitude is 10 cm.

**5** Show that a point in simple harmonic motion will travel from one end of its path to a point half-way to the centre of its path in one-sixth of its period.

**6** A particle moves on the $x$-axis in simple harmonic motion about the origin O. The amplitude is 5 m and the period is $\pi/2$ s. Find, to two significant figures, the time taken by the particle to travel directly from a point 4 m from O to a point 1 m from O.

**7** A point moves in a straight line with simple harmonic motion of amplitude 2 cm. Find the greatest frequency possible if the acceleration must not exceed $200 \text{ m s}^{-2}$.

**8** A point in simple harmonic motion of amplitude 5 m passes through a point X at intervals of 2 seconds and 8 seconds. Show that the distance of X from the centre of the path is $5 \sin 54°$ m.

**9** A particle executing simple harmonic motion in a straight line moves directly from a point A to a point B in two seconds, and passes through B again two seconds later. The speed of the particle at A is equal to its speed at B. If AB $= 10$ cm, find the amplitude of the motion.

**10** A particle P moves on the $x$-axis in such a way that its acceleration is proportional to its distance from the origin O and is always directed towards O. When OP $= 5$ m, the speed of the particle is $24 \text{ m s}^{-1}$ and its acceleration is $20 \text{ m s}^{-2}$ towards O. Find its acceleration when its speed is $10 \text{ m s}^{-1}$.

**11** A point in simple harmonic motion has its maximum speed at a point O. When its distance from O is 3 m its speed is $16 \text{ m s}^{-1}$. When its distance from O is 4 m its speed is $12 \text{ m s}^{-1}$. Show that the amplitude of the motion is 5 m and that the maximum speed is $20 \text{ m s}^{-1}$.

**12** A particle in simple harmonic motion makes 150 complete oscillations per minute. At each end of its path the magnitude of its acceleration is $50 \text{ m s}^{-2}$. Show that the amplitude of the motion is $2/\pi^2$ m.

## 10.5 Particles in simple harmonic motion

Consider a particle P of mass $m$ which is attached to one end of a light elastic string. The other end of the string is fastened to a point O of a smooth horizontal table on which the particle rests.

In Fig. 10.13, OA is the natural length $l$ of the string. When the particle P is at a distance $l + x$ from O, the tension $T$ in the string is given by

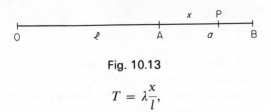

Fig. 10.13

$$T = \lambda\frac{x}{l},$$

where $\lambda$ is the modulus of elasticity of the string.

The force exerted by the string on the particle acts in the direction of $x$ decreasing. Therefore the acceleration of the particle is given by

$$m\ddot{x} = -T = -\lambda x/l,$$

the negative sign indicating that $T$ acts in the negative direction. This gives

$$\ddot{x} = -\left(\frac{\lambda}{ml}\right)x,$$

which is the differential equation of simple harmonic motion with $\omega^2$ replaced by $\lambda/ml$.

Let the particle be released from rest at a point B on the table where $OB = l + a$. The particle will travel towards A in simple harmonic motion of amplitude $a$. The time taken to reach A is one-quarter of a period, i.e. $\pi/2\omega$. Note that this time does not depend on the value of $a$.

The particle will arrive at A with speed $a\omega$, where $\omega$ equals $\sqrt{(\lambda/ml)}$, and then travels towards O at constant speed.

The final kinetic energy of the particle is $\frac{1}{2}m(a\omega)^2$, which equals $\frac{1}{2}\lambda a^2/l$.

When the extension of the string is $a$, the potential energy in the string is $\frac{1}{2}\lambda a^2/l$. This is converted into the kinetic energy of the particle.

*Example 1*
The natural length of a light elastic string is 50 cm and its modulus of elasticity is 20 N. One end of the string is fastened to a point O of a smooth horizontal table, while to the other end is attached a particle of mass 0.4 kg. The particle is held on the table at a distance of 60 cm from O and is then released from rest. Find the time taken by the particle to reach O. Find also the final kinetic energy of the particle.

The situation is the same as that in Fig. 10.13.
Initial extension $a = 10$ cm $= 0.1$ m.
When the extension in the string is $x$ m,

$$\text{tension} = \lambda x/l = 20x/(0.5) = 40x.$$

This force acts in the direction of $x$ decreasing. Hence

$$0.4\ddot{x} = -40x$$
$$\ddot{x} = -100x.$$

This gives $\omega^2 = 100$, $\omega = 10$.

Periodic time $T = 2\pi/\omega = \pi/5$ s.

The string goes slack after time $T/4$, i.e, after $\pi/20$ s.
The speed of the particle is then $a\omega$, which equals $(0.1)(10)$ or $1 \text{ m s}^{-1}$. The particle travels at constant speed from A to O. Since OA equals 50 cm this will take 0.5 s.
Total time taken to reach O from B $= (\pi/20 + 1/2)$ s.

Since there is no loss of energy, the result can be checked by showing that the final kinetic energy of the particle agrees with the initial potential energy in the string.

Final kinetic energy $= \frac{1}{2}(0.4)(1^2) = 0.2$ J.

Initial potential energy $= \frac{1}{2}\lambda a^2/l$
$$= \frac{1}{2}(20)(0.1)^2/(0.5)$$
$$= 0.2 \text{ J.}$$

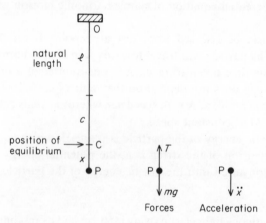

Fig. 10.14

Consider next a particle P of mass $m$ suspended from a fixed point O by a light spiral spring which obeys Hooke's law. Let $l$ be the natural length of the spring and $\lambda$ its modulus of elasticity.
*Step 1*
When the particle hangs in equilibrium at the point C, the tension will equal $mg$ and the extension $c$ of the spring will be such that

$$mg = \lambda c/l. \qquad \qquad \ldots(1)$$

*Step 2*
Let the particle be pulled vertically downwards from C (Fig. 10.14). When it is released, it will oscillate in a vertical line through C.

Let $CP = x$. When P is below C, $x$ is counted as positive, and when P is above C, $x$ is counted as negative.

Extension in the spring $= c + x$.

Tension $= \lambda(c + x)/l = mg + \lambda x/l$. ...(2)

This equation gives the value of the tension throughout the motion, provided that the initial extension is less than $c$.

*Step 3*
The two forces acting on the particle are its weight downwards and the tension in the spring upwards. By the equation $F = ma$,

$$m\ddot{x} = mg - (mg + \lambda x/l) = -\lambda x/l.$$

Hence $\qquad\qquad \ddot{x} = -(\lambda/ml)x.$ ...(3)

Comparison with the equation $\ddot{x} = -\omega^2 x$ shows that the particle executes simple harmonic motion with $\omega^2 = \lambda/ml$. The particle will oscillate about C, the position of equilibrium, with period $2\pi\sqrt{(ml/\lambda)}$.

An alternative approach is provided by consideration of the total energy of the system, which remains constant.

Let the potential energy of the particle be measured from the level of C, the position of equilibrium (Fig. 10.14).

Energy of the particle $= \frac{1}{2}m\dot{x}^2 - mgx$.

Energy in the spring $= \frac{1}{2}\lambda(c + x)^2/l$.

The total energy is constant.

$$\tfrac{1}{2}m\dot{x}^2 - mgx + \tfrac{1}{2}\lambda(c + x)^2/l = \text{constant}.$$

Differentiating with respect to time

$$m\dot{x}\ddot{x} - mg\dot{x} + \lambda(c + x)\dot{x}/l = 0$$
$$m\ddot{x} = mg - \lambda(c + x)/l$$

Since $mg = \lambda c/l$ this gives

$$m\ddot{x} = -\lambda x/l$$
$$\ddot{x} = -(\lambda/ml)x \quad \text{as before.}$$

*Example 2*
A particle of mass 0.08 kg is suspended from a fixed point O by a light elastic string. The natural length of the string is 2 m and its modulus of elasticity is 4 N. The particle is released from rest at a point A vertically beneath O, where OA $=$ 2.6 m. Show that the particle executes simple harmonic motion with period $2\pi/5$ s, and find the speed of the particle when the length of the string is 2.3 m. (Take $g = 10 \text{ m s}^{-2}$.)

Let $c$ be the extension in the position of equilibrium. (Fig. 10.14)

By Hooke's law, $\quad mg = \lambda c/l$ ...(1)
$$c = mgl/\lambda = 0.4.$$

Therefore C, the position of equilibrium, is 2.4 m below O.
When the extension in the string is $c + x$,

$$\text{tension} = \lambda(c + x)/l \qquad \ldots (2)$$

This tension acts upwards on the particle, while its weight acts downwards. By $F = ma$,

$$m\ddot{x} = mg - \lambda(c + x)/l$$
$$0.08\ddot{x} = 0.8 - 4(0.4 + x)/2$$
$$0.08\ddot{x} = -2x$$
$$\ddot{x} = -25x. \qquad \ldots (3)$$

This is the differential equation of simple harmonic motion with $\omega = 5$. Hence the particle executes simple harmonic motion about the point C with period $2\pi/5$ s.

It is important to remember that C, the position of equilibrium, is always the centre of the simple harmonic motion. This is the reason why $x$ is measured from C. The amplitude $a$ of the motion equals the distance CA, which is 0.2 m.

The velocity $v$ of the particle at a distance $x$ from C is given by

$$v^2 = \omega^2(a^2 - x^2)$$
$$= 25(0.04 - x^2).$$

When the length of the string is 2.3 m, $x^2$ equals 0.01 and therefore

$$v^2 = 25(0.04 - 0.01) = 0.75$$
$$v = \pm\sqrt{3/2}.$$

Hence the speed at this point is $\sqrt{3/2}$ m s$^{-1}$.

*Example 3*

A particle of mass $m$ is suspended from a fixed point O by a light elastic string of length $4a$ and modulus of elasticity $2mg$. The particle is held at a point A at a vertical distance $9a$ beneath O, and is then released from rest. Show that the string will go slack for part of the motion. Find the interval of time for which the string is slack, and find the height above A to which the particle rises.

When the particle hangs in equilibrium at the point C, let the extension in the string be $c$ (Fig. 10.15). The tension will then be equal to $mg$.

$$(2mg)(c/4a) = mg$$
$$c = 2a.$$

When the particle is at a distance $x$ below C, the extension is $2a + x$. By $F = ma$,

$$m\ddot{x} = mg - \lambda(2a + x)/4a.$$

Since $\lambda = 2mg$ this gives

$$\ddot{x} = -(g/2a)x.$$

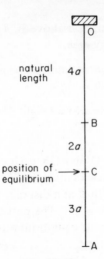

Fig. 10.15

When the particle is released at A, it will begin to execute simple harmonic motion with period $2\pi/\omega$, where $\omega^2 = g/2a$.

Now CA $=$ OA $-$ OC $= 9a - 6a = 3a$. This means that the initial downward displacement from C, the position of equilibrium, is greater than the extension in the position of equilibrium. The particle will rise to a height above C greater than $2a$. Therefore the string will go slack during part of the motion.

When the particle reaches the point B (Fig. 10.15) the tension in the string will be zero. There are two ways in which the speed of the particle at B can be found.
(a) Since $\omega^2 = g/2a$ and the amplitude is $3a$,

$$v^2 = (g/2a)(9a^2 - x^2).$$

Now when $x = $ CB $= 2a$,

$$v^2 = (g/2a)(9a^2 - 4a^2)$$
$$v^2 = 5ga/2.$$

(b) Initial extension of the string $= 5a$.

Initial potential energy in the string $= \frac{1}{2}(2mg)(5a)^2/4a$
$$= 25mga/4.$$

When the particle reaches B, it has risen through a vertical distance $5a$ and gained $5mga$ in potential energy. It has also gained $\frac{1}{2}mv^2$ in kinetic energy.

Energy gained by particle $=$ energy lost by string
$$\frac{1}{2}mv^2 + 5mga = 25mga/4$$
$$v^2 = 5ga/2.$$

After passing through B, the particle will move freely under gravity. It will come to rest after a time $v/g$, and so the string will be slack for a time $2v/g$, i.e. $\sqrt{(10a/g)}$.

Let the particle rise to a height $h$ above B. Then the kinetic energy $\frac{1}{2}mv^2$ is converted into potential energy $mgh$.

$$mgh = \tfrac{1}{2}mv^2 = 5mga/4$$
$$h = 5a/4.$$

Since AB = $5a$, the particle rises to a height $25a/4$ above A.

*Example 4*

Two identical spiral springs PQ and QR each of natural length $2a$ are joined together at Q and carry a particle of mass $m$ at Q. The ends P and R are fixed to two points in a vertical line with P at a distance $8a$ above R. In the position of equilibrium PQ = $5a$ and QR = $3a$. The particle is given a small downward displacement from its position of equilibrium and is then released from rest. Find the period of the oscillations.

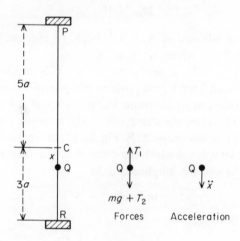

Fig. 10.16

Let $\lambda$ be the modulus of elasticity of each spring.
In the position of equilibrium,

tension in upper spring = $\lambda(5a - 2a)/2a = 3\lambda/2$,

tension in lower spring = $\lambda(3a - 2a)/2a = \lambda/2$.

Therefore
$$3\lambda/2 = mg + \lambda/2$$
$$\lambda = mg.$$

When the particle is at a distance $x$ below C, its position of equilibrium, PQ = $5a + x$ and QR = $3a - x$.

Tension in PQ = $T_1 = mg(3a + x)/2a$,

tension in QR = $T_2 = mg(a - x)/2a$.

The acceleration $\ddot{x}$ of the particle is given by

$$
\begin{aligned}
m\ddot{x} &= mg' - T_1 + T_2 \\
&= mg - mg(3a + x)/2a + mg(a - x)/2a \\
&= -mgx/a
\end{aligned}
$$

i.e. $\qquad\qquad \ddot{x} = -(g/a)x.$

This is the differential equation of simple harmonic motion with $\omega^2 = g/a$. Hence the period of the oscillations is $2\pi\sqrt{(a/g)}$.

*Example 5*

A particle of mass $m$ is suspended from a fixed point O by an elastic string of natural length $l$ and modulus of elasticity $\lambda$. The particle is released from rest at O and first comes to rest at a point A, where OA $= 2l$. Show that $\lambda = 4mg$ and find the maximum speed of the particle.

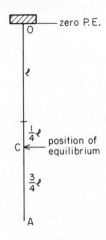

Fig. 10.17

Let the potential energy of the particle be measured from the level of O. When the particle reaches A,

potential energy of the particle $= -2mgl$,

potential energy in the string $= \frac{1}{2}\lambda(l)^2/l = \frac{1}{2}\lambda l.$

Since the total energy is unchanged,

$$
\begin{aligned}
\tfrac{1}{2}\lambda l - 2mgl &= 0 \\
\lambda &= 4mg.
\end{aligned}
$$

Let $c$ be the extension of the string when the particle hangs in equilibrium at the point C. Then the tension in the string will equal the weight of the particle.

$$
\begin{aligned}
(4mg)c/l &= mg \\
c &= l/4.
\end{aligned}
$$

As the particle falls, its speed increases until the extension in the string equals $c$, because the tension is less than $mg$. When the particle is below C, the tension is greater than $mg$ and the speed will decrease. When the extension equals $c$, the speed will have its maximum value $v$. At this instant,

potential energy in the string $= \frac{1}{2}(4mg)c^2/l = mgl/8$,

potential energy of the particle $= -mg(l + c) = -5mgl/4$.

Since the total energy is unchanged,

$$\frac{1}{2}mv^2 - 5mgl/4 + mgl/8 = 0$$
$$v^2 = 9gl/4$$
$$v = \frac{3}{2}\sqrt{(gl)}.$$

Alternatively, when the extension in the string is $c + x$,

$$m\ddot{x} = mg - 4mg(c + x)/l.$$

Since $c = l/4$, this gives

$$\ddot{x} = -(4g/l)x.$$

This is the simple harmonic equation with $\omega^2 = 4g/l$.
Amplitude $a = \text{CA} = 3l/4$.
Greatest speed $= a\omega = (3l/4)\sqrt{(4g/l)} = \frac{3}{2}\sqrt{(gl)}$.

Consider a particle resting on a horizontal platform which is oscillating in simple harmonic motion.

*(i) Vertical oscillations*
In Fig. 10.18, let O be the mid-point of the path AB of the particle P of mass $m$.

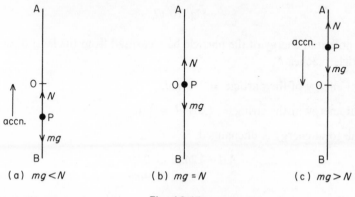

(a) $mg < N$    (b) $mg = N$    (c) $mg > N$

Fig. 10.18

In (a), the particle is below O and its acceleration is upwards towards O. Hence the normal reaction $N$ between the platform and the particle must be greater than $mg$.

In (b), the particle is passing through O and its acceleration is zero, so that $N = mg$.

In (c), the particle is above O and its acceleration is downwards towards O. Therefore $mg$ is greater than $N$. If the downward acceleration of the platform exceeds $g$, $N$ will be zero and the particle will leave the platform.

Let the vertical displacement of the platform at time $t$ be $a \sin(\omega t)$. The downward acceleration of the platform at its highest point is $a\omega^2$. If $a\omega^2 > g$, $N$ will become zero during the motion and the particle will leave the platform.

### (ii) Horizontal oscillations

If the horizontal displacement of the platform at time $t$ is $a \sin(\omega t)$, its greatest acceleration is $a\omega^2$. If $\mu$ is the coefficient of friction between the particle and the platform, the magnitude of the horizontal force on the particle due to friction cannot exceed $\mu$ (normal reaction), i.e. $\mu mg$. This force cannot give the particle an acceleration greater than $\mu g$. Therefore the particle will slide on the platform if $a\omega^2 > \mu g$.

### Example 6

A horizontal shelf makes five complete vertical oscillations per second in simple harmonic motion with amplitude 10 cm. Show that a particle resting on the shelf will lose contact when the shelf is approximately one centimetre above its central position.

Frequency $n = 5$

$$\omega = 2\pi n = 10\pi.$$

Let the vertical displacement $y$ metres of the shelf at time $t$ be given by

$$y = a \sin(\omega t),$$

where $a = 0.1$ and $\omega = 10\pi$.
The acceleration of the shelf is given by

$$\ddot{y} = -a\omega^2 \sin(\omega t) = -\omega^2 y.$$

The particle will leave the shelf when the downward acceleration $\omega^2 y$ of the shelf exceeds $g$, i.e.

$$y > g/\omega^2 = g/(10\pi)^2 \approx 0.01.$$

Therefore the particle leaves the shelf when the shelf is approximately 0.01 m (or 1 cm) above its central position.

### Example 7

A horizontal platform is vibrating horizontally in simple harmonic motion with period $(\pi/10)$ s and amplitude 1 cm. The coefficient of friction between the platform and a particle is 0.3. Find whether the particle can remain on the platform without sliding.

Let the horizontal displacement of the platform at time $t$ be $a \sin(\omega t)$, where $a = 1 \text{ cm} = 0.01 \text{ m}$ and $\omega = 2\pi/T = 20$. The greatest magnitude of the acceleration of the platform is given by

$$a\omega^2 = (0.01) \times 20^2 = 4 \text{ m s}^{-2}.$$

If $m$ is the mass of the particle, the normal reaction between the particle and the platform is $mg$ newtons. The limiting force of friction is $(0.3)mg$ newtons, which can give the particle an acceleration of $(0.3)g$ m s$^{-2}$. Since this is less than $4$ m s$^{-2}$, the particle will not stay at rest relative to the platform.

### Exercises 10.5

1   A particle P of mass 0.5 kg moving on the $x$-axis is acted on by a force towards the origin O of magnitude (2 OP) newtons, where OP is measured in metres. The speed of the particle at O is 10 m s$^{-1}$. Find the amplitude and the period of the simple harmonic motion.

2   A particle of mass 0.5 kg is in simple harmonic motion in a straight line with amplitude 4 cm and period $\pi/10$ s. Find the maximum force exerted on the particle.

3   The displacement $x$ at time $t$ seconds of a particle of mass 0.1 kg is given in metres by the equation $x = 10 \sin(2t)$. Find the force exerted on the particle when its speed is (a) zero (b) 10 m s$^{-1}$ (c) 12 m s$^{-1}$.

4   One end of a light elastic string of natural length 40 cm and modulus of elasticity 2 N is fastened to a point O of a smooth horizontal table. A particle of mass 0.2 kg is attached to the other end A of the string. The particle is released from rest at a point B of the table where OB = 50 cm. Find
(a) the greatest speed of the particle
(b) the time taken to reach O.

5   A particle of mass 4 kg is suspended from a fixed point O by a light elastic string of natural length 50 cm and modulus of elasticity 50 N. Find the amplitude and the period of the oscillations if the particle is released from rest when the length of the string is (a) 95 cm (b) one metre. (Take $g = 10$ m s$^{-2}$.)

6   A particle of mass $m$ is suspended from a fixed point O by a light elastic string of natural length $l$. When the particle hangs in equilibrium the length of the string is $2l$.
(a)  Find the period and amplitude of the oscillations if the particle is pulled vertically downwards and released from rest when the length of the string is $5l/2$.
(b)  The particle is released from rest at a point A vertically beneath O such that OA = $4l$. Find the kinetic energy of the particle at the instant when the string becomes slack, and show that when the particle reaches O its speed is $\sqrt{(gl)}$.

7   A particle is suspended from a fixed point by a light elastic string, and in the

position of equilibrium the extension of the string is $c$. The particle is raised through a vertical distance $c/2$ and then released from rest. Find its greatest speed and its greatest acceleration.

8    A particle of mass $m$ is suspended from a fixed point by a light elastic string and in the position of equilibrium the extension of the string is $c$. When the particle is in equilibrium it is given a vertical velocity of magnitude $\sqrt{(gc)}$. Find the amplitude of the oscillations, and the greatest and least values of the tension in the string.

9    A particle of mass $m$ is suspended by a spiral spring and makes $n$ complete vertical oscillations per second. Find the frequency of the oscillations if the particle is replaced by one of mass $4m$.

10   Two particles of equal mass are together suspended from a fixed point by an elastic string. When they are hanging at rest, one particle falls off. Show that the other particle will rise until the string is just taut.

11   When a particle is suspended by a light elastic string, the period of its vertical oscillations is $T$. One end of the string is fixed at a point on a smooth horizontal table and the particle is held on the table with the string extended. If the particle is released from rest, find the time that elapses before the string becomes slack.

12   A particle of mass $m$ is attached to one end of a light elastic string of natural length $l$ and modulus of elasticity $mg$. The other end of the string is fixed at a point O on a smooth horizontal table. The particle is released from rest at a point on the table at a distance $3l$ from O. Find the time it takes to reach O.

13   The natural length of a light elastic string is $2l$ and its modulus of elasticity is $2mg$. A particle of mass $m$ is attached to the mid-point of the string, and the ends of the string are fixed at two points A and B on a smooth horizontal table where $AB = 3l$. The particle is moved towards B through a distance less than $l/2$ and is then released from rest. Show that it will execute simple harmonic motion with period $\pi\sqrt{(l/g)}$.

14   A particle of mass $m$ is attached to one end of a light elastic string of natural length $l$ and modulus of elasticity $\lambda$. The other end of the string is fastened to a point on a smooth plane inclined at $30°$ to the horizontal. When the particle rests in equilibrium the extension of the string is $l/2$. Show that $\lambda = mg$.
   Find the period and amplitude of the oscillations if the particle is pulled downwards along a line of greatest slope from its position of equilibrium and is released from rest when the length of the string is (a) $2l$ (b) $7l/4$.

15   A particle suspended by a spiral spring is moving in a vertical line, making five complete oscillations per second. Show that when the particle hangs in equilibrium, the extension of the spring is approximately one centimetre.

16   When a scale-pan of mass $m$ is suspended by a spiral spring the extension of the spring is $b$. A particle of mass $3m$ is placed in the pan which is then released. Show that the period of the oscillations is $4\pi\sqrt{(b/g)}$.

17 A horizontal shelf is oscillating vertically in simple harmonic motion with amplitude 5 m and period 5 s. A body of mass 10 kg rests on the shelf. Show that the reaction between the body and the shelf varies between 20 N and 180 N approximately.

18 A particle rests on a horizontal table which is vibrating horizontally in simple harmonic motion with amplitude $10^{-3}$ m and frequency 10 oscillations per second. Show that the particle will not move relative to the table if the coefficient of friction between the particle and the table is 0.5.

19 The vertical displacement $x$ at time $t$ seconds of a horizontal shelf is given in metres by the equation $4x = \sin(10t)$. Show that a particle on the shelf loses contact with the shelf when the shelf is approximately 15 cm below its highest position.

20 A horizontal platform is performing simple harmonic motion horizontally, making five complete oscillations per second. Find, in centimetres, the greatest amplitude possible if a particle on the platform is to remain at rest relative to the platform, given that the coefficient of friction between the particle and the platform is 0.1. (Take $g = 10\ \mathrm{m\,s^{-2}}$.)

## 10.6  Dimensions

Let the units of mass, length and time be denoted by M, L and T respectively. Since unit speed equals unit length divided by unit time, the dimensions of speed are said to be $[L/T]$, or $[LT^{-1}]$.

Since unit acceleration is unit speed divided by unit time, the dimensions of acceleration are $[L/T^2]$, or $[LT^{-2}]$.

Since unit force equals unit mass times unit acceleration, the dimensions of force are $[MLT^{-2}]$.

The radian is defined as the angle subtended at the centre of a circle of unit radius by an arc of unit length. In circular measure an angle is measured by arc/radius. Hence an angle has zero dimensions.

Since angular velocity is angle/time, the dimensions of angular velocity are $[T^{-1}]$.

In any equation all terms must have the same dimensions, numerical factors being regarded as having zero dimensions.

Consider the differential equation of simple harmonic motion

$$\frac{d^2 x}{dt^2} = -(\text{constant})x.$$

The left-hand side represents an acceleration with dimensions $[LT^{-2}]$, and on the right-hand side the dimensions of $x$ are $[L]$. Therefore the dimensions of the constant must be $[T^{-2}]$. This will be true if the constant is taken to be $\omega^2$, the square of an angular velocity.

### Exercises 10.6

1 Find the dimensions of (a) volume (b) power (c) angular acceleration.

2   Show that impulse and momentum have the same dimensions.
3   Show that kinetic energy, potential energy and work have the same dimensions.
4   Check the dimensions of the terms in the formulae
    (a) $s = ut + \frac{1}{2}at^2$          (b) $v^2 = u^2 + 2as$
    (c) $v^2 = \omega^2(a^2 - x^2)$          (d) $T = 2\pi\sqrt{(ml/\lambda)}$.
5   The speed $v$ of a particle is given by the equation $v^2 = 2a^m g^n$ where $a$ is a length and $g$ is an acceleration. Show that $m = n = 1$.


**Miscellaneous exercises 10**

1   A particle of mass $m$ is attached to one end of an elastic string of unstretched length $a$, the other end being attached to a fixed point on a smooth horizontal table. When the particle moves in a circle on the table with constant speed $\sqrt{(ag)}$, the length of the string is $3a/2$. Find the modulus of the string.

2   The depth of water at high tide at the end of a pier at 3 a.m. on a certain day is 9 m and at low tide, 6 hours 15 minutes later, it is 4 m. Show that the greatest rate at which the water level changes is approximately 2 cm per minute. Find to the nearest minute the time that same morning at which the depth of water was 8.75 m.

3   A light string ABCD carries a particle of mass $m$ at B, where AB $= 2a$, and another particle of the same mass at C, where CD $= a$. The ends of the string are fastened to two fixed points A and D in the same vertical line with A above D. The system rotates with constant angular velocity $\omega$ about AD with the string taut and with BC vertical. Show that the tension in AB is twice the tension in CD. Given that B is at a vertical depth $x$ below A and that C is at a vertical height $y$ above D, show that

$$\omega^2 = 2g/(x - y).$$

4   A particle is moving in a straight line with simple harmonic motion. When the particle is 7 m from its mean position its speed is 48 m s$^{-1}$, and when it is 15 m from its mean position its speed is 40 m s$^{-1}$. Find the maximum acceleration of the particle.

5   A small ring of mass $m$, free to slide on a thin smooth vertical rod, is attached by a light inelastic string of length $2a$ to a point on the rod. A particle of mass $2m$ is attached to the mid-point of the string. Prove that the system can rotate in steady motion about the rod, with each part of the string inclined to the rod, provided that the angular velocity exceeds $\sqrt{(2g/a)}$.

6   A particle moves in a straight line with simple harmonic motion, the centre of oscillation being A. When the particle is 1.5 m from A, its speed is 4 m s$^{-1}$ and its acceleration is 6 m s$^{-2}$. Find the amplitude and the period of the motion.

**7** A particle attached to a fixed point O by a string moves in a horizontal circle with uniform speed. Prove that the period of the motion is $2\pi\sqrt{(h/g)}$, where $h$ is the depth of the circle below O.

The string is elastic, with natural length $a$ and modulus of elasticity twice the weight of the particle. Show that if $h = a$, the speed of the particle is $\sqrt{(3ga)}$.

**8** A rough horizontal circular disc is rotating with uniform angular velocity $\omega$ about a fixed vertical axis through its centre. Two particles each of mass $m$ lie on the disc and are at rest relative to it. The particles are on the same radius at distances $a$ and $2a$ from the centre. They are connected by a light inextensible string of length $a$, and $\mu$ is the coefficient of friction between each particle and the disc. Show that both particles are on the point of slipping outwards if

$$3a\omega^2 = 2\mu g,$$

and find an expression for the corresponding tension in the string.

**9** The depth of water over a harbour bar varies in simple harmonic motion, the depth at low tide being 1.5 m and at high tide 5 m. On a certain day low tide occurs at 9.15 a.m. and high tide at 3.30 p.m.; find the times on that day between which a ship drawing 3 m of water can cross the bar.

**10** A horizontal plate is oscillating in its own plane in simple harmonic motion. The period of the oscillation is 4 seconds and the greatest speed is $2\,\mathrm{m\,s^{-1}}$. Show that the amplitude of the motion is $(4/\pi)$ m, and find the time taken by the plate to move a distance $(2/\pi)$ m from its central position.

Show that a particle resting on the plate will slip if the coefficient of friction between the particle and the plate is 0.3.

**11** Two particles of mass $m$ and $4m$ are connected by a light inextensible string of length $3a$, which passes through a small smooth fixed ring. The heavier particle hangs at rest at a distance $2a$ beneath the ring, while the lighter particle describes a horizontal circle with constant speed. Find the distance of the plane of this circle below the ring and the number of revolutions per second made by the lighter particle.

**12** An elastic string of natural length $3a$ is stretched between two points $6a$ apart on a smooth horizontal plane. A particle of mass $m$ is attached to the string at a point of trisection. Given that the tension in the position of equilibrium is $2mg$, find the period of oscillation of the particle along the length of the string.

**13** A particle is moving in simple harmonic motion between two points A and B four metres apart. The greatest speed of the particle is $2\pi\,\mathrm{m\,s^{-1}}$. Find the least time taken by the particle to travel from A to a point one metre from B.

**14** A particle moving in a straight line passes through a point O, and is then acted on by a force towards O and proportional to its distance from O. In three seconds after passing through O the particle travels 0.72 m, and in the next three seconds it travels 0.36 m. Show that the particle returns to O after approximately 13 seconds.

**15** One end of a light elastic string of natural length 2 m is attached to a fixed point O. From the other end is suspended a particle of mass 2 kg which hangs in equilibrium at a point A where OA = 2.25 m. Another particle of mass 3 kg is attached and the combined particles are released from rest at A. Show that the motion is simple harmonic, and find the period and the greatest value of the acceleration. (Take $g = 10 \text{ m s}^{-2}$.)

**16** One end of an elastic string of natural length 0.5 m is fixed at a point O on a smooth horizontal table. To the other end is attached a particle of mass 2 kg which is released from rest from a point on the table 0.7 m from O. The initial tension in the string is 10 N. Find the maximum speed of the particle and show that it reaches O after 0.81 s approximately.

**17** A point P moves on a circle with centre O and radius 2 m, starting from rest. The angular acceleration of OP is constant and equals 3 rad s$^{-2}$. Find the speed of P at the instant when its acceleration is of magnitude 10 m s$^{-2}$.

**18** A light inextensible string of length 0.2 m joins a particle of mass 0.5 kg to a fixed peg on a smooth horizontal table. The particle is moving on the table with constant speed 2 m s$^{-1}$ with the string taut. Find the tension in the string.

If the particle strikes and coalesces with another particle of mass 0.75 kg which is lying at rest on the table, find the tension in the string after the impact.

**19** Two particles P and Q, connected by light inextensible strings AP and AQ to a fixed point A, are moving in the same sense in two horizontal circles. The strings AP and AQ are each of length 0.8 m and they make angles of 45° and 60° respectively with the downward vertical. At a certain instant the strings are in the same vertical plane. Show that they will next both be in the same vertical plane after an interval of nearly 4 seconds.

**20** Two particles P and Q of equal mass are connected by a light inextensible string of length 2a threaded through a small hole O in a smooth horizontal table. The particle P describes a circle on the table so that OP rotates with constant angular velocity $2\sqrt{(g/a)}$. The particle Q describes a horizontal circle below the table with the same angular velocity. Show that OP = a and find the angle made by OQ with the vertical.

**21** A particle of mass $m$ is attached to the middle point of a light elastic string of natural length $a$ and modulus $mg$. The ends of the string are attached to two fixed points A and B, A being at a distance 2a vertically above B. Prove that the particle can rest in equilibrium at a depth 5a/4 below A, and find the period of oscillation if it is displaced slightly in the vertical direction.

[L]

**22** A particle of mass $m$ is at rest suspended from a fixed point by a light elastic spring of modulus $2mg$ and natural length $l$. Find the period and the amplitude of the oscillations performed if the particle is projected vertically from this position with speed $\frac{1}{2}\sqrt{(gl)}$.

When the particle is at its lowest position it is held there whilst another identical particle is affixed to it. If the particles are then released to oscillate together, find the period and the amplitude of the new oscillations.   [L]

**23** A particle of mass $m$ is attached to one end Q of an elastic string PQ of modulus $3mg$ and natural length $l$. If the string will break when its extension is greater than $2l$, find the maximum constant angular speed of the particle when describing a horizontal circle with the end P of the string fixed. Find the ratio of the kinetic energy of the particle during this motion to the potential energy stored in the string PQ.   [L]

**24** A particle moves in a straight line. At time $t$ its displacement from a fixed point O on the line is $x$, its velocity is $v$ and its acceleration is $k \sin \omega t$, where $k$ and $\omega$ are constant. The displacement, velocity and acceleration are all measured in the same direction. Initially the particle is at O and is moving with velocity $u$. Show that

$$v = \frac{k}{\omega}(1 - \cos \omega t) + u$$

and hence find $x$ in terms of $t$, $k$, $\omega$ and $u$.

(i) Show that if $u = 0$, the particle first comes to instantaneous rest after travelling a distance $2\pi k/\omega^2$.

(ii) If $u \neq 0$, find the relation which holds between $u$, $k$ and $\omega$ if the motion is simple harmonic.   [JMB]

**25** A particle moves on the line Ox so that after time $t$ its displacement from O is $x$, and $d^2x/dt^2 = -9x$. When $t = 0$, $x = 4$ and $dx/dt = 9$. Find

(i) the position and velocity of the particle when $t = \pi/6$

(ii) the maximum displacement of the particle from O.   [JMB]

**26** A particle moves in a straight line with simple harmonic motion of period $6\pi$ s about a centre O. Find the magnitude of its acceleration when it is distant 6 m from O.

At this instant the speed of the particle is $2\frac{2}{3}$ m s$^{-1}$. Find the amplitude of the motion.

**27** A particle P is attached by a light inextensible string of length $l$ to a fixed point O. The particle is held with the string taut and OP at an acute angle $\alpha$ to the downward vertical, and is then projected horizontally at right angles to the string with speed $u$ chosen so that it describes a circle in a horizontal plane. Show that $u^2 = gl \sin \alpha \tan \alpha$.

The string will break when the tension exceeds twice the weight of P. Find the greatest possible value of $\alpha$.   [JMB]

**28** The ends of a light inextensible string ABC of length $3l$ are attached to fixed points A and C, C being vertically below A at a distance $\sqrt{3l}$ from A. At a distance $2l$ along the string from A, a particle B of mass $m$ is attached. When both portions of the string are taut, B is given a horizontal velocity $u$, and then continues to move in a circle with constant speed. Find the tensions in the two portions of the string and show that the motion is possible only if $u^2 \geqslant \frac{1}{3}gl\sqrt{3}$.  [JMB]

29  A particle of mass $m$ is attached to one end of a light elastic string of natural length $l$ and modulus $2mg$. The other end of the string is fixed at a point A. The particle rests on a support B vertically below A, with $AB = 5l/4$. Find the tension in the string and the reaction exerted on the particle by the support B.

The support B is suddenly removed. Show that the particle will execute simple harmonic motion and find
(i) the depth below A of the centre of oscillation
(ii) the period of the motion.
(Standard formulae for simple harmonic motion may be quoted without proof.)  [JMB]

30  A light inextensible string of length $5a$ has one end fixed at a point A and the other end fixed at a point B which is vertically below A and at a distance $4a$ from it. A particle P of mass $m$ is fastened to the mid-point of the string and moves with speed $u$, and with the parts AP and BP of the string both taut, in a horizontal circular path whose centre is the mid-point of AB. Find in terms of $m$, $u$, $a$ and $g$ the tensions in the two parts of the string, and show that the motion described can take place only if $8u^2 \geqslant 9ga$.  [JMB]

31  A smooth ring of mass 6 kg is free to slide on a light inextensible string of length 8 m. One end of the string is attached to a fixed point B on a smooth horizontal table and the other to a fixed point A, 4 m vertically above B. The ring describes a horizontal circle as a conical pendulum at a constant angular velocity of $n$ revolutions per minute with both portions of the string taut. Find in terms of $n$ the reaction between the ring and the table when the ring moves on the table.

Hence or otherwise find the largest value of $n$ in order that the ring may remain on the table. With this value of $n$, find
(i) the force on the ring due to the tension in the string
(ii) the resultant force acting on the ring.
(Take the acceleration due to gravity to be $10 \text{ m s}^{-2}$.)  [AEB]

32  The ends of the prongs of a tuning fork are vibrating with amplitude 1 mm and with frequency 512 complete oscillations per second. Show that the magnitude of their maximum acceleration exceeds $10 \text{ km s}^{-2}$.

33  A particle A is performing simple harmonic oscillations about a point O with amplitude 2 m and period $12\pi$ s. Find the least time from the instant when A passes through O until the instant when
(i) its displacement is 1 m
(ii) its velocity is half that at O
(iii) its kinetic energy is half that at O.  [AEB]

**34** A particle moves in a straight line with simple harmonic motion of period $4\pi$ about a point D. Its speeds at two points A and B, which are distant 8 m and 10 m from a fixed point C on the line of motion are $4\,\mathrm{m\,s^{-1}}$ and $3\,\mathrm{m\,s^{-1}}$ respectively. The points lie in the order C, D, A, B. Prove that the point C is a distance 2 m from the centre of oscillation and find the amplitude of the oscillations.

An impulse is applied to the particle when it is at the point D so that the particle in the subsequent motion comes to rest at the point C. If the particle is of mass 2 kg, find the possible magnitudes and directions of the impulse. [AEB]

**35** A light elastic spring, of modulus $8mg$ and natural length $l$, has one end attached to a ceiling and carries a scale pan of mass $m$ at the other end. The scale pan is given a vertical displacement from its equilibrium position and released to oscillate with period $T$. Prove that $T = 2\pi\sqrt{(l/8g)}$.

A weight of mass $km$ is placed in the scale pan and from the new equilibrium position the procedure is repeated. The period of oscillation is now $2T$. Find the value of $k$. Find also the maximum amplitude of the latter oscillations if the weight and the scale pan do not separate during the motion. [AEB]

**36** A light elastic string of natural length $l$ has a particle of mass $m$ attached at one end B and the other end A is fixed. If $\mathrm{AB} = 3l/2$ when the particle hangs freely at rest, show that the modulus of elasticity of the string is $2mg$.

When hanging at rest the particle is suddenly given a downward vertical velocity $v$ so that it describes simple harmonic motion of amplitude $l/2$. Find in terms of $l$ and $g$ the period of this motion and the value of $v$.

Find the speed of the particle and the tension in the string when $\mathrm{AB} = 5l/4$. [AEB]

**37** If $\mathrm{d}^2x/\mathrm{d}t^2 = -\omega^2 x$, show that $(\mathrm{d}x/\mathrm{d}t)^2 = \omega^2(a^2 - x^2)$, where $\omega$ and $a$ are constants.

A particle is moving with simple harmonic motion in a straight line between the extreme points A and B; O is the centre of the oscillations. When the particle is 4 m from O, its speed is $4\,\mathrm{m\,s^{-1}}$, and when the particle is 2 m from O, its speed is $8\,\mathrm{m\,s^{-1}}$. Calculate the distance AB and the time taken by the particle for one oscillation.

If the particle is of mass 2 kg, find its kinetic energy when it is 2 m from A.

Find the time taken by the particle to travel directly from A to M, where M is the mid-point of OB. [AEB]

**38** If a particle describes a circle of radius $r$ with constant speed $u$, show that the acceleration of the particle is $u^2/r$ directed towards the centre of the circle.

A car enters a circular bend of radius $a$ moving horizontally with speed $v$. If the track is banked at an angle $\theta$ to the horizontal, show that, if there is no side-thrust on the car, then $v^2 = ga\tan\theta$.

The car enters the bend a second time travelling at speed $3v/2$. If the mass of the car is $M$ and no slipping occurs, show that the side-thrust on the car is $(5/4)Mgv^2/\sqrt{(v^4 + g^2a^2)}$. [AEB]

**39** A particle P moves in a circle, centre O and radius $r$, so that OP has constant angular speed $\omega$. Prove that the acceleration of the particle is $r\omega^2$ and is directed towards O.

One end of a light inextensible string of length $5a$ is tied at a fixed point A which is at a distance $3a$ above a smooth horizontal table. A particle of mass $m$, which is tied to the other end of the string, rotates with constant speed in a circle on the table. If the reaction between the particle and the table is $R$, find the tension in the string when (i) $R = 0$ (ii) $R = 3mg/4$. Show that the respective times of one revolution for these two values of $R$ are in the ratio 1:2. [AEB]

**40** The motion of a particle is defined by the differential equation $d^2x/dt^2 = -\omega^2 x$, where $\omega$ is a constant and $x$ is the distance of the particle from a fixed point in its path at time $t$. Show that the speed $v$ of the particle is given by $v^2 = \omega^2(a^2 - x^2)$, where $a$ is a constant.

A cubical box of mass $M$ stands on a horizontal platform with one square face of the box in contact with the platform. The platform moves up and down with simple harmonic motion of amplitude $a$ and frequency $n$. Throughout the motion the platform remains horizontal and the box remains in contact with the platform. Show that $n \leqslant \sqrt{(g/4a\pi^2)}$. Find the speed of the box and the reaction between the box and the platform when the latter is at a height $\frac{1}{2}a$ above its lowest position. Find also the least time taken to reach this position from the lowest position. [AEB]

# 11  Probability

## 11.1  Permutations

Suppose that three playing cards, an ace, a king and a queen are to be placed in line. The first place can be filled in three ways, with the ace or the king or the queen. For each of these three ways, the second place can be filled in two ways, and then the last card takes the third place. The cards can therefore be arranged in $3 \times 2 \times 1$ ways, and these six ways are shown in Fig. 11.1.

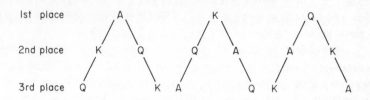

Fig. 11.1

Each arrangement of $n$ objects in a definite order is called a permutation. When the $n$ objects are all different, the first object can be chosen in $n$ ways and the second object in $(n - 1)$ ways. This gives $n(n - 1)$ ways of choosing the first two objects. The third object can then be chosen in $(n - 2)$ ways and so on. The number of permutations of $n$ different objects, denoted by $^nP_n$, is therefore

$$n(n - 1)(n - 2) \ldots 3 \times 2 \times 1,$$

i.e. $^nP_n = n!$ (factorial $n$).

If only two objects are to be included in each arrangement, the number of permutations will be $n(n - 1)$.

If only $r$ objects are to be included in each arrangement, the number of permutations will be the product of the $r$ factors

$$n(n - 1)(n - 2) \ldots (n - r + 1)$$

This is denoted by $^nP_r$, so that

$$^nP_r = \frac{n!}{(n - r)!}$$

Consider now the case when the objects are not all different. An ace, a king and a queen can be arranged in six ways, but an ace and two kings can be arranged in only three ways – AKK, KAK, KKA – if the suits of the kings are ignored.

An ace, a king and four queens can be arranged in 6! ways when the suits of the queens are taken into account. Once the positions of the ace and the king are chosen, the queens can be arranged in 4! ways. If the suits of the queens are ignored, the number of permutations is reduced to 6!/4!, i.e. 30.

In the general case, when $r$ of $n$ objects are identical and the rest are all different, the $n!$ arrangements of all $n$ objects can be divided into groups each containing $r!$ identical arrangements. Therefore the number of permutations of $n$ objects of which $r$ are identical is $n!/r!$.

### Example 1
Find the number of three-figure integers that can be formed from the numbers 2, 3, 5, 7, 8 (a) if no number is used twice (b) if any number may be used more than once.

(a) The first number can be chosen in five ways, the second number in four ways and the third number in three ways. The number of permutations is therefore $5 \times 4 \times 3$, i.e. 60.
(b) The first number can be chosen in five ways. The second can also be chosen in five ways, and so can the third. Therefore $5 \times 5 \times 5$ integers can be formed, i.e. 125.

### Example 2
Find the number of permutations of the letters of the word PUPPETS.

If the seven letters in the word were all different, there would be 7! permutations. As three letters are identical, the number of permutations is 7!/3!, i.e. 840.

### Example 3
Find the number of permutations of the letters of the word FOOTBALL
(a) which begin with L and end in A
(b) which begin with F and end with T.

(a)

| L |  |  |  |  |  |  | A |
|---|---|---|---|---|---|---|---|

Each empty box in the diagram has to be filled with one of the six remaining letters F, O, O, T, B, L. Since the two O's are indistinguishable, the number of ways in which this can be done is 6!/2!, i.e. 360.

(b)

| F |  |  |  |  |  |  | T |
|---|---|---|---|---|---|---|---|

The empty boxes have to be filled using the letters O, O, B, A, L, L. If the six letters were all different this could be done in 6! ways. This number must be divided by 2! since the two O's are indistinguishable, and again by 2! since the two L's are indistinguishable. The required number is therefore 6!/(2!2!), i.e. 180.

*Example 4*

Find the number of permutations of the letters of the word SWEET in which the two E's are not adjacent.

The total number of permutations of the five letters is 5!/2!, i.e. 60.

When the two E's are next to one another, they can be regarded as a single letter – a double E. The number of permutations of the letters S, W, T and a double E is 4!, so that the number of permutations in which the E's are adjacent is 24.

The number of permutations in which the two E's are not adjacent will be 60 − 24, i.e. 36.


*Example 5*

Find the number of permutations of the letters of the word MONOPOLISES in which the consonants and the vowels come alternately.

The six consonants M, N, P, L, S, S can be arranged in 6!/2! ways, i.e. in 360 ways.

The five vowels O, O, O, I, E can be arranged in 5!/3! ways, i.e. in 20 ways.

Each arrangement of the vowels can be fitted into each arrangement of the consonants. Therefore the number of such permutations is 360 × 20, i.e. 7200.


**Exercises 11.1**

1   Find the number of permutations (a) of three letters (b) of four letters, chosen from the six letters A, B, C, X, Y, Z.

2   Find the number of permutations (a) of six letters (b) of five letters, taken from the word SLOPING.

3   Ten athletes run in a race. Find the number of ways in which the first three places can be filled.

4   Find the number of four-figure integers that can be formed from the digits 2, 3, 4, 5 (a) if each digit may be used more than once (b) if no digit may be used more than once.

5   Find the number of odd three-figure integers that can be formed from the digits 2, 3, 4, 6, 7, 8 if no digit is to be used more than once.

6   Find the number of permutations of the letters of the word DISTINCT.

7   Find the number of permutations of the letters ABABABA.

8   Find the number of permutations of the five letters XXXYY in which the two Y's are not adjacent.

9   Find the number of permutations of the letters of the word MONDAY that begin either with A or with O.

10   Find the number of permutations of the letters of the word AROUND in which the vowels and the consonants come alternately.

11 Find the number of permutations of the letters of the word MINIMUM which begin with M and end in I.

12 Find the number of permutations of the letters of the word TATTOO which begin and end with a vowel.

13 A forecast is made of the results of twelve football matches, each result being given as a home win, a draw or an away win. Find the number of possible different forecasts.

Find also the number of possible different forecasts (a) containing only one error (b) containing only two errors.

14 In a competition, eight safety features of a car have to be placed in order of importance. Find the number of possible different entries.

If the features are labelled A, B, C, D, E, F, G, H, find the number of different entries in which D precedes A, B and C while E is last.

15 Find the number of permutations of the letters of the word ARREARS. Find also the number of these permutations in which no two R's are adjacent.

## 11.2 Combinations

The number of permutations of two letters from the four letters a, b, c, d is $4 \times 3$, i.e. 12. They are listed below.

$$
\begin{array}{cccccc}
ab & ac & ad & bc & bd & cd \\
ba & ca & da & cb & db & dc
\end{array}
$$

Each of these twelve permutations can be paired with another containing the same two letters, e.g. ab with ba.

If the order of the letters does not matter, only six different selections of two letters can be made i.e. the number of combinations of four letters taken two at a time is 6.

The number of permutations of three letters from the four letters a, b, c, d is $4 \times 3 \times 2$, i.e. 24. If the order of the letters does not matter, only four different selections of three letters can be made – abc, abd, acd, bcd. The three letters in each selection can be rearranged in six ways. This shows that the number of combinations of four letters taken three at a time is 4.

In the general case, the number of ways in which $r$ objects can be selected from $n$ different objects, irrespective of their order, is called the number of combinations of $n$ objects taken $r$ at a time. This number can be denoted either by $^nC_r$ or by $\binom{n}{r}$. (Do not confuse this with a vector.)

Each combination of $r$ different objects can be permuted in $r!$ ways. Therefore the number of permutations of $n$ different objects $r$ at a time equals the number of combinations multiplied by $r!$, i.e.

$$\binom{n}{r} \times r! = {}^nC_r \times r! = \frac{n!}{(n-r)!}.$$

It follows that

$$\binom{n}{r} = {}^{n}C_r = \frac{n!}{r!(n-r)!}.$$

For each set of $r$ objects selected, a corresponding set of $(n-r)$ objects is left. Therefore the number of combinations of $n$ objects taken $r$ at a time equals the number of combinations of $n$ objects taken $(n-r)$ at a time, i.e.

$$\binom{n}{r} = \binom{n}{n-r}$$

*Example 1*
Find the number of combinations of the letters a, b, c, d, e taken two at a time.

The number of permutations of two letters of the five letters is $5 \times 4$, i.e. 20. Each permutation can be paired with another containing the same two letters, e.g. bd and db. Therefore the number of combinations will be 10.
   Alternatively

$$\binom{5}{2} = {}^{5}C_2 = \frac{5!}{2!3!} = \frac{5 \times 4}{2!} = 10.$$

The ten combinations are ab, ac, ad, ae, bc, bd, be, cd, ce, de.

*Example 2*
Find the number of ways in which a working party of three members can be chosen from eight people.

The number of permutations of three objects from eight is $8 \times 7 \times 6$, i.e. 336.
   Each choice of three objects can be permuted in 3! ways. Hence the number of ways in which three members can be chosen is 336/6, i.e. 56.
   Alternatively

$$\binom{8}{3} = {}^{8}C_3 = \frac{8!}{3!5!} = \frac{8 \times 7 \times 6}{3 \times 2 \times 1} = 56.$$

*Example 3*
Ten points lie in a plane. Given that no three points are collinear, find the number of straight lines each of which passes through two of the points.

The number of combinations of two points from ten points is needed.

$$\binom{10}{2} = \frac{10!}{2!8!} = \frac{10 \times 9}{2 \times 1} = 45.$$

*Example 4*
A team of two boys and two girls is to be chosen from six boys and five girls. Find the number of possible teams.

Two boys can be chosen in $\binom{6}{2}$ ways, i.e. 15 ways.

Two girls can be chosen in $\binom{5}{2}$ ways, i.e. 10 ways.

Each choice of two boys can be taken with each choice of the two girls. This gives $15 \times 10$, i.e. 150 possible teams.

*Example 5*
Find the number of ways in which six boys can be divided into two teams of three.

Select any boy, and then choose two others from the remaining five boys. This can be done in $\binom{5}{2}$ ways, i.e. in 10 ways.

**Exercises 11.2**
1  Find the number of ways in which (a) two letters (b) six letters, can be chosen from eight different letters.
2  Find the number of ways in which (a) four books (b) five books, can be selected from ten different books.
3  Find the number of ways in which three cards can be selected from a pack of 52 cards.
4  Find the number of ways in which a group of eight boys can be split into two groups of four.
5  Find the number of ways in which a team of three boys and three girls can be chosen from a group of five boys and six girls.
6  Six straight lines are drawn in a plane, each line crossing each other line. Find the maximum number of points of intersection.
7  Six points are marked on a circle. Find the number of chords each passing through two points, and the number of triangles having three of the points as vertices.
8  Verify that

$$\binom{8}{3} + \binom{8}{2} = \binom{9}{3}$$

By considering the number of combinations of $(n + 1)$ different objects taken $r$ at a time, prove the general result

$$\binom{n}{r} + \binom{n}{r-1} = \binom{n+1}{r}.$$

**9** Three out of ten houses are to be painted white and the rest green. Find the number of ways in which the choice can be made.

**10** A team of 11 players is to be chosen from 15 boys. Find the number of ways in which this can be done if there are two boys who refuse to play in the same team.

## 11.3 Possibility spaces

Consider an experiment which can be repeated as many times as required, such as throwing a die, tossing a coin or choosing a card from a pack. Each performance of the experiment is called a *trial*. The result of any trial is known as its *outcome*. The set of possible outcomes of an experiment is called the *possibility space* or *sample space* for the experiment. Each outcome is called a *sample point* of this space.

For the experiment consisting of tossing a coin there are only two possible outcomes to any trial, a head $H$ or a tail $T$. The possibility space can be represented by $\{H, T\}$, where $H$ and $T$ denote the two sample points.

For the experiment which consists of throwing a die, there are six possible outcomes. The possibility space is the set $\{1, 2, 3, 4, 5, 6\}$.

Any subset of the possibility space (apart from the empty set) is known as an *event*. A simple event is a subset containing only one sample point. When the experiment is performed and the outcome is known, any event containing this outcome is said to have occurred.

For the experiment in which a die is thrown, the simple event $\{4\}$ occurs when the outcome of a trial is 4. The event $\{2, 4, 6\}$ occurs when the outcome of a trial is 2 or 4 or 6, but not when the outcome is 1 or 3 or 5.

Two events $E_1$ and $E_2$ are said to be complementary if $E_2$ contains all the sample points which are not contained in $E_1$. Then if the event $E_1$ occurs in any trial, the event $E_2$ does not, and vice versa. The complement of an event $E$ can be denoted either by $E'$ or by $\bar{E}$.

*Example 1*
An experiment consists of selecting a card from the four aces of a pack. State the sample points and the possibility space of this experiment. Give examples of complementary events.

The outcome of any trial will be a spade $S$, a heart $H$, a diamond $D$ or a club $C$. The sample points can be denoted by $S, H, D, C$ and then the possibility space is the set $\{S, H, D, C\}$.

The simple event $\{S\}$ occurs when the outcome of a trial is the ace of spades. The complementary event will be $\{H, D, C\}$.

The event $\{H, C\}$ occurs when the outcome is the ace of hearts or the ace of clubs. The complementary event will be $\{S, D\}$.

The event $\{S, D, C\}$ occurs when the spade or the diamond or the club is chosen. The complementary event will be $\{H\}$.

*Example 2*

A coin is tossed twice in succession. Make a list of the possible outcomes of this experiment and find the complement of the event $\{(HT), (TT)\}$.

There are four possible outcomes.
(a) A head followed by a head, denoted by $(HH)$.
(b) A head followed by a tail, denoted by $(HT)$.
(c) A tail followed by a head, denoted by $(TH)$.
(d) A tail followed by a tail, denoted by $(TT)$.

The possibility space is the set $\{(HH), (HT), (TH), (TT)\}$. Therefore the complement of the event $\{(HT), (TT)\}$ is the event $\{(HH), (TH)\}$.

## Exercises 11.3

1  Find the number of sample points in the possibility space for each of the following experiments.
   (a) A die is thrown twice.
   (b) A coin is spun twice.
   (c) A die is thrown and a coin is spun.

2  A die is thrown twice. Express in terms of sample points the simple events such that
   (a) the total score is 12
   (b) the total score is less than 4
   (c) the total score is 10.

3  A pack of cards is shuffled and then the top two cards are taken. Show that there are 2652 sample points in the sample space. Find the number of outcomes such that
   (a) both cards are aces
   (b) both cards are hearts
   (c) both cards are red
   (d) the two cards are of different colours.

4  An even positive integer less than 50 is chosen. Find the number of sample points in the events such that
   (a) the integer is a perfect square
   (b) the integer is a multiple of 7
   (c) the integer is not a factor of 12!

5  Three discs numbered 1, 2 and 3 respectively are placed in a bag. A boy draws a disc from the bag with his left hand, and another disc with his right hand. Make a list of the sample points of the possibility space. Find the event which occurs whenever the disc numbered 2 is not drawn, but does not occur when this disc is drawn.

## 11.4 Probabilities

The probability of an event which is certain to occur is taken to be 1, and the probability of an impossible event is taken to be 0. The probability of an event $X$ is denoted by $P(X)$, where $0 \leqslant P(X) \leqslant 1$.

When a coin is spun, it is intuitive that the chance that it will come down heads $(H)$ is equal to the chance that it will come down tails $(T)$. It is felt that the two simple events $X$ and $Y$, where $X = \{H\}$ and $Y = \{T\}$, are equally likely and the probability 1/2 is assigned to each of these events, i.e.

$$P(X) = 1/2, \quad P(Y) = 1/2.$$

Consider a possibility space $S$ containing $n$ sample points. The sum of the probabilities assigned to the $n$ sample points must be 1, since the event $S$ is bound to occur in any trial. If it is assumed that the $n$ outcomes are all equally likely, the probability of each outcome will be $1/n$. The probability of an event $A$ which contains $r$ of the $n$ sample points will be $r/n$, i.e.

$$P(A) = \frac{\text{number of favourable outcomes}}{\text{number of possible outcomes}}$$

The complementary event $A'$ contains the remaining $(n - r)$ sample points, so that $P(A') = (n - r)/n$.
It follows that

$$P(A') = 1 - P(A).$$

*Example 1*
A playing card is to be drawn at random from a pack. Find the probability that
(a) it will be red
(b) it will be a heart
(c) it will be a king.

The phrase 'at random' implies that the possible outcomes are all equally likely.
(a) There are 52 possible outcomes, of which 26 are favourable. Hence the probability that the card will be red is 26/52, i.e. 1/2.
(b) Since there are 13 favourable outcomes out of 52 outcomes, the probability that the card will be a heart is 13/52, i.e. 1/4.
(c) Only 4 of the 52 outcomes are favourable, so that the probability that the card will be a king is 4/52, i.e. 1/13.

*Example 2*
In the experiment in which a die is thrown, $X$ is the event $\{2, 3\}$ and $Y$ is the event $\{3, 4, 5\}$. Find $P(X)$, $P(Y)$, $P(X \cup Y)$ and $P(X \cap Y)$.

The assumption is made that the six possible outcomes are all equally likely. Since there are six sample points, the probability assigned to each one is 1/6. The event $X$ contains two sample points, giving $P(X) = 2/6 = 1/3$. The event $Y$ contains three sample points, giving $P(Y) = 3/6 = 1/2$.

The union of $X$ and $Y$ is the event $\{2, 3, 4, 5\}$. This contains four sample points and so $P(X \cup Y) = 4/6 = 2/3$.

The intersection of $X$ and $Y$ is the simple event $\{3\}$, and so $P(X \cap Y) = 1/6$.

Notice that $$P(X) + P(Y) = 1/3 + 1/2 = 5/6,$$

and that $$P(X \cup Y) + P(X \cap Y) = 2/3 + 1/6 = 5/6,$$

and so in this example

$$P(X \cup Y) = P(X) + P(Y) - P(X \cap Y).$$

It will now be shown that this relation holds for any experiment having a possibility space $S$ with $n$ equally likely sample points.

Let the event $X$ contain $r$ sample points and let the event $Y$ contain $s$ sample points. Then

$$P(X) = r/n, \quad P(Y) = s/n.$$

If $X$ and $Y$ have $c$ sample points in common, the probability that both $X$ and $Y$ occur in a given trial is $c/n$, i.e.

$$P(X \cap Y) = c/n.$$

The union of the events $X$ and $Y$ will contain $r + s - c$ sample points, and so

$$\begin{aligned} P(X \cup Y) &= (r + s - c)/n \\ &= r/n + s/n - c/n \\ &= P(X) + P(Y) - P(X \cap Y). \end{aligned}$$

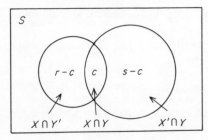

Fig. 11.2

In the Venn diagram in Fig. 11.2 the set $X \cup Y$ contains $r + s - c$ sample points.

Of these, $r - c$ belong to $X$ but not to $Y$, i.e. to $X \cap Y'$,
$s - c$ belong to $Y$ but not to $X$, i.e. to $X' \cap Y$,
and $c$ belong to both $X$ and $Y$, i.e. to $X \cap Y$.

$P(X \cup Y)$ will equal the sum of $P(X)$ and $P(Y)$ only when $X$ and $Y$ have no sample points in common. This case is considered in the next section.

*Example 3*

A die is to be thrown twice. Find the probability that

(a) the sum of the scores will be less than 4
(b) the product of the scores will be at least 20
(c) no 6 will be thrown
(d) at least one six will be thrown.

Let $m$ be the score on the first throw and $n$ the score on the second throw. Then the possibility space is the set of ordered pairs $(m, n)$ where $m$ and $n$ each take the values 1, 2, 3, 4, 5 and 6. The 36 sample points are shown in the table below.

$$
\begin{array}{cccccc}
(1, 1) & (1, 2) & (1, 3) & (1, 4) & (1, 5) & (1, 6) \\
(2, 1) & (2, 2) & (2, 3) & (2, 4) & (2, 5) & (2, 6) \\
(3, 1) & (3, 2) & (3, 3) & (3, 4) & (3, 5) & (3, 6) \\
(4, 1) & (4, 2) & (4, 3) & (4, 4) & (4, 5) & (4, 6) \\
(5, 1) & (5, 2) & (5, 3) & (5, 4) & (5, 5) & (5, 6) \\
(6, 1) & (6, 2) & (6, 3) & (6, 4) & (6, 5) & (6, 6)
\end{array}
$$

On the assumption that the outcomes are all equally likely, each sample point will have the probability 1/36.

(a) There are three sample points such that the sum of the scores is less than 4, namely (1, 1), (1, 2) and (2, 1). Hence the probability of this event is 3/36, i.e. 1/12.

(b) There are eight sample points for which the product of the scores is at least 20, namely (5, 4), (4, 5), (6, 4), (5, 5), (4, 6), (6, 5), (5, 6) and (6, 6). Hence the probability of this event is 8/36, i.e. 2/9.

(c) There are 25 sample points in which neither $m$ nor $n$ takes the value 6. Therefore the probability that no 6 is thrown is 25/36.

(d) There are 11 sample points containing at least one 6, and so the required probability is 11/36.

Note that the events described in (c) and (d) are complementary events, and that the sum of their probabilities is 1.

*Example 4*

A card is to be drawn at random from a pack and replaced. Then another card is to be drawn at random. Find the probability that

(a) both cards will be spades
(b) neither card will be a spade
(c) one card will be a spade and the other will not.

Let $S$ denote 'a spade is drawn', and similarly $H$, $D$ and $C$ for the other suits. The sixteen sample points in the possibility space are shown in the table below.

$$
\begin{array}{cccc}
SS & HS & DS & CS \\
SH & HH & DH & CH \\
SD & HD & DD & CD \\
SC & HC & DC & CC
\end{array}
$$

If the 16 outcomes are assumed to be equally likely, the probability of each sample point is 1/16.

(a) The event $\{SS\}$ contains only one sample point and so the probability of this event is 1/16.

(b) There are 9 outcomes in which neither card is a spade. Therefore the probability of this event is 9/16.

(c) There are 3 sample points in the first column of the table and 3 in the top row such that one card is a spade and the other is not. Hence the probability of this event is 6/16, i.e. 3/8.

## Exercises 11.4

1  Perform the experiment described in the last example one hundred times, counting the number of times that neither card is a spade.

2  Spin two coins and record whether the result is two heads, two tails or a head and a tail. Repeat this experiment a hundred times.

3  Draw a square grid on paper, the length of the side of each square being twice the diameter of a coin. Place the paper on a table and roll the coin on the grid. Record whether the coin comes to rest across a line or not. Repeat the experiment forty times. Prove that the probability that the coin will rest across a line is 3/4.

4  A pencil of length $L$ is thrown in a random manner on to a horizontal sheet of ruled paper, the distance between adjacent lines being $d$. The number $n$ of lines crossed by the pencil when it comes to rest is recorded. Repeat this experiment fifty times, observing whether $n$ is greater or less than $L/2d$.

5  Two coins are to be tossed. Find the probability that at least one of them will come down heads.

6  Two dice are to be thrown simultaneously. Find the probability that the total score will be (a) 1 (b) 5 (c) 11.

7  A card is to be drawn at random from a pack. Find the probability that (a) it will be an ace (b) it will not be a club.

8  A positive integer less than 100 is to be chosen at random. Find the probability that it will be a multiple of 11.

9  A card is to be drawn at random from a pack. Find the probability that it will be (a) a king, queen or jack (b) none of these.

10  A die is to be thrown twice. Find the probability that the second score will be greater than the first.

11  Three dice are to be thrown. Find the number of sample points in the probability space. Find also the probability that the total score will not exceed 4.

12  A coin is to be tossed and a die thrown. List the sample points for this experiment, and find the probability that the coin will show heads and the die will show an even number.

## 11.5 Mutually exclusive events

Two events $X$ and $Y$ are said to be mutually exclusive if, in any trial, $X$ cannot occur when $Y$ occurs and $Y$ cannot occur when $X$ occurs.

When a die is thrown, the event $\{1, 3\}$ occurs when either 1 or 3 is scored, and the event $\{2, 4, 6\}$ occurs when 2 or 4 or 6 is scored. These two events are mutually exclusive since they cannot both occur in the same trial. The events $\{1, 3\}$ and $\{2, 3, 5\}$ are not mutually exclusive since they both occur when the outcome of a trial is 3.

Let an experiment have $n$ outcomes, all equally likely. Suppose that the event $X$ contains $r$ sample points and that none of these sample points is contained in the event $Y$. Suppose that $Y$ contains $s$ sample points none of which is contained in $X$. Then $X$ and $Y$ are mutually exclusive. Now $P(X) = r/n$ and $P(Y) = s/n$. But there are $(r + s)$ outcomes for which either $X$ or $Y$ will occur. It follows that

$$P(\text{either } X \text{ or } Y) = (r + s)/n,$$

i.e. $$P(\text{either } X \text{ or } Y) = P(X) + P(Y).$$

This is the *addition law* for mutually exclusive events.

*Example 1*
A card is to be drawn at random from a pack. Find the probability
(a) that it will be either a spade or a red king
(b) that it will be either a spade or a black king (or both).

(a) The probability that the card is a spade is $13/52$.
The probability that it is a red king is $2/52$.
Since the card cannot be both a spade and a red king, the two events are mutually exclusive and the addition law applies.
Hence the required probability $= 13/52 + 2/52 = 15/52$.
(b) Since the card may be the king of spades, the two events are not mutually exclusive.
There are 13 outcomes in which a spade is chosen and 2 outcomes in which a black king is chosen, but there is one outcome which belongs to both sets. Hence there are only 14 favourable outcomes, and the required probability is $14/52$, i.e. $7/26$.

*Example 2*
A die is to be thrown twice. Find the probability that either the product of the scores will be 12 or that the sum of the scores will be 10.

From the table of sample points in Example 3 of Section 11.4, it can be seen that the sample points for which the product of the scores is 12 are $(2, 6)$, $(3, 4)$, $(4, 3)$, $(6, 2)$.
Let $X$ be the event $\{(2, 6), (3, 4), (4, 3), (6, 2)\}$.

The sample points for which the sum of the scores is 10 are $(4, 6)$, $(5, 5)$, $(6, 4)$.
Let $Y$ be the event $\{(4, 6), (5, 5), (6, 4)\}$.

The events $X$ and $Y$ have no sample points in common, and so they are mutually exclusive events. By the addition law,

$$P(\text{either } X \text{ or } Y) = P(X) + P(Y)$$
$$= 4/36 + 3/36 = 7/36.$$

**Exercises 11.5**

1  A bag contains 3 red discs, 4 green discs, 5 yellow discs and 6 blue discs. If a disc is drawn at random, find the probability that it will be (a) blue (b) green (c) either green or blue (d) neither green nor blue.

2  A die is to be thrown. Find the probability that the score will be either less than four or greater than five.

3  A card is to be drawn from a pack. Find the probability that it will be either the ace of clubs or else not an ace.

4  Find whether, when a die is thrown, the following pairs of events are mutually exclusive or not.
   (a) $\{1, 3, 5\}, \{4, 5\}$
   (b) $\{1, 2, 3\}, \{4, 5, 6\}$
   (c) $\{4, 5\}, \{6\}$
   (d) $\{3\}, \{2, 4, 6\}$.

5  A card is to be drawn at random from a pack. Find the probability that it will be
   (a) the queen of hearts or an ace
   (b) a two, a red four or a black six
   (c) neither a king nor a red card.

6  Two dice are to be thrown. $X$ is the event described by 'the sum of the scores is 10' and $Y$ is the event described by 'the difference between the scores is 3'. Show that the events $X$ and $Y$ are mutually exclusive. Find $P(X)$, $P(Y)$, $P(X \cup Y)$ and $P(X \cap Y)$.

## 11.6  Independent events

Consider an experiment in which a red die and a green die are thrown. Let the score on the red die be denoted by $m$ and the score on the green die be denoted by $n$. The possibility space will have 36 sample points, which are shown in the table of Example 3 in Section 11.4.

Let $A$ be the event '$m = 1$ or 2' and let $B$ be the event '$n = 6$'. 12 sample points belong to $A$ and 6 sample points belong to $B$. Therefore.

$$P(A) = 12/36 = 1/3, \quad P(B) = 6/36 = 1/6.$$

This gives $P(A) \times P(B) = 1/18$.

Only two sample points are favourable to the event $A \cap B$, namely $(1, 6)$ and $(2, 6)$, and so $P(A \cap B) = 2/36 = 1/18$. In this case

$$P(A \cap B) = P(A) \times P(B).$$

In contrast to this, consider the event $C$ described by $mn = 6$. Four sample points $(1, 6)$, $(2, 3)$, $(3, 2)$ and $(6, 1)$ belong to this event, and so

$$P(C) = 4/36 = 1/9.$$

Only the outcome $(1, 6)$ is favourable to both $B$ and $C$, and so $P(B \cap C)$ equals $1/36$. In this case, $P(B \cap C)$ does not equal $P(B) \times P(C)$.

When two events $X$ and $Y$ are such that

$$P(X \cap Y) = P(X) \times P(Y),$$

$X$ and $Y$ are said to be independent. This is the *multiplication law* for independent events.

In the experiment considered above, $A$ and $B$ are independent events, whereas $B$ and $C$ are not independent events.

Do not confuse independent events with mutually exclusive events. Independent events can occur together, but mutually exclusive events cannot occur together.

*Example*
A card is to be drawn at random from a pack. Show that the event $X$, 'a king is drawn', is independent of the event $Y$, 'a heart is drawn'.

There are four kings in the pack and so four of the 52 outcomes are favourable to $X$, giving $P(X) = 4/52 = 1/13$.
There are 13 hearts in the pack, giving $P(Y) = 13/52 = 1/4$.

Only one outcome (when the king of hearts is drawn) produces the event $X \cap Y$. Therefore

$$P(X \cap Y) = 1/52 = P(X) \times P(Y),$$

showing that the events $X$ and $Y$ are independent.

**Exercises 11.6**
1  A card is drawn at random from a pack. Find which of the following pairs of events are independent.
   (i) An ace is drawn; the ace of clubs is drawn.
   (ii) A spade is drawn; a red card is drawn.
   (iii) The queen of clubs is drawn; the two of hearts is drawn.
   (iv) A red six is drawn; a heart is drawn.
2  Two dice are thrown. Show that the events 'the sum of the scores is 6' and 'the difference between the scores is 2' are not independent.
3  A die is thrown. Show that the events $\{1, 2, 3\}$ and $\{3, 4\}$ are independent. Show also that the events $\{1, 2, 3, 4\}$ and $\{1, 2, 6\}$ are independent.
4  A boy picks a number $m$ at random from the integers $1, 2, 3$ and $4$. A girl also picks a number $n$ at random from the same four integers. Show that the events '$m = 1$' and '$m = n$' are independent. Show also that the events '$m + n = 5$' and '$mn = 4$' are not independent.

5    A bag contains 4 red balls numbered 1, 2, 3 and 4 respectively and 2 yellow balls numbered 5 and 6 respectively. A ball is drawn at random from the bag. Show that the events 'yellow ball' and 'even number' are independent. Show also that the events 'red ball' and 'ball number 1' are not independent.

6    In a class of 35 girls, 21 have fair hair and 25 have blue eyes, while 4 have neither. A girl is chosen at random from the class. Show that the event 'she has fair hair' and the event 'she has blue eyes' are independent.

7    In a class of 30 boys, 18 are 16 years old and 12 are 15 years old. Five boys in the class are left-handed. When a boy is chosen at random, the events 'he is 16 years old' and 'he is left-handed' are independent. Find the number of boys who are 16 years old and left-handed.

8    A football team plays 45 matches, winning 36 and losing 9. A certain goalkeeper $X$ plays in 25 of the matches, of which 5 were lost. If a match is chosen at random, show that the events 'a win' and '$X$ did not play' are independent.

## 11.7   Conditional probability

When a die is thrown, the probability that the outcome will be a 6 is 1/6. If it is known that in a given trial the outcome was an even number, the possibility space is reduced to three sample points. Given that the outcome is an even number, the probability that the outcome is a 6 is 1/3. This is known as the conditional probability of the event $\{6\}$, given that the event $\{2, 4, 6\}$ occurs.

The conditional probability of an event $X$, given the event $Y$, is denoted by $P(X \mid Y)$ and is defined by

$$P(X \mid Y) = \frac{P(X \cap Y)}{P(Y)}.$$

When all outcomes are equally likely, this gives

$$P(X \mid Y) = \frac{\text{number of outcomes favourable to both } X \text{ and } Y}{\text{number of outcomes favourable to } Y}.$$

The conditional probability of the event $Y$, given the event $X$, is defined by

$$P(Y \mid X) = \frac{P(Y \cap X)}{P(X)}$$
$$= \frac{\text{number of outcomes favourable to both } X \text{ and } Y}{\text{number of outcomes favourable to } X}.$$

*Example 1*
A die is thrown. $X$ is the event $\{1, 3, 5, 6\}$ and $Y$ is the event $\{1, 2\}$. Find $P(X \mid Y)$ and $P(Y \mid X)$.

Since the possibility space is $\{1, 2, 3, 4, 5, 6\}$, $P(X) = 2/3$ and $P(Y) = 1/3$. The intersection $X \cap Y$ of the events $X$ and $Y$ is the event $\{1\}$, and so $P(X \cap Y)$ equals 1/6. Then

$$P(X \mid Y) = \frac{P(X \cap Y)}{P(Y)} = (1/6)/(1/3) = 1/2,$$

$$P(Y \mid X) = \frac{P(X \cap Y)}{P(X)} = (1/6)/(2/3) = 1/4.$$

Suppose that the probability of an event $X$ is the same whether the event $Y$ occurs or not, i.e. suppose that

$$P(X \mid Y) = P(X).$$

This means that

$$\frac{P(X \cap Y)}{P(Y)} = P(X),$$

i.e. that

$$P(X \cap Y) = P(X) \times P(Y).$$

This is the multiplication law for independent events. When this is true, $X$ and $Y$ are independent events. It follows that

$$P(Y \mid X) = \frac{P(Y \cap X)}{P(X)} = P(Y),$$

i.e. the probability of $Y$ is the same whether $X$ occurs or not.

This proves that when the probability of $X$ is the same whether $Y$ occurs or not, the probability of $Y$ is the same whether $X$ occurs or not. When this is the case,

$$P(X \cap Y) = P(X) \times P(Y),$$

and $X$ and $Y$ are independent events.

*Example 2*
A die is to be thrown twice. Find the probability that
(a) two sixes will be thrown
(b) no sixes will be thrown
(c) at least one six will be thrown.

The outcome of the second throw is not affected by the outcome of the first throw, and so the multiplication law applies.
(a) For the first throw, $P(A) = P(\{6\}) = 1/6$.
    For the second throw, $P(B) = P(\{6\}) = 1/6$.
The events $A$ and $B$ are independent, and so

$$P(A \cap B) = P(A) \times P(B) = 1/36.$$

(b) For the first throw, $P(X) = P(\{1, 2, 3, 4, 5\}) = 5/6$.
    For the second throw, $P(Y) = P(\{1, 2, 3, 4, 5\}) = 5/6$.
Since $X$ and $Y$ are independent,

$$P(X \cap Y) = P(X) \times P(Y) = 25/36.$$

(c) Let $Z = X \cap Y$, i.e. let $Z$ be the event 'no sixes are thrown'. The complement of $Z$ will be the event 'at least one six is thrown'. This is denoted by $Z'$, and

$$\begin{aligned} P(Z') &= 1 - P(Z) \\ &= 1 - P(X \cap Y) \\ &= 1 - 25/36 \\ &= 11/36. \end{aligned}$$

*Example 3*
The letters a, b, c, d are arranged in a random order, all the permutations being equally likely. Find the probability that the letter a will come first, given that d comes last.

Let $X$ be the event 'the first letter is a', and let $Y$ be the event 'the last letter is d'.
  Only two permutations, abcd and acbd, are favourable to both $X$ and $Y$. With d last, there are six possible permutations.

$$P(X \mid Y) = \frac{\text{number of outcomes favourable to both } X \text{ and } Y}{\text{number of outcomes favourable to } Y}$$

$$= 2/6 = 1/3.$$

*Example 4*
A box contains two red discs and three green discs. A disc is taken from the box at random and is replaced. Then a second disc is taken at random. Find the probability that one and only one of the two discs is red.

The probability that the first disc is red is 2/5.
The probability that the second disc is not red is 3/5.

These two events are independent, and so the probability that the first disc is red and the second is green is given by (2/5)(3/5), i.e. 6/25.
  In the same way, the probability that the first disc is not red and the second disc is red equals (3/5)(2/5), i.e. 6/25. These two events are mutually exclusive. Now if two events $A$ and $B$ are mutually exclusive, $P(A \cap B) = 0$ and

$$P(A \cup B) = P(A) + P(B).$$

It follows that the probability that one and only one disc is red is 6/25 + 6/25, i.e. 12/25.
  Alternatively, it can be argued that the probability of drawing two red discs is (2/5)(2/5), i.e. 4/25, and the probability of drawing two green discs is (3/5)(3/5), i.e. 9/25. These two events are mutually exclusive, and so the probability of drawing two discs of the same colour is given by 4/25 + 9/25, i.e. 13/25.
Now the event 'two discs of the same colour' is the complement of the event 'one and only one red disc'. The sum of the probabilities of two complementary

events is 1. Hence the probability of drawing one and only one red disc is
$1 - 13/25$, i.e. 12/25 as before.

### Exercises 11.7

1  A card is drawn at random from a pack, and is replaced. A second card is then drawn. Given that the first card is red, find the probability that the second card is also red.

2  A coin is tossed three times in succession. Find the probability that it comes down heads each time. Given that this event occurs, state the probability that a fourth toss will give tails.

3  A coin is tossed and a die is thrown. Find the probability
   (a) of a head and a six
   (b) of a tail and an odd number.

4  The five letters a, b, c, d, e are arranged in a random order, all the permutations being equally likely. Find the probability that the letters a and b will come next to one another, given that the letter c comes in the middle place.

5  Find the probability that when two dice are thrown they will show the same score.

6  Two coins are to be tossed and two dice thrown. Find the probability that there will be a head, a tail, a two and a four.

7  Given that $P(A) = 1/4$, $P(B) = 1/3$ and $P(A \cup B) = 1/2$, find the values of $P(A \mid B)$ and $P(B \mid A)$.

8  Two dice are thrown. Find the probability that the sum of the numbers thrown is greater than 9, given that at least one 6 is thrown.

9  When two dice are thrown, show that the events 'throwing a 4 on at least one throw' and 'throwing the same number twice' are not independent.

10  A box contains three red discs, two green discs and a blue disc. A disc is drawn at random and is replaced. Then a second disc is drawn at random. Show that the probability that the two discs are the same colour is 7/18. Given that the two discs are not the same colour, find the probability that one of the discs drawn is blue.

11  Seven cards marked 1 to 7 respectively are placed in a box. Three cards are drawn at random and are not replaced. Find the probability that their sum is 10, given that the card marked 5 is not drawn.

12  Two dice are thrown. Find the probability that the total score will be 6, given that neither die shows an even number.

13  Three dice are thrown. Find the probability that the total score is 10, given that no sixes are thrown.

14  A pack of cards is divided into four hands each containing thirteen cards of one suit. A card is chosen at random from one hand. Then another card is chosen at random from one of the other hands. Given that the first card is the ace of hearts, find the probability that the second card is (a) the ace of diamonds (b) the king of hearts.

**15**  Ten cards marked from 1 to 10 respectively are placed in a box and a card is drawn at random. A is the event $\{1, 3, 5, 7\}$ and B is the event $\{4, 5, 6\}$. Find the values of $P(A \mid B)$, $P(B \mid A)$ and $P(A' \mid B)$.

**16**  A boy spins a coin three times and a girl spins a coin twice. Show that the probability that the girl gets more heads than the boy is 3/16.

Given that the girl gets more heads than the boy, show that the probability that the boy got no heads is 1/2.

## 11.8  Tree diagrams

The purpose of a tree diagram is to assist in finding the various probabilities of the outcomes of a sequence of trials. Consider the tree diagram shown in Fig. 11.3. The first trial has two possible outcomes, represented by the branches OA and OB.

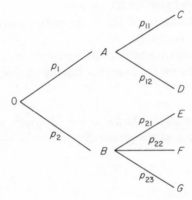

Fig. 11.3

The event $A$ has probability $p_1$ and the event $B$ has probability $p_2$. Since these are the only possible outcomes,

$$p_1 + p_2 = 1.$$

When $A$ is the outcome of the first trial, there are two possible outcomes $C$ and $D$ of the second trial, with probabilities $p_{11}$ and $p_{12}$ respectively. Note that $p_{11} + p_{12} = 1$ (Fig. 11.3).

When $B$ is the outcome of the first trial, the diagram shows three outcomes $E$, $F$ and $G$ for the second trial, with probabilities $p_{21}, p_{22}$ and $p_{23}$ respectively. For this trial,

$$p_{21} + p_{22} + p_{23} = 1.$$

The probability that $A$ will be the outcome of the first trial and then $C$ will be the outcome of the second trial is given by

$$P(A \cap C) = P(A) \times P(C \mid A)$$
$$= p_1 \times p_{11}.$$

The probability that $A$ will be the outcome of the first trial and then $D$ will be the outcome of the second trial is given by

$$P(A \cap D) = P(A) \times P(D \mid A)$$
$$= p_1 \times p_{12}.$$

In the same way,

$$P(B \cap E) = p_2 \times p_{21},$$
$$P(B \cap F) = p_2 \times p_{22},$$
$$P(B \cap G) = p_2 \times p_{23}.$$

The sum of these five probabilities equals

$$p_1(p_{11} + p_{12}) + p_2(p_{21} + p_{22} + p_{23}) = p_1 + p_2 = 1.$$

There are two essential facts to remember.
(a) At any junction, the sum of the probabilities marked on the branches leaving the junction must equal 1.
(b) The probability of any particular final outcome is equal to the product of the probabilities marked on the branches leading to that outcome.

*Example 1*
A coin is tossed three times. Draw a tree diagram showing the possible outcomes, and use it to find the probability of two heads and one tail.

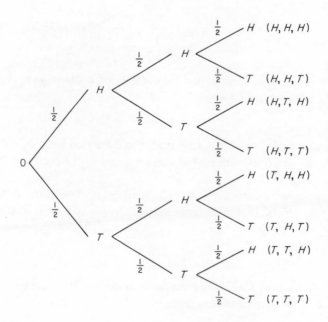

Fig. 11.4

The eight sample points are shown on the right-hand side of the diagram (Fig. 11.4). The probability of each outcome is $\frac{1}{2} \times \frac{1}{2} \times \frac{1}{2}$, i.e. 1/8. The three sample points $(H, H, T)$, $(H, T, H)$ and $(T, H, H)$ give 'two heads and one tail', and therefore the probability of this event is 3/8.

*Example 2*
Eight playing cards, two hearts and six spades are face downwards on a table. A card is drawn at random and is not replaced. Another card is then drawn. Find the probability that the two cards drawn are of the same suit.

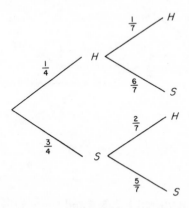

**Fig. 11.5**

At the first draw, $P(H) = 1/4$ and $P(S) = 3/4$ (Fig. 11.5). If the first card is a heart, only one heart is left. Therefore at the second draw, $P(H) = 1/7$ and $P(S) = 6/7$. Hence the probability of drawing two hearts is $(1/4)(1/7)$, i.e. 1/28.

If the first card is a spade, five spades are left. The probability of drawing a second spade is 5/7, and therefore the probability of drawing two spades is $(3/4)(5/7)$, i.e. 15/28. Since the events 'two hearts' and 'two spades' are mutually exclusive, the probability that one or other will occur is the sum of their probabilities. This gives $1/28 + 15/28$, i.e. 4/7.

*Example 3*
Box A contains 5 balls:1 red, 3 green and 1 blue. Box B contains 5 balls:2 red, 1 green and 2 blue. A die is thrown and if the throw is a 6, box A is chosen; otherwise box B is chosen. A ball is drawn at random from the chosen box. Find the probability that this ball is (a) red (b) green. Given that a green ball is drawn, find the probability that it came from box A.

$$P(\text{red}) = P(A \cap \text{red}) + P(B \cap \text{red})$$
$$= (1/6)(1/5) + (5/6)(2/5) = 11/30.$$

$$P(\text{green}) = P(A \cap \text{green}) + P(B \cap \text{green})$$
$$= (1/6)(3/5) + (5/6)(1/5) = 4/15.$$

$$P(A \mid \text{green}) = \frac{P(A \cap \text{green})}{P(\text{green})} = \frac{(1/6)(3/5)}{(4/15)} = 3/8.$$

These results can be verified from Fig. 11.6.

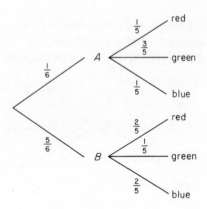

Fig. 11.6

## Exercises 11.8

1  From five tickets numbered 1, 2, 3, 4 and 5 respectively two tickets are to be drawn without replacement. Draw a tree diagram and use it to find the probability that
   (a) the two even numbers will be drawn
   (b) two odd numbers will be drawn
   (c) one even and one odd number will be drawn.

2  The three top cards are to be dealt from a pack. Find the probability that (a) they will all be red (b) two will be red and one will be black.
   Illustrate with a tree diagram.

3  Two blue balls and two yellow balls are placed in a box. Two balls are to be taken out at random without replacement. Draw a tree diagram and find the probability that they will be of different colours.

4  Five playing cards, a king and four aces, are placed face downwards on a table. A card is taken at random without replacement, and this action is repeated until the king is taken. Draw a tree diagram and find the probability that the king will be the last card taken.

5  Three playing cards, a king, a queen and a jack, are placed face downwards on a table. A card is chosen at random and then the three cards are shuffled and replaced on the table. This is done three times. Draw a tree diagram and find the probability that
   (a) the king is chosen each time
   (b) the king is chosen twice and the queen once
   (c) each card is chosen once.

## 11.9   Repeated trials

Consider an experiment in which a coin is tossed four times. The probability of four heads will be $\frac{1}{2} \times \frac{1}{2} \times \frac{1}{2} \times \frac{1}{2}$, i.e.

$$P(4 \text{ heads}) = 1/2^4.$$

The outcome can be 3 heads and 1 tail in four ways, represented by $HHHT$, $HHTH$, $HTHH$ and $THHH$.

$$P(3 \text{ heads, 1 tail}) = 4/2^4.$$

Since there are six permutations of the four letters $H$, $H$, $T$, $T$,

$$P(2 \text{ heads, 2 tails}) = 6/2^4.$$

The outcome can be 1 head and 3 tails in four ways.

$$P(1 \text{ head, 3 tails}) = 4/2^4.$$

Finally

$$P(4 \text{ tails}) = 1/2^4.$$

Note that the five probabilities are the terms in the binomial expansion

$$(\tfrac{1}{2} + \tfrac{1}{2})^4 = 1/2^4 + 4/2^4 + 6/2^4 + 4/2^4 + 1/2^4.$$

In the general case, suppose that in each trial the outcome is either a success with probability $p$, or a failure with probability $q$, where $p + q = 1$. Consider the probability of $r$ successes in $n$ independent trials. The number of permutations of $r$ successes and $n - r$ failures is $\dfrac{n!}{r!(n-r)!}$, i.e. $\binom{n}{r}$. Therefore

$$P(r \text{ successes in } n \text{ trials}) = \binom{n}{r} p^r q^{n-r}.$$

Thus the terms in the binomial expansion of $(p + q)^n$ give in turn the probability of $n, n-1, n-2, \ldots$ successes in $n$ trials.

*Example 1*
A coin is tossed seven times. Find the probability of four heads and three tails.

Consider a head as a success and a tail as a failure. In this case, $p = q = \frac{1}{2}$ and $n = 7, r = 4$.

$$\binom{7}{4} = \frac{7!}{4!3!} = 35.$$

$P(4 \text{ heads, 3 tails}) = 35(\tfrac{1}{2})^4(\tfrac{1}{2})^3 = 35/128.$

*Example 2*
Six cards are drawn in turn from a pack with replacement. Find the probability that exactly four will be spades.

The probability that a card will be a spade is 1/4. In this case, $p = 1/4, q = 3/4$, $n = 6$ and $r = 4$. The probability that exactly four cards will be spades is given by

$$\binom{6}{4} \left(\frac{1}{4}\right)^4 \left(\frac{3}{4}\right)^2 = \frac{135}{4096}.$$

*Example 3*
A die is thrown repeatedly until a six is thrown. Find an expression for the probability that the first six is thrown at the $n$th attempt.

Let P($n$th throw) denote the probability that the first six comes on the $n$th throw.
P(1st throw) = 1/6.
Since the probability of failure on the first throw is 5/6,
P(2nd throw) = (5/6)(1/6).
Since the probability of failure on the first two throws is $(5/6)^2$
P(3rd throw) = $(5/6)^2(1/6)$.
By considering the probability of $(n - 1)$ consecutive failures followed by a success,
P($n$th throw) = $(5/6)^{n-1}(1/6) = 5^{n-1}/6^n$.

**Exercises 11.9**
1  A coin is to be tossed three times. Find the probability that two heads and one tail will be obtained.
2  A die is to be thrown four times. Find the probability of two even numbers and two odd numbers being thrown.
3  Three cards are to be drawn in turn from a pack with replacement. Find the probability that exactly one card will be a club.
4  A coin is to be tossed ten times. Find the probability that the result will be five heads and five tails.
5  A die is to be thrown ten times. Show that the probability that a six will be thrown at least twice is $1 - 3(5/6)^{10}$.
6  Five cards, an ace and four kings, are placed face downwards on a table. A card is drawn at random and is then replaced. This is done five times. Find the probability that
   (a) the ace is picked once only
   (b) the ace is picked three times.
7  The probability that a certain marksman will score a bull is 1/6. Find the probability that, if he fires repeatedly until he scores a bull, he will succeed at the third attempt.
8  Four coins are to be tossed simultaneously. Find the probability that all but one will fall the same way up.
9  In a batch of 1000 switches it is known that ten are faulty. Five switches are selected at random from this batch. Show that, to three decimal places, the

probability that exactly one will be faulty is 0.048.

10   It is claimed that the probability that a letter posted on Monday will be delivered on Tuesday is 0.9. Assuming that this claim is true, calculate to two significant figures the probability that
(a) if six letters are posted on Monday, all will be delivered on Tuesday
(b) if six letters are posted on Monday, only two will be delivered on Tuesday.

## Miscellaneous exercises 11

1   Find the number of permutations of the letters of the word CROSSES. Find also the number of permutations
(a) which begin with S
(b) which begin and end with S
(c) which neither begin nor end with S.

2   Three integers are to be chosen from the integers from 1 to 9 inclusive. Find the number of ways in which this can be done (a) if they must be even numbers (b) if they must be odd numbers (c) if two must be odd and one even.

3   A tennis club is to select a team of three pairs, each pair consisting of a man and a woman, for a match. The team is to be chosen from 7 men and 5 women. In how many different ways can the three pairs be selected?

[L]

4   A committee of three people is to be chosen from four married couples. Find in how many ways this committee can be chosen
(a) if all are eligible
(b) if the committee must consist of one woman and two men
(c) if all are equally eligible except that a husband and wife cannot both serve on the committee.                                                                    [L]

5   Find the number of permutations of the letters of the word COMMITTEE in which no two vowels are adjacent.

6   A letter is to be chosen at random from the word DEED and a letter is to be chosen at random from the word SEED. Make a list of the possible outcomes and find the probability that the two letters will be different.

7   An experiment has six outcomes, all equally likely. Two of the outcomes produce the event $A$, three produce the event $B$ and two produce both $A$ and $B$. Evaluate $P(A \mid B)$, $P(B \mid A)$, $P(A \mid B')$ and $P(B \mid A')$.

8   Three coins are tossed. Show that the events 'two heads and one tail' and 'at least one tail' are not independent.

9   Four cards are to be drawn in succession from a pack with replacement. Find the probability that
(a) four hearts will be drawn
(b) four cards of the same suit will be drawn
(c) the cards drawn will be all of different suits.

10   A bag contains two white balls and eight black balls. Two balls are to be

drawn from the bag without replacement. Use a tree diagram to find the probability that one will be white and one black.

11 The random events $A$, $B$ and $C$ are defined on a finite sample space $S$. The events $A$ and $B$ are mutually exclusive and the events $A$ and $C$ are independent. $P(A) = 1/5$, $P(B) = 1/10$, $P(A \cup C) = 7/15$ and $P(B \cup C) = 23/60$. Evaluate $P(A \cup B)$, $P(A \cap B)$, $P(A \cap C)$ and $P(B \cap C)$, and state whether events $B$ and $C$ are independent. [L]

12 A gold urn contains 3 red balls and 4 white balls, and a silver urn contains 5 red balls and 2 white balls. A die is rolled and, if a six shows, one ball is selected at random from the gold urn. Otherwise a ball is selected at random from the silver urn. Find the probability of selecting a red ball.

The ball selected is not replaced and a second ball is selected at random from the same urn. Find the probability that both balls are white. [L]

13 A box contains 9 discs, of which 4 are red, 3 are white and 2 are blue. Three discs are to be drawn at random without replacement from the box. Calculate

(i) the probability that the discs, in the order drawn, will be coloured red, white and blue respectively

(ii) the probability that one disc of each colour will be drawn

(iii) the probability that the third disc drawn will be red

(iv) the probability that no red disc will be drawn

(v) the most probable number of red discs that will be drawn

(vi) the expected number of red discs that will be drawn, and state the probability that this expected number of red discs will be drawn. [JMB]

14 Two events $A$ and $B$ are such that $P(A) = 0.4$ and $P(A \cup B) = 0.7$.

(i) Find the value of $P(A' \cap B)$.

(ii) Find the value of $P(B)$ if $A$ and $B$ are mutually exclusive.

(iii) Find the value of $P(B)$ if $A$ and $B$ are independent. [JMB]

15 A trial consists of selecting a card at random from a pack and then replacing it. Find the probability that in 12 trials a red card will be selected more than 6 times.

16 Two different integers are to be chosen at random from the integers 1 to 9 inclusive. Given that the sum of the two integers is even, find the probability that

(a) they will differ by 4

(b) they will both be odd.

17 Find the number of ways in which a committee of 4 can be chosen from 4 boys and 6 girls

(a) if it must contain at least one boy and one girl

(b) if two girls who are sisters cannot both be chosen.

18 Eight cards bear the numbers 1, 1, 2, 2, 2, 3, 4, 4. Two cards are drawn without replacement. Find the probability that the numbers on the two cards will add up to (i) exactly 5 (ii) more than five. [AEB]

19 Find the number of different arrangements of all the letters of the word

GEOMETRIC. Find also the number of arrangements in which
(i) the vowels occur together
(ii) no two vowels are together.
(Answers may be left in factorial form.) [AEB]

20 The probabilities of two events $A$ and $B$ are 2/3 and 3/4. Show that the probability of them both occurring lies between 5/12 and 2/3. Describe the relationships between the events for these limiting values to be taken. What is the probability of $A$ or $B$ (or both) occurring if the events are independent? [AEB]

21 Two cards are drawn without replacement from a pack of playing cards. Using a tree diagram, or otherwise, calculate the probability
(a) that both cards are aces
(b) that one (and only one) card is an ace
(c) that the two cards are of different suits.
  Given that at least one ace is drawn, find the probability that the two cards are of different suits. [L]

22 A card is drawn at random from a pack of playing cards. Find the probability that it is
(a) a court card (i.e. king, queen or jack)
(b) a red card
(c) a red card or a court card. [L]

23 Two cards are taken, without replacement, from a well-shuffled pack of playing cards. Calculate the probability that
(a) neither card is an ace
(b) neither card is a spade.
  Given that neither card is an ace, find the probability that neither card is a spade. [L]

24 In a game a turn consists of drawing a card from a pack of playing cards and rolling one of two dice. If the card is a heart, the die rolled is a red one numbered 2, 4, 6, 8, 10, 12; if not, a blue die numbered 1, 2, 3, 4, 5, 6 is rolled. Find the probability of obtaining a total score of 6 on the dice in two turns. [L]

25 In a competition, six items have to be chosen from eight items labelled A, B, C, D, E, F, G, H and arranged in order of preference. Find the number of different entries that can be made. In how many of these will the items A and B be found in adjacent positions?

26 Find the number of ways in which ten children can be divided into two groups of five. In how many of these ways are the two oldest children in the same group?

27 The following information is known about the three events $A$, $B$ and $C$:
$P(A) = \frac{1}{3}$, $P(B) = \frac{1}{4}$, $P(C) = \frac{1}{2}$, $P(A \cup B \cup C) = 1$,
$P(B \cap C) = 0$, $P(C \cap A) = 0$.
Show that
(a) $P(A \cap B) = P(A) \times P(B)$,

(b) $P(A \cup B) = P(C)$,

(c) $P(A \cup C) = P(A) + P(C)$.

Write down two of the events $A$, $B$, $C$ which are independent.　　[L]

**28**　Three cards are drawn at random without replacement from a pack of ten cards which are numbered from 1 to 10 respectively. Calculate

(i) the probability that the numbers drawn consist of two even numbers and one odd number

(ii) the probability that at least one of the numbers drawn is a perfect square greater than 1

(iii) the probability that the smallest number drawn is the 5.　　[JMB]

**29**　In a certain gambling game, a player nominates an integer $x$ from 1 to 6 inclusive and he then throws three fair cubical dice. Calculate the probabilities that the number of $x$'s thrown will be 0, 1, 2, 3. The player pays 5 pence per play of the game and he receives 48 pence if the number of $x$'s thrown is three, 15 pence if the number of $x$'s thrown is two, 5 pence if only one $x$ is thrown and nothing otherwise. Calculate the player's expected gain or loss per play of the game.　　[JMB]

**30**　The events $A$ and $B$ are such that $P(A) = 0.6$, $P(B) = 0.2$, $P(A \mid B) = 0.1$. Calculate the probabilities that

(i) both of the events occur

(ii) at least one of the events occurs

(iii) exactly one of the events occurs

(iv) $B$ occurs given that $A$ has occurred.　　[JMB]

**31**　Nine counters are identical except for their colour. Two are blue, two red, two green, two yellow and one is black. How many distinguishable sets of three counters can be selected from the nine?

　Three counters are to be selected at random from the nine counters. Find the probability that they will be of three different colours.

**32**　In a school, 40% of the boys have fair hair, 25% have blue eyes, 15% have both fair hair and blue eyes. A boy is selected at random.

(a) Given that he has fair hair, find the probability that he has blue eyes.

(b) Given that he has blue eyes, find the probability that he does not have fair hair.

(c) Find the probability that he has neither fair hair nor blue eyes.　[L]

**33**　A boy and a girl spin a coin in turn, and the first to get a 'head' is the winner. The girl spins first. Find the probability that the boy wins.　[L]

**34**　Tom, Dick and Harry throw a die in turn, and the first to throw a six is the winner. Find the probability that each boy has of winning.

**35**　Four cards are drawn at random from a pack, one at a time with replacement. Find the probability that

(a) no heart is drawn

(b) four hearts are drawn

(c) two hearts and two diamonds are drawn (in any order)

(d) one card from each suit is drawn.　　[L]

**36** A box contains only four red balls and five white balls. Three balls are drawn successively at random from the box without replacement. Draw a tree diagram illustrating the various possibilities. Hence, or otherwise, find the probability that three white and three red balls are left in the box.

[L]

**37** Nine discs numbered from 1 to 9 are placed in a bag and three discs are then drawn at random without replacement. The number on the first disc drawn is denoted by $n$, and the sum of the numbers on the three discs is denoted by $S$.
(a) Find the probability that $S = 10$.
(b) Find the probability that $n = 2$ and $S = 10$.
(c) Given that $n = 2$, find the probability that $S = 10$.
(d) Given that $n \neq 2$, find the probability that $S = 10$.     [L]

**38** A man rolls a die to select one of three boxes. If he rolls a 6, he selects the red box; if he rolls a 5 or a 4, he selects the blue box; and if he rolls a 3 or a 2 or a 1, he selects the yellow box. He opens the box he has chosen and selects a coin at random from it. The red box contains three gold coins, the blue box two gold coins and one silver coin, and the yellow box, one gold coin and two bronze coins. Using a tree diagram, or otherwise, find the probability that
(a) he selects a silver coin
(b) he selects a gold coin
(c) having selected and retained a gold coin, if he now selects at random a second coin from the same box, it will also be gold.     [L]

**39** One of three coins is biased so that the probability of obtaining a head is twice as great as the probability of obtaining a tail. The other two coins are fair. One of the three coins is chosen at random and tossed three times, showing a head on each occasion. Using a tree diagram, or otherwise, find the probability that the chosen coin is biased.     [L]

**40** In a building programme, the event that all the materials will be delivered at the correct time is $M$, and the event that the building programme is completed on time is $F$. Given that $P(M) = 0.8$ and $P(M \cap F) = 0.65$, explain in words the meaning of $P(F \mid M)$ and calculate its value.

If $P(F) = 0.7$, find the probability that the building programme will be completed on time if all the materials are not delivered at the correct time.     [L]

**41** Given that $P(A) = 1/2$, $P(B) = 1/3$, $P(C) = 1/4$ and that events $A$ and $B$ are independent, events $A$ and $C$ are independent and events $B$ and $C$ are mutually exclusive, find
(a) $P(A \cup C)$ (b) $P(A \cap B)$ (c) $P(A' \cap B' \cap C')$
giving in each case sufficient explanation to show how your results are obtained.     [L]

**42** A trial consists of rolling a red die and a blue die. The result $R$ of the trial is defined as the sum of the numbers showing when the numbers on the red and blue dice are the same but as the product of these two numbers when they are different. Find

(a) $P(R = 6)$ (b) $P(R = 8)$ (c) $P(R = 12)$.

If the results of two trials are added together, what is the probability of getting more than 45? [L]

43 A drawer contains four black, six brown and two blue socks. Two socks are taken at random from the drawer, without replacement. Find the probability that they are the same colour. Given that they are of the same colour, find the probability that they are blue.

# Formulae

## Numerical methods

*Newton-Raphson method* $\quad x_2 = x_1 - \dfrac{f(x_1)}{f'(x_1)}$ 

*Trapezium rule*

$$\int_a^b y\,dx \approx h(\tfrac{1}{2}y_0 + y_1 + y_2 + \cdots + y_{n-1} + \tfrac{1}{2}y_n)$$

where $\quad h = (b - a)/n$ 

*Simpson's rule for two strips*

$$\int_a^b y\,dx \approx \frac{h}{3}(y_0 + 4y_1 + y_2) \quad \text{where} \quad h = (b - a)/2.$$

*Simpson's rule for n strips (n even)*

$$\int_a^b y\,dx \approx \frac{h}{3}(y_0 + 4y_1 + 2y_2 + 4y_3 + \cdots + 4y_{n-1} + y_n)$$

where $\quad h = (b - a)/n.$ 

## Geometry

*Vector equations*

Equation of the line through the point with position vector **a** parallel to the vector **b** is

$$\mathbf{r} = \mathbf{a} + t\mathbf{b}$$

Equation of the plane at right angles to the vector **a** passing though the point with position vector **b** is

$$\mathbf{r} \cdot \mathbf{a} = \mathbf{b} \cdot \mathbf{a}$$

Equation of the plane at right angles to the unit vector **n** at a distance $p$ from the origin is

$$\mathbf{r} \cdot \mathbf{n} = p$$

*Cartesian equations*

Equations of the line through the point $(\alpha, \beta, \gamma)$ with direction ratios $a{:}b{:}c$ are

$$\frac{x - \alpha}{a} = \frac{y - \beta}{b} = \frac{z - \gamma}{c}$$

or $\quad x = \alpha + at, y = \beta + bt, z = \gamma + ct$

The direction ratios of the normal to the plane $ax + by + cz = d$     36
are $a{:}b{:}c$.

The perpendicular distance from the point $(x_1, y_1, z_1)$ to the plane
$ax + by + cz = d$ is

$$\left| \frac{ax_1 + by_1 + cz_1 - d}{(a^2 + b^2 + c^2)^{1/2}} \right|$$     49

## Mechanics

*Uniform acceleration*

$$s = \tfrac{1}{2}(u + v)t \qquad s = ut + \tfrac{1}{2}at^2$$
$$v = u + at \qquad v^2 = u^2 + 2as$$     57

*Variable acceleration*

$$v = \frac{ds}{dt} \qquad a = \frac{dv}{dt} = \frac{d^2s}{dt^2} = v\frac{dv}{ds}$$     56

$$\mathbf{v} = \frac{d\mathbf{r}}{dt} = \dot{\mathbf{r}} \qquad \mathbf{a} = \frac{d\mathbf{v}}{dt} = \frac{d^2\mathbf{r}}{dt^2} = \ddot{\mathbf{r}}$$     89

*Projectiles*

$$\text{Time of flight} = \frac{2u \sin \alpha}{g}$$     92

$$\text{Horizontal range} = \frac{u^2 \sin 2\alpha}{g}$$     93

$$\text{Maximum height} = \frac{u^2 \sin^2 \alpha}{2g}$$     92

Equation of path

$$y = x \tan \alpha - \frac{gx^2}{2u^2}(1 + \tan^2 \alpha)$$     93

*Newton's second law*

$$\mathbf{F} = \mathbf{ma} \qquad F = ma$$     120

*Energy of a particle*

Kinetic energy $= \tfrac{1}{2}mv^2$     201

Potential energy $= mgh$     201

*Elastic strings*

$$\text{Tension} = \left(\frac{\lambda}{l}\right)x$$ 146

$$\text{Potential energy} = \frac{1}{2}\left(\frac{\lambda}{l}\right)x^2$$ 208

*Simple harmonic motion*

| | | |
|---|---|---|
| Acceleration | $\ddot{x} = -\omega^2 x$ | 339 |
| Velocity | $v^2 = \omega^2(a^2 - x^2)$ | |
| Displacement | $x = a \sin(\omega t)$  if $x = 0$ and $\dot{x} = a\omega$ when $t = 0$. | |
| Periodic time | $T = 2\pi/\omega$ | |
| Frequency | $n = 1/T$ | |

## Probability

Number of permutations of $n$ different things $r$ at a time

$$^nP_r = \frac{n!}{(n-r)!}$$ 362

Number of combinations of $n$ different things $r$ at a time

$$^nC_r = \binom{n}{r} = \frac{n!}{r!(n-r)!}$$ 366

For any events $X$ and $Y$,

$$P(X \cup Y) = P(X) + P(Y) - P(X \cap Y)$$ 371

*Addition law for mutually exclusive events*
If $X$ and $Y$ are mutually exclusive,

$$P(\text{either } X \text{ or } Y) = P(X) + P(Y)$$ 374

*Multiplication law for independent events*
If $X$ and $Y$ are independent,

$$P(X \cap Y) = P(X) \times P(Y)$$ 376

*Complementary events*
The complement of the event $X$ is denoted by $X'$ (or by $\bar{X}$)

$$P(X') = 1 - P(X)$$ 370

*Conditional probability*

$$P(X \mid Y) = \frac{P(X \cap Y)}{P(Y)}$$ 377

# Answers

## Chapter

**1.1**
**1** 0.5, $-8$, 1.5
**2** (a) $s = 3t + 1$
(b) $2v = 3u + 5$
(c) $4q = 3p - 10$
(d) $y = 9 - 5x$
**3** $L = 20 + 0.8T$
**4** $u = 6.0, f = -1.5$
**5** 19.3 for 20.3

**1.2**
**1** (a) $X = x^2, Y = y/x$
(b) $X = u, Y = \ln v$
(c) $X = \log (1 + t), Y = \log s$
(d) $X = \sqrt{x}, Y = \sqrt{y}$
**2** (a) $a, b$ (b) $\log a, -\log b$
(c) $a, b + 1$ (d) $-k, \log a$
**3** 0.20, 0.25
**4** 2.4, 3.5
**5** 100, 4
**6** 31, $-1.9$
**7** 500, $-3$
**8** $-3$, 20
**9** 2.0, 1.2
**10** 1.9, 0.24
**11** 1.73, $-0.5$
**12** 4, $-2$

**1.3**
**1** $-2.97, 0.96, 7.02$
**3** (a) 2.7 (b) $-2.2, 0, 2.2$ (c) 1.6
(d) 1.4 (e) 2.9 (f) $-1.4$
**4** (a) 0.62, 1.51 (b) $-0.26$
**5** (a) 2.55 (b) 1.17
**6** 1.22
**7** 2.09
**8** $-1.88$
**9** 0.64
**10** 1.13
**11** 0.96
**12** 0.739
**13** 2.05

**1.4**
**1** (a) 1.89 (b) $-1.92$ (c) 0.54
(d) 0.67
**2** 1.44
**3** 0.26, 2.54
**5** (a) 5.76 (b) 5.75
**6** 0.242
**7** 2.786
**10** 0.7391

**1.5**
**1** 3.16
**1.6**
**1** (a) $n$ (b) $1/2^{n-1}$ (c) $1/n!$
(d) $n^2$
**3** 1
**4** 1
**5** 1.796
**7** 2.080
**8** (a) 0.618 (b) 1.618
**9** 1.8556

**1.7**
**1** 14, $14\frac{2}{3}$
**2** 0.168
**3** 0.693
**5** 65.33
**6** 16.7, 16.7
**7** 7.13
**8** 13960 m³
**10** (a) 1.47 (b) 1.47

## Miscellaneous exercises 1

**1** $a = 1/2, b = 12$
**2** $a = 10, k = 7.4$
**3** $a = 2, b = 36$
**4** $a = 6.0, b = 1.2$
**5** $T_0 = 95, k = 0.0115$
**6** $a = 20, k = 0.5$
**7** $a = 0.76, b = -0.62$
**8** 2:1:2
**9** 1.32
**10** $a = 0.33, b = -0.50$
**12** $A = 17.6, m = 2.30, n = 0.61$
**13** 1.466
**15** 2.104
**17** 1.1025, $-0.36$
**18** 2.21
**19** 0.453
**20** 0.709, 1.492
**21** 0.893 ($a = 5, b = 6, c = -2$)
**22** 2.481, 0.689, $-1.170$
**23** 3.111, 0.100, $-3.211$
**24** (a) $-1.1$ (b) $-0.9, 1.3, 2.6$
**25** 1.35
**26** (a) 1 (b) 2 (c) 3
**27** 2.99
**28** 0.3
**29** $(b - a)^3/24$
**30** 1.20106

33   0.46

34   0.436

35   1.29, $-1.90$

37   250

38   2.3

39   all 0.764

40   2.407; $a = 9, b = -2$; 2.41

41   (a) 16.699   (b) 16.678; 16.778

42   4.78, 1.14; 4.781

43   7.51, 30.0

44   $a = 4, n = 3$ (a) 4.826
(b) 4.8224

45   1.18

46   0.89

47   1.7

48   1.52

49   2.0, 8.1

50   1.083

52   $a = 2, b = 5$

53   (c)

54   $n = 1, x_2 = 1.260, x_3 = 1.312$

55   2.714, 5.848

## Chapter 2

2.1  **1**  (a) $\mathbf{r} = \mathbf{i} + t\mathbf{j}$
(b) $\mathbf{r} = \mathbf{i} - \mathbf{j} + t(2\mathbf{i} + \mathbf{j} + \mathbf{k})$
(c) $\mathbf{r} = 4\mathbf{i} + 3\mathbf{j} - 2\mathbf{k} + t(\mathbf{j} + \mathbf{k})$
(d) $\mathbf{r} = t(2\mathbf{i} + 3\mathbf{k})$
**2**  (a) $\mathbf{r} = \mathbf{i} + \mathbf{k} + t(-\mathbf{i} + \mathbf{j})$
(b) $\mathbf{r} = \mathbf{i} + \mathbf{j} + \mathbf{k} + t\mathbf{i}$
(c) $\mathbf{r} = 2\mathbf{i} - \mathbf{j} + 3\mathbf{k} + t(\mathbf{j} - \mathbf{k})$
(d) $\mathbf{r} = t(\mathbf{j} - \mathbf{k})$
**3**  2:4:1;
$(x - 2)/2 = (y - 3)/4 = (z - 4)$
**4**  (a) $(x - 1)/3 = (y + 1)/2 = (z - 2)/2$  (b) $x/4 = (y - 1)/3 = z + 2$
(c) $(x - 2)/2 = y + 3 = z/(-1)$
(d) $(x + 2)/3 = y/4 = (z - 1)/2$
**5**  (a) $\mathbf{i} + 2\mathbf{j} + \mathbf{k}$  (b) $2\mathbf{i} + 4\mathbf{j}$
(c) $\mathbf{i} + 2\mathbf{j}$  (d) $\mathbf{i}$
**6**  (a) $1/\sqrt{3}, -1/\sqrt{3}, -1/\sqrt{3}$
(b) 2/3, 1/3, $-2/3$  (c) $-6/11, 7/11, 6/11$  (d) $-6/7, 3/7, 2/7$
**7**  $x = y = -z$
**9**  $x = 2 + t, y = 1 + 3t,$
$z = 3 - 2t$
**10**  $(2, 2, 2); \cos^{-1}(8/9)$

2.2  **1**  (a) $\mathbf{r} \cdot \mathbf{i} = 1$
(b) $\mathbf{r} \cdot (\mathbf{i} + 2\mathbf{j})/\sqrt{5} = \sqrt{5}$
(c) $\mathbf{r} \cdot \mathbf{k} = 0$
(d) $\mathbf{r} \cdot (2\mathbf{i} + 2\mathbf{j} + \mathbf{k})/3 = 2$
**2**  (a) $\sqrt{2}$  (b) 1  (c) 2  (d) 3
**3**  (a) 45°  (b) 30°  (c) 45°  (d) 30°
**4**  (a) 0°  (b) 45°  (c) 30°  (d) 60°
**5**  (a) 1  (b) 3  (c) 2  (d) 4

**6**  $\mathbf{r} \cdot (2\mathbf{i} + 4\mathbf{j} - 5\mathbf{k}) = 0$
**7**  $3/\sqrt{22}$
**8**  $\mathbf{r} \cdot (\mathbf{i} - \mathbf{j}) = 0, \mathbf{r} \cdot (\mathbf{i} + \mathbf{j} + \mathbf{k}) = 1,$
$\mathbf{r} = \mathbf{k} + t(\mathbf{i} + \mathbf{j} - 2\mathbf{k})$
**9**  $\mathbf{r} \cdot (\mathbf{i} - \mathbf{j} - \mathbf{k}) = 0$
**10**  $\mathbf{r} \cdot (\mathbf{i} - \mathbf{j}) = 3$

2.3  **1**  2; 1
**2**  $m = -2, n = 1, d = 2$
**3**  60°
**4**  1; 2
**5**  $3x + 3y - 4z = 1$
**7**  $x + y + z = 0$
**8**  $4x - y - z = 2$
**9**  $4x + y + 3z = 14$
**10**  $(x - 2)/2 = (y + 1)/3 = (z - 3)/4$

## Miscellaneous exercises 2

**1**  $\mathbf{r} \cdot \mathbf{i} = \mathbf{r} \cdot \mathbf{j} = \mathbf{r} \cdot \mathbf{k} = 0,$
$\mathbf{r} \cdot \mathbf{i} = \mathbf{r} \cdot \mathbf{j} = \mathbf{r} \cdot \mathbf{k} = 1,$
$\mathbf{r} = t(\mathbf{i} + \mathbf{j} + \mathbf{k})$
**2**  14
**3**  45°
**4**  $\mathbf{r} \cdot (\mathbf{i} + \mathbf{j} + \mathbf{k}) = 4$
**5**  $\mathbf{r} = \mathbf{i} + \mathbf{k} + t\mathbf{j}$
**6**  $(\mathbf{i} + \mathbf{j} + 2\mathbf{k})/3$
**7**  $11\mathbf{i} + 8\mathbf{j} + 11\mathbf{k}$
**8**  $\sqrt{(2/3)}$
**9**  45°, 45°, 90°
**10**  $x + 2y + 3z = 4$
**11**  $x = y = z/2$
**12**  $\mathbf{i} + 3\mathbf{j} + 6\mathbf{k}$
**13**  $(3\mathbf{i} - 2\mathbf{j} - \mathbf{k})/2$
**14**  7/3, $-2/3$, 4/3
**15**  $-10/27$
**17**  $(29\mathbf{i} - 11\mathbf{j} + 54\mathbf{k})/4$
**18**  (i) 30°; $p = 1, q = -2$
**19**  1:6; $\mathbf{r} = \mathbf{a} + t(\mathbf{a} + 6\mathbf{b})$; $-6\mathbf{b}$
**20**  $\mathbf{r} \cdot (\mathbf{i} + \mathbf{j} + \mathbf{k}) = 3$
**21**  $4t\mathbf{i} + (2t - 4)\mathbf{j} + 3\mathbf{k}$; $4\mathbf{i} + 2\mathbf{j}$,
$(43/35)(\mathbf{i} - 3\mathbf{j} + 2\mathbf{k}),$
$(-1/7)(\mathbf{i} - 3\mathbf{j} + 2\mathbf{k})$
**22**  (i) 7/5; 7/5 (ii) $3x + 4y = 0$,
$3x + 4y + 14$
(iii) $\mathbf{r} = \mathbf{i} + \mathbf{j} + t(8\mathbf{i} - 6\mathbf{j} - 11\mathbf{k})$
**23**  $x + 3y - 2z = 13$,
$\mathbf{r} \cdot (\mathbf{i} + 3\mathbf{j} - 2\mathbf{k}) = 13$
**24**  (i) 1/2 (ii) 60°
**25**  $3x + y - 2z = -7$;
$x/3 = y - 2 = (z - 1)/(-2)$;
$-3/2, 3/2, 2$; $-1/2, 11/6, 4/3$; $\sqrt{14}$
**26**  $y + z = 0$; $\sqrt{6}$
**27**  2
**28**  $\mathbf{r} = \mathbf{i} - 3\mathbf{k} + t(\mathbf{i} + \mathbf{j} - 7\mathbf{k})$;
$\mathbf{r} = 3\mathbf{i} - 3\mathbf{j} + s(5\mathbf{i} - 5\mathbf{j} - \mathbf{k})$;

$r = (i − j + k)/2 + t(18i + 17j + 5k)$
**29**  $r = i + j + t(i + k); 1/\sqrt{3}$
**30**  $(3i + 5j + 4k)/(5\sqrt{2})$;
$3x + 5y + 4z = 30; 3\sqrt{2}$
**31**  $r \cdot (i + 2j − k) = −2$;
$r \cdot (i + 2j − k) = 22$
**33**  $3x + 4y − z = 5$
**34**  (a) 5  (b) $\sqrt{5}$
**35**  $r = (i − j + k) + t(i − 4j + 6k)$

*Chapter 3*

**3.1  1**  (a) $a = 1.5, s = 36$  (b) $a = −3$,
$s = 14$  (c) $a = 2.8, t = 30/7$
(d) $a = −2\frac{3}{4}, t = 40/7$  (e) $s = 150$,
$t = 10$  (f) $s = 72, t = 4$
(g) $s = 32, v = 20$  (h) $s = 0$,
$v = −8$  (i)  $u = 5, a = 1\frac{1}{4}$
(j)  $u = 26/3, a = −31/9$
(k) $v = −18, a = −4$  (l)  $v = 20$,
$a = 3\frac{3}{4}$  (m) $u = −12, v = 28$
(n) $u = 3, v = −27$  (o) $v \approx 21.6$,
$t \approx 3.14$  (p) $v \approx 22.3, t \approx 3.28$
(q) $u = 12, s = 64$  (r)  $u = 12$,
$s = 0$  (s) $u \approx \pm17.4, t \approx 2.43, 8.24$
(t) $u \approx 22.4, t \approx 6.18$
**2**  (a) 120 m  (b) 30 m
**3**  (a) 2400 m  (b) 600 m  (c) 150 m
**4**  3 s, 7.5 m
**5**  $9/40 \text{ m s}^{−2}, 133\frac{1}{3}$ s
**6**  $6 \text{ m s}^{−2}, 16 \text{ m s}^{−1}, 4 \text{ m s}^{−1}$
**3.2  1**  180 m, 12 s, at height 160 m,
$20 \text{ m s}^{−1}$
**2**  (a) $20 \text{ m s}^{−1}$  (b) after 0.586 s and
3.414 s
**3**  (a) $2\sqrt{3}$ s  (b) $20\sqrt{3} \text{ m s}^{−1}$,
$10\sqrt{6} \text{ m s}^{−1}$
**4**  $3.43 \text{ s}, 46.3 \text{ m s}^{−1}$  (a) $33.8 \text{ m s}^{−1}$
(b) $29.2 \text{ m s}^{−1}$
**5**  (a) 81.25 m  (b) $40.3 \text{ m s}^{−1}$
(i) $5 \text{ m s}^{−1}$ after 2 and 3 s,
(ii) $32.02 \text{ m s}^{−1}$ after 5.7 s
**3.3  1**  (a) $181 \text{ m s}^{−1}, 66 \text{ m s}^{−2}$
(b) $18 \text{ m s}^{−1}, 2 \text{ m s}^{−2}$
**2**  (a) $24 \text{ m s}^{−2}, 114 \text{ m}$  (b) $31 \text{ m s}^{−2}$,
$93\frac{3}{4}$ m
**3**  (a) $21 \text{ m s}^{−1}, 276 \text{ m}$
(b) $23\frac{5}{6} \text{ m s}^{−1}, 575\frac{1}{3}$ m
**4**  (a) $0.9 \text{ m s}^{−2}$  (b) 466 m
**5**  (a) $4 \text{ m s}^{−2}$  (b) 159 m
**6**  0.14, 0.72, 2.01, 4.06, 6.69,
$9.66 \text{ m s}^{−1}, 1.92 \text{ m}, 16.6 \text{ m}$
**7**  (a) 7 km  (b) 280 s  (c) $5/12 \text{ m s}^{−2}$
**8**  $400/13 \text{ m s}^{−1}, 80$ s
**9**  10

**10**  $30 \text{ m s}^{−1}, 100$ s
**11**  3/52, 200 m
**12**  (a) $3 \text{ m s}^{−2}$  (b) $8 \text{ m s}^{−1}$
(c) 21.5 m
**13**  (a) $2 \text{ m s}^{−2}$  (b) $40 \text{ m s}^{−1}$
(c) 20 s
**3.4  1**  $3.12 \text{ s}, 11.2 \text{ m s}^{−2}$
**2**  $11.4 \text{ m s}^{−1}$
**3**  (a) 16.1 s  (b) 15.6 s
**4**  220 s
**5**  $12/\pi$ s
**6**  $26.96 \text{ m s}^{−1}$
**7**  $34.4 \text{ m s}^{−1}$
**8**  (a) $\ln(7/4)$ s  (b) $6 − 2\ln(7/4)$
**9**  (a) $\pi/4$ s  (b) $\frac{1}{2}\ln 2$
**3.5  1**  $5 \text{ cm s}^{−1}$ at $36° 52'$ to AB
**2**  $50\sqrt{17} \text{ m min}^{−1}$ at $14° 2'$
to the banks
**3**  154.9 knots on bearing $138° 52'$
**4**  upstream at $36° 52'$ to the banks,
$8\frac{1}{3}$ s; 40 m downstream of A
**5**  $250° 6', 19° 54'$; 8 h 29 min
**6**  (a) $3\frac{3}{4}$ s, 22.5 m downstream
(b) upstream at $41° 25'$ to the banks,
5.67 s
**3.6  1**  (a) $i + 2j, −i − 2j$
(b) $2i − 4j, −2i + 4j$  (c) $−4i + 4j$,
$4i − 4j$  (d) $−i − 7j, i + 7j$
(e) $−i + 7j − 3k, i − 7j + 3k$
(f) $−4i − j − k, 4i + j + k$
**2**  (a) $i + j$  (b) $6i + 8j$
(c) $−8i + 5j$
**3**  (a) $20j, 15i$  (b) $−15i + 20j$;
25 knots, $323° 8'$
**4**  $9i + 9\sqrt{3}j, −12\sqrt{3}i − 12j$,
40.6 knots, $047° 11'$
**5**  $345° 58'$
**6**  17.83 knots, $007° 29'$
**7**  (a) $−\frac{1}{2}i + 4\frac{1}{2}j$  (b) $4.53 \text{ m s}^{−1}$, in
direction $353° 40'$
**8**  (a) 20.5 knots, $046\frac{1}{2}°$(b) 29 knots,
$042°$  (c) 38 knots, $097\frac{1}{2}°$
(d) 32.5 knots, $333°$
**9**  (a) 36 knots, $304°$  (b) 31 knots,
$060°$  (c) 22 knots, $076\frac{1}{2}°$
(d) 31 knots, $275°$
**10**  21.6 knots, $146°$; 33.3 nautical
miles, 1102 h
**11**  32.5 knots, $258°$; 58.7 nautical
miles, 0823 h
**3.7  1**  (a) $4i + 12j, 2i + 12j$  (b) $8i + j, 2i$
(c) $39i + 6j, 22i + 2j$
(d) $ei + 2e^2j, ei + 4e^2j$
(e) $12i + 4j + k, 12i + 2j$

(f) $(-\pi/6)\mathbf{j} + 6\mathbf{k}$, $(-\pi^2/36)\mathbf{i} + 2\mathbf{k}$

**2** (a) $2\sqrt{2}$, $2\mathbf{i} + \frac{3}{2}\mathbf{j}$

(b) $3\sqrt{5}$, $-5\mathbf{i} + \frac{10}{3}\mathbf{j}$

(c) $\frac{1}{2}\sqrt{757}$, $-\frac{7}{2}\mathbf{i} + 21\mathbf{j}$

(d) $\left[\frac{1}{4}(e^8 - 1)^2 + \frac{1}{9}(e^{12} + 2)^2\right]^{1/2}$,

$\frac{1}{4}(e^8 - 5)\mathbf{i} + \frac{1}{9}(e^{12} + 23)\mathbf{j}$

(e) $\frac{1}{\pi}\left[\pi^2 + 16\sqrt{2}\pi + 144\right]^{1/2}$,

$\left[\frac{\pi^2 + 8\pi - 16}{\pi^2}\right]\mathbf{i}$

$+ \left[\frac{2\pi^2 + 128 - 64\sqrt{2}}{\pi^2}\right]\mathbf{j}$

(f) $\sqrt{1630}$, $28\mathbf{i} - 51\mathbf{j} + 13\frac{1}{2}\mathbf{k}$

(g) $\sqrt{3521}$, $73\mathbf{i} - 60\mathbf{j} + 8\mathbf{k}$

**3.8** **1** (a) 4 s, $80\sqrt{3}$ m, 20 m (b) $6\sqrt{3}$ s, $180\sqrt{3}$ m, 135 m (c) $5\sqrt{2}$ s, 250 m, 62.5 m (d) 9.6 s, 614.4 m, 115.2 m

(e) 9.6 s, 345.6 m, 115.2 m

(f) 4 s, 192 m, 20 m (g) 14.4 s, 432 m, 259.2 m

**2** (a) $+16°6'$ (b) $-30°$ (c) $-16°6'$

(d) $+18°26'$ (e) $-16°16'$

**3** after (a) 1 s and 3 s (b) 2 s

(c) 1.5 s and 2.5 s (d) 2 s and 7.6 s

(e) 4 s and 5.6 s (f) 4.6 s and 5 s

**4** (a) $60\sqrt{3}$ m, 40 m (b) $120\sqrt{3}$ m, 40 m (c) 64 m, 28 m (d) 96 m, 27 m

**5** (a) $30°$ (b) $35°16'$ (c) $34°27'$

**6** (a) $24°18'$ or $65°42'$ (b) $19°20'$ or $70°40'$ (c) $20°54'$ or $69°6'$

**7** (a) $45°$ or $71°34'$ (b) $63°26'$ or $65°13'$ (c) $18°26'$ or $66°48'$

**3.9** **1** (a) 240 m, 720 m (b) $156\frac{1}{4}$ m, 625 m (c) 462.2 m, 1040 m

(d) 187.2 m, 4680 m (e) 125 m, 222.2 m

**2** (a) $60°$, $30°$ (b) $63°26'$, $26°34'$

(c) $56°18'$, $33°42'$ (d) $78°42'$, $11°18'$

(e) $53°8'$, $36°52'$, to the horizontal

**3** (a) 219.7 m, 419.4 m (b) 336 m, 1200 m (c) 114.4 m, 218.4 m

(d) $166\frac{2}{3}$ m, $333\frac{1}{3}$ m (e) 175.7 m, 655.7 m

## Miscellaneous exercises 3

**1** $+40 \text{ m s}^{-2}$, $-10 \text{ m s}^{-1}$

**3** $10/u$, $14/u$, $6/u$ (i) 5 (ii) 5/2, 25/6

**4** (i) 80 s (ii) 90 km h$^{-1}$ (iii) 37 s

**5** (a) $192/(u + 6)$, $60/u$ (i) 10

(ii) $\frac{1}{3}$ m s$^{-2}$, 5/3 m s$^{-2}$ (b) 14 m s$^{-1}$, $051°47'$

**7** (a) $-1$ m s$^{-2}$ (b) 28.4 m

**8** (a) 6.95 m s$^{-2}$ (b) 2.4 m s$^{-2}$

(c) 36.2 m (d) 77.6 m

**9** 87 s

**10** 87 s

**11** (i) after 15 s

**13** (i) 28/9 m

**14** 22.4

**16** $2u$

**17** (i) 10 s (ii) $-32/2197$ m s$^{-2}$

(iii) 1 m

**18** 4; 5, $u - 2\lambda$

**19** (i) 20 m s$^{-1}$, 40 m

(ii) $\tan^{-1}(4/3)$ (iii) $10\sqrt{13}$ m s$^{-1}$ at $\tan^{-1}(2/3)$

**20** 2 h 11 min

**21** AB:30 min, BC:50.6 min, CA:50.6 min

**23** (a) 17.03 knots (b) 1 h 34 min

**24** (i) 100 s, 583 m (ii) at $\tan^{-1}(3/4)$ to banks, 125 s, 500 m

**25** 3, $\sqrt{134}$ m

**26** $3\mathbf{i} - 2\mathbf{j} - 3\mathbf{k}$,

$(a - 7 + 3t)\mathbf{i} + (5 - 2t)\mathbf{j} - (2 + 3t)\mathbf{k}$;

$\left[(a - 7 + 3t)^2 + 13t^2 - 8t + 29\right]^{1/2}$,

$-19/3$

**27** $8\mathbf{i} - 7\mathbf{j} + 6\mathbf{k}$, $6\mathbf{i} - 6\mathbf{j} + 2\mathbf{k}$,

$16\mathbf{i} + 12\mathbf{j} + 8\mathbf{k}$

**28** $(3 - 2T)\mathbf{i} + (4 - 5T)\mathbf{j}$

$+ (-2 + T)\mathbf{k}$; 1.693,

$\frac{17}{15}\mathbf{i} - \frac{2}{3}\mathbf{j} - \frac{16}{15}\mathbf{k}$

**29** $25\mathbf{i} + 10\mathbf{j}$

**30** $7\mathbf{i} + 2\mathbf{j}$, 12.30 p.m., $4\frac{1}{2}\mathbf{i} + \mathbf{j}$

**31** $\frac{1}{4}u\sqrt{5}$, west $\tan^{-1} 2$ south

**32** $\frac{10}{3}\mathbf{i} - \frac{2}{3}\mathbf{j}$, 14/13

**33** $\frac{1}{2}a(u^2 - v^2)^{-1/2}$

**34** 3.2 m s$^{-1}$, $5\frac{1}{3}$ m s$^{-1}$ perpendicular to the banks, $5\frac{1}{3}$ m s$^{-1}$

**35** 50 m s$^{-1}$ at $\tan^{-1}(4/3)$ below $Ox$, 32 m s$^{-1}$

**36** 12.13.5 p.m., due east, 12.21.8 p.m.

**37** $36°52'$, 54 min

**38** 2.828 km, 1224 p.m., $019°28'$

**39** $53°8'$, (a) 4.7 nautical miles

(b) 0902 hours

**40** 1448 hours

**41** 30 knots due west, $225°$

**42** (i) 19.8 knots, $108°21'$

(ii) 1214.4 h, 15.5 min

**43** (i) 6.4 m s$^{-1}$ (ii) $\tan^{-1}(4/3)$

(iii) 6.4 m s$^{-1}$, $\tan^{-1}(4/5)$

(iv) $100/(\sqrt{41})$ m

**44** 271 s, $050°16'$; 173 s, 153.6 m; 13 m s$^{-1}$ at $247°23'$

**45** (a) $6\sqrt{2}$ knots from NW
(b) 1.1 nautical miles
**46** $5\sqrt{7}$, $\tan^{-1}(2\sqrt{3}/3)$; 69.4 km h$^{-1}$, 351.7°
**47** $\mathbf{r} = 2t\mathbf{i} + (1 - t)\mathbf{j} + t^2\mathbf{k}$, $t = \frac{1}{2}$
**48** $\sqrt{2}:\sqrt{3}$, $(\sqrt{5})/(2\sqrt{2})$
**49** $\mathbf{r} = t^2\mathbf{i} + t(\mathbf{i} + \mathbf{j})$,
$\mathbf{v} = (\mathbf{i} + \mathbf{j}) + 2t\mathbf{i}$; 6 m s$^{-1}$, 2 s
**50** $y^2 + z^2 = a^2$, $x = 0$;
$y^2 = 4ax + a^2$, $z = 0$; $t = n/4$,
$\sqrt{(27/40)}$
**51** $x^2 + y^2 = a^2$, $x - y - a = 0$;
$a\mathbf{i}$, $-a\mathbf{j}$, 0, $\pm\frac{1}{10}$
**52** $a\mathbf{i} \pm (a\sqrt{3})\mathbf{j}$, $(2n + 1)\pi/(2\omega)$
**53** $x^2 + y^2 = 4$, $x^2 = 4y$
(a) $|2\omega - 2|$  (b) $2\sqrt{(\omega^4 + 1)}$
(c) $(2n - 1)\pi/\omega$  (d) $2n\pi/\omega$
**54** (i) $\mathbf{i}(3 + 4t) + \mathbf{j}(-2 - 3t + t^2)$
$+ \mathbf{k}(-4 + 4t - t^2)$,
$\mathbf{i}(8 - t) + \mathbf{j}(-2 + 4t) + \mathbf{k}(-3 + 3t)$
(ii) $6\mathbf{i} + 6\mathbf{j} + 3\mathbf{k}$
**55** (i) $-15\mathbf{i} + 70\mathbf{k}$, $5\mathbf{i} + 30\mathbf{k}$
(ii) $\sqrt{(100t^2 - 600t + 925)}$,
$10\mathbf{i} + 95\mathbf{k}$, $25\sqrt{2}$ m, $t = 37/12$
**56** $\frac{1}{2}V\sqrt{3}$ at $\tan^{-1}(\frac{1}{2}\sqrt{2})$ to the horizontal
**57** $22\frac{1}{2}$
**58** $26°34' \leqslant \alpha \leqslant 63°26'$, 98 m, 22.05 m
**59** $P(Vt \cos\alpha, Vt \sin\alpha - \frac{1}{2}gt^2)$,
$Q(Vt \sin\alpha, Vt \cos\alpha - \frac{1}{2}gt^2)$,
$\left(\dfrac{V^2 \sin\alpha \cos\alpha}{g}, \dfrac{V^2 \sin^2\alpha}{2g}\right)$
**60** $3u^2/(8g)$, $\pm50°46'$
**61** 26.2 m s$^{-1}$, 13.2 m
**62** $\tan^{-1}(1/12)$ with horizontal
**63** $\sqrt{(10gh/3)}$
**64** 32 m s$^{-1}$ at $\tan^{-1}(4/3)$, $20\sqrt{5}$ m s$^{-1}$
**68** $bg/V$
**69** (i) $\tan^{-1}\frac{1}{2}$ (ii) $\sqrt{5}$ s (iii) $\tan^{-1} 2$ with horizontal
**70** $\tan^{-1}(b/2h)$, $b^2/(4h)$
**71** $x = ut \cos\alpha$, $y = ut \sin\alpha - \frac{1}{2}gt^2$,
$\tan\alpha = 7/24$, $u = 22.8$ m s$^{-1}$, 22.4 m s$^{-1}$
**72** 30°, $30\sqrt{3}$ m, $\tan^{-1}(2/\sqrt{3})$, $(\sqrt{3} + 1)$ s
**74** 0, $\tan^{-1} 5$  (a) 240 m  (b) 2.9 s
**77** (a) $\pi/4$  (b) $\tan^{-1} 3$
**78** $2V^2 \sin\alpha \cos(\alpha + \beta)/(g \cos^2\beta)$
**79** $\dfrac{2V}{(\sqrt{5})g \cos\alpha}$; $1, \frac{1}{3}$
**80** $\sin^{-1}\frac{1}{3}$

## Chapter 4

**4.2** **1** (a) 10 N at 53°8′ to AB at A
(b) 20 N at 53°8′ to AB at B
(c) 13 N at 67°23′ to CD at C
(d) 25 N at 16°16′ to AD at D
(e) 26 N at 22°37′ to CB at C
(f) 20.3 N at 24°39′ to AB at A
(g) 21.4 N at 29°41′ to AD at D
(h) 25.6 N at 6°20′ to AB at centre of square ABCD
**2** (a) 9.17 N at 10°54′ to AD at A
(b) 10 N at 36°52′ to BF at point of intersection of AD and BF
(c) $10\sqrt{3}$ N along AD  (d) $6\sqrt{5}$ N at 3°26′ to AD at D  (e) $10\sqrt{7}$ N at 79°6′ to AD at centre of hexagon
(f) $10\sqrt{7}$ N at 19°6′ to CF at C
(g) 50 N at 36°52′ to AE at E
**4.3** **1** 17.3 N, 34°58′
**2** 27.1 N, 32°7′
**3** 19.4 N, 76°7′
**4** 4.64 N, 29°59′
**5** 82.6 N, 156°57′
**6** 10.1 N, 61°20′
**7** 40.2 N, 46°4′
**8** 25.8 N, 60°2′
**9** 43.7 N, 54°7′
**10** 26.0 N, 142°14′
**11** 53.5 N, 158°21′
**12** 43.6 N, 23°25′
**13** 31.9 N, 6°36′
**14** 52.4 N, 20°29′
**15** 12.7 N, $-36°17'$
**4.4** **1** 10, 10
**3** 12, $12\sqrt{3}$
**4** 12, 9
**5** 14.1 N, 8° away from DC
**6** $57\frac{1}{2}$ N, 175°
**7** 28.7 N, $96\frac{1}{2}°$
**4.5** **1** $W/\sqrt{2}$
**2** 13 N, 15 N
**3** $3mg/5$, $4mg/5$
**4** $15\sqrt{3}$ N, 15 N
**5** $10\sqrt{3}$ N, $10\sqrt{3}$ N
**6** 20.5 N, 16 N
**8** 0.58 $W$, 0.58 $W$
**9** (a) $\frac{1}{2}W$, $\frac{1}{2}W\sqrt{3}$
(b) $W/\sqrt{3}$, $2W/\sqrt{3}$
**10** $\frac{1}{2}W$, 30°
**11** $10\sqrt{3}$ N, 10 N outwards along radius
**12** 5 N, 5 N outwards along radius
**13** $4\sqrt{2}$ N, 4 N outwards along radius
**14** (a) BC 120 N, tension

AB $120\sqrt{2}$ N, compression  
(b) BC $100\sqrt{3}$ N, tension  
AB $50\sqrt{3}$ N, compression  
(c) CD 400 N, tension  
BC $200\sqrt{2}$ N, tension  
AB 200 N, compression  
AC $200\sqrt{2}$ N, compression  
(d) AD 600 N, tension  
CD $300\sqrt{3}$ N, tension  
BC 600 N, compression  
BD 300 N, compression  

4.6   1   0.3  
     2   0.4  
     3   720  
     4   600  
     5   0.3  
     6   0.1  
     7   9.375  
     8   7.2  
     9   161.7  
    10   8/11  

4.7   1   (a) 3 N   (b) 2.69 N   (c) $6/\sqrt{5}$ N  
     2   (a) 5.72 N   (b) 5.67 N  
     (c) 7.65 N  
     3   (a) $\frac{1}{2}$   (b) $1/\sqrt{3}$  
     4   (a) $7.5 \text{ N} \leqslant P \leqslant 16.5 \text{ N}$  
     (b) $7.5 \text{ N} \leqslant P \leqslant 82.5 \text{ N}$  
     5   0.45  

## Miscellaneous exercises 4

     1   (a) 13 N, 22° 37′   (b) 35.0 N, 59° 2′   (c) 44.7 N, −169° 42′  
(d) 31.8 N, 76° 34′   (e) 30.2 N, 116° 18′   (f) 74.8 N, 142° 4′  
     2   (a) 58.9 N, 351° 55′   (b) 45.2 N, 136° 32′   (c) 17.7 N, 205° 40′  
(d) 15.9 N, 199° 6′  
     3   32, 24  
     4   41.8 N at 35° 30′ to BA  
     5   (a) $10, 10\sqrt{3}$   (b) $10\sqrt{3}$ in direction DO, 50  
     6   (a) $100/\sqrt{3}$ N in every rod (b) 100 N  
     7   $W/\sqrt{3}$  
     8   168 N, 576 N  
     9   (a) $W\sqrt{3}$   (b) $\frac{1}{2}W\sqrt{3}$  
   10   $\frac{3}{4}W$ perpendicular to OA  
   11   (a) $2W/\sqrt{13}$   (b) $9W/(13\sqrt{13})$  
   12   $\frac{3}{4}$; (a) $7W/24$   (b) $7W/25$  
   13   (i) $P = 815$ N, $Q = 455$ N (ii) 355 N in the same direction as the force of 600 N  
   14   $k = 7/25$; $T_{AB} = 24W/25$, $T_{BC} = 7W/25$, $T_{CD} = 56W/125$  
   15   $W = 30$, $T_{BC} = 30$ N,  

$T_{AB} = T_{CD} = 30\sqrt{3}$ N  
   16   (i) BC:$5mg$, AB:$4mg\sqrt{2}$  
(ii) BC:$\tan^{-1} (4/3)$, AB:$45°$  
   17   $(\sqrt{3} - \frac{3}{2})W$  
   18   $\frac{1}{3}a, \frac{2}{3}a; 3m$  
   19   $T_{AB} = 3W\sqrt{3}$, $T_{BC} = 3W$, $T_{CD} = 3W\sqrt{3}$; $CD = 3a$  
   20   (i) $\frac{3}{4}W$, $5l/(4K)$ (ii) $3W/5$, $4l/(5K)$  
   21   (a) $30°$ (b) $W\sqrt{3}$, $30°$  
   22   $W(19\sqrt{3} - 4)/22$, $W(75 + 4\sqrt{3})/22$  
   23   $\dfrac{l(2k\lambda + k + 1)}{k\lambda}$; $\dfrac{kl}{1 + k}$, $\dfrac{l}{1 + k}$; $\dfrac{2\lambda k\sqrt{3}}{1 + k}$  

## Chapter 5

5.1   1   12.8 m, 1.6 s  
     2   72.8 m, 3.432 s, 1.768 s  
     3   25 m, $2\frac{1}{2}$ s, $10\sqrt{2}$ m s⁻¹  
     4   $4\sqrt{6}$ m s⁻¹, $\frac{2}{3}\sqrt{6}$ s  
     5   2.27 s, 29.5 m  
     6   65 m, $4\frac{1}{3}$ s, 10 m s⁻¹  
     7   $30°$  
     8   $\sin^{-1} (3/5)$  
     9   $\frac{1}{2}$  
   10   $\tan^{-1} (5/12)$  
   11   75 s, 1125 m  
   12   19.4 s, 290 m  
   13   (a) 225 s, 1800 m   (b) 22.5 s, 180 m  
   14   (a) $3/200$ m s⁻²  
(b) $43/200$ m s⁻²; 30 km, 124 s  
   15   (a) 260 N   (b) 120 N  
   16   (a) 50 N   (b) 60 N   (c) 42.5 N  

5.2   1   $g/3$  
     2   $g/2$  
     3   $7g/23$  
     4   $17/15$ m s⁻², 440 N  
     5   $8/7$ m s⁻², 537 N  
     6   $5/7$ m s⁻², 671 N  
     7   (a) 4 m s⁻²   (b) 18 N (c) $18\sqrt{3}$ N  
     8   (a) $50/9$ N   (b) $80/9$ N   (c) 14.5 N  
     9   (a) $3\frac{3}{4}$ m s⁻²   (b) $12\frac{1}{2}$ N   (c) 19.8 N  
   10   (a) 3.56 m s⁻²   (b) 25.8 N  
   11   (a) 6.07 m s⁻²   (b) 11.8 N  
   12   (a) 4.77 m s⁻²   (b) 10.5 N  
   13   $9/20$ m s⁻², 875 N  

5.3   1   (a) 56 m s⁻¹, $117\frac{1}{3}$ m from O  
(b) 536 m s⁻¹, 734 m from O  
(c) 11.73 m s⁻¹, 81.7 m from O  
     2   (a) 11 m s⁻¹   (b) $3(e^6 - 1)$ s  
     3   1.86 m s⁻¹

**4** (a) $\frac{1}{2}(e^8 - 1)\,\text{m s}^{-1}$
(b) $\frac{1}{4}(e^8 - 9)\,\text{s}$
**5** (a) $\frac{1}{2}\ln 51$  (b) $(2e^{60} - 2)^{1/2}$
**6** (a) 1.7, 7.3, 21, 46.7, 88.3,
150 m s$^{-1}$  (b) 240 m
**7** (a) 6.8, 9.5, 11.6 m s$^{-1}$  (b) 12.7 m
**8** (a) 16.2, 30.4, 44.5, 58.7 m s$^{-1}$
(b) 3.07 s
**9** 2.2 s
**10** 34.6 s

## Miscellaneous exercises 5

**1** 450 N  (a) 100 s  (b) 0.015
**2** 42.35 s, 423.5 m, 4050 m
**3** 300 N, 228.6 m
**4** 200 N, 1000 N, 900 m, 3200 N
**5** 0.2625, 1 m s$^{-1}$
**6** (i) (a) 2500 N  (b) 500 N
(ii) (a) 3000 N  (b) 600 N
(iii) (a) 2000 N  (b) 400 N
**8** $5u^2/(8g)$, $u/\sqrt{2}$
**9** $10/u$, $14/u$, $6/u$, $u = 5$, $x = 5/2$,
$y = 25/6$, $R_1 = 375$ N, $R_2 = 300$ N,
$R_3 = 175$ N
**10** $mg + mf$
**11** (a) 47.04 N  (b) 1.96 m s$^{-2}$
(c) 15.68 N  (d) 94.08 N
**12** 1 m s$^{-2}$, 36 N, 22 N; 1 s
**13** (i) $4g/9$ (ii) $\frac{4}{3}\sqrt{(2ga)}$
(iii) $3\sqrt{(2a/g)}$
**14** $\frac{1}{2}(x + y)$ (i) $3g/5$, $3mg/5$
(ii) $9g/17$, $12mg/17$
**15** $g(\frac{3}{4} - \frac{1}{3}\sin\beta)$
**16** $3M/5$, $2g/3$, $5Mg/3$
**17** $\frac{1}{2}g(\cos\theta - \sin\theta)$,
$\frac{1}{2}mg(\cos\theta + \sin\theta)$,
$\frac{1}{2}g(2M + 3m + m\sin 2\theta)$
**18** $\sqrt{(8d/3g)}$; $8mg/11$, $4mg/11$
**19** $g/9$, $40mg/9$, $5mg/3$
**20** $\frac{1}{4}g$, $\frac{1}{2}g$, $3mg/2$; $5g/28$, $33mg/14$,
$6mg/7$
**21** $(k - 4)g/(k + 8)$
(a) $6kmg/(k + 8)$  (b) $3kmg/(k + 8)$
**22** (a) $3/2$, $-\frac{1}{3}g$, $\frac{2}{3}g$  (b) 3, $-\frac{1}{2}g$, $\frac{1}{2}g$
**23** (i) $g/7$, $48g/7$, $24g/7$ (ii) 3.6
**24** 1.0, 1.1, 1.25, 1.45, 1.75, 2.2,
3.0 m s$^{-2}$, 85 km h$^{-1}$
**25** 27.4 m s$^{-1}$, 24.5 m s$^{-1}$
**26** 12 s, 210 m
**27** 24.5 m s$^{-1}$
**28** (i) $\frac{1}{2} < t < 5\frac{1}{2}$, $t > 6$ (ii) $-g$
**29** $U\sqrt{3}$
**30** (a) 2.2  (b) $u\sqrt{(85/7)}$

## Chapter 6

**6.1** **1** 480 J
**2** 0
**3** 210 J
**4** 0
**5** 300 J
**6** 320 J
**7** 48 J
**8** 180 J
**9** 250 J
**10** $250\sqrt{3}$ J
**11** 600 J
**12** 450 J
**13** 8 J
**14** 7 J
**15** $-35$ J
**16** (a) 0  (b) 120 J  (c) $-120$ J
(d) 120 J
**17** (a) 300 J  (b) 192 J  (c) 108 J
(d) 246 J
**18** (a) 43.2 J  (b) 76.8 J  (c) $-33.6$ J
**19** 800 000 J
**20** (a) $30\,000\sqrt{3}$ J  (b) $40\sqrt{2}$ kJ
(c) 56 000 J
**21** (a) $3200\sqrt{3}$J  (b) 4800 J
(c) 4431 J
**22** (a) 400 J  (b) $200\sqrt{2}$ J  (c) 369 J
**6.2** **1** 12 kW
**2** 42.5 kW
**3** 49 kW
**4** 87 kW
**5** (a) 25 m s$^{-1}$  (b) 0.6 m s$^{-2}$
**6** (a) 30 m s$^{-1}$  (b) $\frac{1}{3}$ m s$^{-2}$
**7** (a) 40 m s$^{-1}$  (b) $\frac{1}{4}$ m s$^{-2}$
**8** (a) 16 m s$^{-1}$  (b) 0.9 m s$^{-2}$
**9** (a) 20 m s$^{-1}$  (b) 1.5 m s$^{-2}$
**10** (a) 20 m s$^{-1}$  (b) 0.175 m s$^{-2}$
**11** (a) 18.66 s  (b) 296 m
**12** (a) 6.39 s  (b) 61.3 m
**6.3** **1** 10 m s$^{-1}$
**2** 16.1 m s$^{-1}$
**3** 15.5 m s$^{-1}$
**4** $4\sqrt{3}$ m s$^{-1}$
**5** 6.32 m s$^{-1}$
**6** 3.16 m s$^{-1}$
**7** 7.21 m s$^{-1}$
**8** $2\sqrt{2}$ m s$^{-1}$
**9** 4.12 m s$^{-1}$
**10** 3.1 m s$^{-1}$
**11** (a) 10 m  (b) $5\sqrt{2}$ m s$^{-1}$
**12** (a) 23.1 m  (b) $10\sqrt{2}$ m s$^{-1}$
**13** (a) 48 m  (b) $12\sqrt{2}$ m s$^{-1}$
**14** (a) 48.75 m  (b) $15\sqrt{2}$ m s$^{-1}$
**15** 7.75 m s$^{-1}$
**16** $2\sqrt{5}$ m s$^{-1}$

**17** $90°$

**18** $60°$

**6.4** **1** $6.32\,\text{m s}^{-1}$
 **2** $17.8\,\text{m s}^{-1}$
 **3** $34.6\,\text{m s}^{-1}$
 **4** $2.72\,\text{kW}$
 **5** $32\,000$
 **6** $11.1\,\text{m s}^{-1}$
 **7** $16.2\,\text{m s}^{-1}$
 **8** $34.2\,\text{m s}^{-1}$
 **9** (a) $5.77\,\text{m s}^{-1}$ (b) $20.8\,\text{m s}^{-1}$

**6.5** **1** $1\,\text{J}$
 **2** $3\,\text{J}$
 **3** $\frac{1}{2}\,\text{J}$
 **4** $20\,\text{J}$
 **5** $6\,\text{m}$
 **6** $6\,\text{m}$
 **7** $3.73\,\text{m}$
 **8** $3\,\text{m}$
 **9** $1.805\,\text{m}$

*Miscellaneous exercises 6*

**1** $\sqrt{(2gr)}, \frac{3}{4}mgr$
**2** (i) $\sqrt{[ga(2 - \sqrt{3})]}$ (ii) $\sqrt{(ga)}$
(iii) $\sqrt{(2ga)}$
**3** $-9(5x - 3)/5$,
$18(6x - 5x^2 - 1)/5$
**4** (a) $10.95\,\text{m s}^{-1}$ (b) $8.16\,\text{m s}^{-1}$
**5** (i) (a) $52\,\text{m}$ (b) $20\,\text{m s}^{-1}$
(ii) (a) $32.5\,\text{m}$ (b) $10\,\text{m s}^{-1}$
**6** (i) (a) $29\,400\,\text{J}$ (b) $45\,000\,\text{J}$;
$74.4\,\text{kW}$ (ii) $81\,\text{kW}$
**7** $144\,000\,\text{kg}$
**8** $76\,\text{kW}, 40\,\text{m}$
**9** $15\,\text{kW}, 40\,\text{m}$
**10** (i) $8\,\text{kW}$ (ii) $0.6\,\text{m s}^{-2}$ (iii) $150$
**11** $5/6\,\text{m s}^{-2}, 86.2\,\text{km h}^{-1}$
**12** $25\,\text{kW}$ (i) $43.2\,\text{km h}^{-1}$
(ii) $\frac{1}{3}\,\text{m s}^{-2}$
**13** $735\,\text{kW}, 0.27\,\text{m s}^{-2}$
**14** $45\,\text{m s}^{-1}, 1.715\,\text{m s}^{-2}$
**15** $3W/n, 2g/n$
**16** (a) $(1000H - MVg \sin \alpha)/(MV)$
(b) $(1000H + MVg \sin \alpha)/(MV)$,
$(1000H)/(3MVg), \frac{3}{4}V$
**17** (a) $Hnv/(Hn + Mgv)$
(b) $Hnv/(Hn - Mgv)$
**18** $R = 5\,Mg \sin \theta, 2.4u$
**19** $75\,\text{kW}, 3/35\,\text{m s}^{-2}$
**20** $73\frac{1}{3}\,\text{kW}, 50\,\text{km h}^{-1}$
**21** $A = 2655, B = 1.8; 67.5\,\text{kW}$
**22** (i) $32.4\,\text{kW}$ (ii) $\frac{1}{4}\,\text{m s}^{-2}$
**23** $5 \times 10^4\,\text{N}, 2 \times 10^4\,\text{N}, 41\,666\frac{2}{3}\,\text{N}$
**24** $u/3, \frac{1}{2}\,\text{m s}^{-2}$
**25** $3270\,\text{W}$ (a) $40\,\text{km h}^{-1}$

(b) $24\,\text{km h}^{-1}$
**26** $60\,\text{km h}^{-1}, 0.24\,\text{m s}^{-2}, 106\,\text{N}$;
$1.45$ megajoules
**27** $\frac{3}{2}\sqrt{(ga)}, 5a/4$
**28** $\sqrt{(5ag/2)}$
**29** $a/12$ (a) $5a/12$ (b) $5\sqrt{(ag/12)}$,
$5g$
**30** $3a/2, 2\sqrt{(ga/3)}$
**31** $\lambda b^2/(2a), 17g/15$

*Chapter 7*

**7.1** **1** $16\sqrt{5}\,\text{N m}, -8\sqrt{5}\,\text{N m}$
 **2** All $Pa\sqrt{3}$
 **3** $-34, 22, 0$
 **4** $2x + y = 5$
 **5** $2x - y = 2, \sqrt{5}$
 **6** (a) $-9$ (b) $-20$ (c) $2$ (d) $20$
 **7** $6, \sqrt{2}$
 **8** (a) $-20\,\text{N m}$ (b) $30\,\text{N m}$
 **9** $0.7\,\text{N m}, 1.4\,\text{N m}, 1.6\,\text{N m}$
 **10** $\mathbf{r} = 3t\mathbf{i} + t\mathbf{j}$

**7.2** **1** $3\,\text{N}$
 **2** (a) $9\,\text{N}$ (b) $9/\sqrt{2}\,\text{N}$ (c) $3\sqrt{3}\,\text{N}$
(d) $4\frac{1}{2}\,\text{N}$
 **4** $5\,\text{N}$ along BC, $5\,\text{N}$ along BA,
$5\,\text{N}$ along AC
 **5** $W/(2\sqrt{3}), W/4$
 **6** $75\,\text{N}$
 **7** $5\frac{1}{2}\,\text{N}, 8\frac{1}{2}\,\text{N}$
 **8** $20\,\text{cm}$
 **9** $35\,\text{cm}$
 **10** $9\,\text{N}$

**7.3** **1** (a) $2$ (b) $18$ (c) $6$ (d) $6$
 **2** (a) $8$ (b) $1$ (c) $2$ (d) $2$
 **3** $p = 8, q = 2$

**7.4** **2** $x = 4, y = 5; 18\,\text{N m}$ in the sense
ABCD
 **3** $12\,\text{N m}$
 **4** $10$
 **6** $5\,\text{N}$ along AB, $5\,\text{N}$ along BC,
$5\sqrt{2}\,\text{N}$ along CA

**7.5** **5** (a) $x - y\sqrt{3} = 0$
(b) $x - y\sqrt{3} = 12$

**7.6** **1** (a) $2\,\text{N}$ along AD (b) A couple
of moment $2\,\text{N m}$ in the sense ADCB
(c) $9\sqrt{2}\,\text{N}$ along AC
 **2** $2P$ along CD
 **3** $4\,\text{N}$ along BC
 **6** $5a, 4a$
 **7** $8\,\text{N}, 2\sqrt{10}\,\text{N}; 6a$
 **9** $\sqrt{2}, 2\sqrt{2}$ and $2\,\text{N}$
 **10** $4, 2$ and $4\,\text{N}$

**7.7** **1** (a) $5$ (b) $4$ (c) $3$ (d) $0$
 **2** $(3, 4)$
 **3** (a) $3\mathbf{i} + \mathbf{j}$ (b) $3.5\mathbf{i} + 6.5\mathbf{j}$

(c) $3\mathbf{i} + 2\mathbf{j}$  (d) $6\mathbf{i}$
  4  $(2\mathbf{i} + \mathbf{j})/3$
  5  $(4, -6)$
  6  6.5 cm, 4.5 cm
  7  $(-3, 3)$
  8  2.5 cm
  9  19 cm
 10  $(3.5, 0)$
 11  $2a/(3\sqrt{3})$
 12  4.15 cm, 5.38 cm
 13  6.88 cm, 1.64 cm
 14  2.69 cm, 6.85 cm
7.8  1  $(3/8, 2/5)$
  2  $(3/4, 3/10)$
  3  $(6/5, 3\sqrt{2}/4)$
  4  $2a/5\pi$
  5  $[1/(e - 1), (e + 1)/4]$
  6  $(\pi/2, \pi/8)$
  7  $13a/3\pi$
  8  $3/5$
  9  $(\pi - 2, \pi/8)$
 10  $(0, 7/10)$
 11  $(2, 0)$
 12  $5/8$
 13  $2 \ln 2$
 14  $2 \ln 2 - 1$
 15  $\frac{1}{2}(e^2 + 1)/(e^2 - 1)$
7.9  1  5, 8 and 8 cm
  2  8 cm, 7 cm
  3  2 cm
  4  (a)  3.8 cm, 3 cm   (b) $2\frac{1}{3}$ cm, $1\frac{5}{6}$ cm   (c) $2\frac{1}{2}$ cm, $1\frac{7}{12}$ cm
  5  $\sqrt{2}$ cm
  6  $3a/2\pi$
 10  $5a/3, 22a/27; 3a/2, 13a/18$
 11  $\sqrt{6}{:}1$
 12  $2\frac{1}{3}$ cm

## Miscellaneous exercises 7

  1  39 cm, 29 cm
  2  3 N in the direction $\overrightarrow{BC}$; 12 cm
  3  $\sqrt{5}$ N, $\tan^{-1} 0.5$, 14 N m
  4  $11\frac{1}{2}$ cm
  5  40 cm; 3 N
  7  $\sqrt{3}P; 2\sqrt{3}Pa$
  8  $3\frac{6}{13}$ cm
  9  $23\frac{3}{4}$ cm, $11\frac{3}{4}$ cm
 10  $a/\ln 2$; $a \ln 4$
 11  $\sqrt{13}/2$
 12  3.9 cm, 5.9 cm
 13  $2a/3$
 14  $(7\sqrt{3} - 2)/13, (6\sqrt{3} - 3)/11$
 15  (i) $2P$ along MN, $P$ along NA, $P$ along AM  (ii) 6 units along OB

 16  $19\frac{1}{2}$ cm, $13\frac{1}{2}$ cm
 18  $-11, 14, -9$; $6Pa\sqrt{3}$; $P\sqrt{19}$, $\tan^{-1} (\sqrt{3}/4)$
 19  $a, 2a$
 20  (a) $P = R, Q = S$
     (b) $P = R, Q = S, 2Q + R = 0$
     (c) $P = R, Q + R + S = 0$; $Q = S = P/2, R = -P$
 21  7 N
 22  9 cm
 23  4.9 cm
 24  25 N in the direction $\overrightarrow{AC}$; 2.4 m
 25  $4\frac{3}{16}$ cm
 26  G lies on AC, AG $= 4a/3$
 27  (a) 4, 5  (b) 4, 4
 29  1, 4; $5Pa$ in the sense DCBA
 30  $13P$ at $\tan^{-1} 2.4$ to OA; $140Pa$ in the sense OABC
 31  (i) 2, 6 (iii) $x = 3\sqrt{3}a$; $\sqrt{31}P$, $5\sqrt{3}Pa$
 32  (i) $\sqrt{7}/2$, 27 m (ii) 3, $4\frac{1}{2}$; $9\sqrt{7}/2$ N m in the sense ABC
 33  $24\mathbf{i} - 12\mathbf{j} - 6\mathbf{k}, -\mathbf{j} + 1.5\mathbf{k}$
 34  (i) $2P$ in the direction $\overrightarrow{CA}$; $3d$ (ii) $2K/(a\sqrt{3})$
 35  $m = 1, n = 2$
 36  (i) $3\sqrt{21}P$ (ii) $4W$
 38  $(1\frac{3}{4}, 4\frac{3}{4})$; $(1\frac{19}{28}, 5\frac{13}{28})$
 40  $P\sqrt{5}/3, 2Pa/3$
 41  $11/(6 + \sqrt{2})$ cm
 42  $20/(9\pi - \sqrt{3})$ cm
 44  $a = 1, b = -2, c = 1$; $(18\mathbf{i} + 19\mathbf{j})/13$; $4\sqrt{10}$ in the direction $3\mathbf{i} + \mathbf{j}$; $\mathbf{r} = (11/4)\mathbf{j} + t(3\mathbf{i} + \mathbf{j})$
 45  (i) $3\sqrt{2}$ N; $\mathbf{r} = (-\mathbf{i} - \mathbf{j} + 2\mathbf{k})$ $+ t(\mathbf{i} - \mathbf{j})$ (ii) $3\mathbf{i} - 3\mathbf{j}$; $6\sqrt{3}$ N m

## Chapter 8

8.1  1  $\tan^{-1} (1/2)$
  2  $\tan^{-1} (1/3)$
  3  $69.4°$
  4  $a/3, 2a/3$; $\tan^{-1} (1/2)$
  5  $\tan^{-1} (2/9)$
  6  yes
  7  $97a/132$
  8  yes
  9  $2a$
8.2  1  10 N, 30 N
  2  350 N, 150 N
  3  $10g$ N
  4  0.45 m
  5  15 N
  6  40, 120 and 80 N
  7  100 N; 35 N downward

**8** 4 N, 2 N

**9** 2:1

**10** $\sqrt{3}$:1:1

**12** $4a/3$

**8.3** **1** (a) 73 N, 52 N; 20° (b) 68 N at
70° to the vertical; 22° (c) 53°, 44°; 8°

   **2** 5 N, $5\sqrt{3}$ N

   **3** $W$, $\sqrt{2}W$

   **4** 8 N, $4\sqrt{3}$ N

   **7** 7.2 N, 30 cm

**8.4** **1** (a) $W/4$ (b) $W$ (c) $\sqrt{5}W/3$

   **2** $W\sqrt{3}/4$, $W/4$, $W/2$

   **3** 80 N, 60 N; 100 N; $\tan^{-1}(4/3)$

   **4** $W$, $W/2$, $W/2$

   **5** $W/4$, $3W/4$

   **6** 34°, 63°; $\sqrt{5}W$, $\sqrt{2}W$

   **7** 20 N, $10\sqrt{3}$ N, 20 N

   **8** 0.8 N m

   **9** $3W/5$, $9W/20$, $W/4$

   **10** $3W/10$, $2W/5$

**8.5** **1** (a) 1/2 (b) 68° (c) 14°

   **2** (a) $a/2$, $W\sqrt{2/3}$ (b) $\sqrt{3}/4$, $2W/5$

   (c) $\tan^{-1}(3/4)$, 1/2

   **7** (a) 10 N, $\mu \geqslant 1/6$ (b) $15\sqrt{2}$ N,

   $\mu \geqslant 1/5$

   **10** (a) 4 N (b) 5 N

*Miscellaneous exercises 8*

   **1** $\tan^{-1}(7/18)$

   **2** $2a$

   **4** $5W$, $\sqrt{19}W$

   **5** $\sqrt{2}P$ in the direction $\overrightarrow{BD}$

   **6** 10 N, 19°

   **7** $3W/4$, $W/4$

   **8** $W\sqrt{(31/12)}$, $W\sqrt{(7/12)}$

   **12** 13 N, 10 N, 7 N

   **15** 14 N

   **16** 3 N, 3 N, $\sqrt{10}$ N

   **17** $W$, $W$

   **18** $4W\sqrt{10}$

   **19** 11°

   **20** 2 N, 9 N

   **21** 20 N

   **23** $5W/3$, $a/(2\sqrt{2})$

   **24** 45°, $W/2$; halfway; to the top

   **25** 3/46

   **26** $2W/\sqrt{3}$

   **27** $W/\sqrt{3}$

   **28** $\sqrt{5}W/2$; $7l/12$

   **29** 1/3

   **31** (a) $W/4$ (b) $3W/4$

   **33** 22°

   **37** $(1 - 2\mu^2)/2\mu$; $(1 + 6\mu^2):(4 + 8\mu^2)$

   **38** (i) $r/\pi$ (ii) $h/4$; $\sqrt{\pi}:2$

**39** $1/(3\sqrt{3})$; $\frac{1}{2}W(2 + \sqrt{3})$;
by slipping at C

**40** BC, 45°, $\frac{1}{2}W\sqrt{5}$ at $\tan^{-1}\frac{1}{2}$ to the
horizontal

**42** $7a/2$; $9wa/2$, $21wa/2$; 15

**43** 1/2; 7/8

**44** (i) $35\sqrt{2}$ N (ii) $5\sqrt{2}$ N

**45** 20/7

**46** (i) $W\cos\theta$ (ii) $W\sin\theta$ along
bisector of $\angle$ BAC; $\cos^{-1}(3/4)$

**47** $9a/7$, $13a/7$ (i) 49° 46′ (ii) 5/7

**48** $(4a^2 + 2ax + x^2)/(6a + 3x)$

**49** $13a/8$, $11a/8$

**50** $69r/11$

**51** $W/\sqrt{3}$; $2W/\sqrt{3}$ along AB; $W/\sqrt{3}$
in each rod; AB, AC in compression

**52** (ii) $5W/6$ along CA (iii) AB,
CD $2W/3$; BC, AD $W/2$; BD $5W/6$;
AB, AD, CD in compression

**53** 14 N; in tension AD, CD $14\frac{7}{12}$ N
in compression AB $4\frac{1}{12}$ N, BC $52\frac{1}{12}$ N,
BD $17\frac{1}{2}$ N

**54** (i) $50\sqrt{3}$ N (ii) $50\sqrt{7}$ N along BD
(iii) AB 50 N; AC 100 N; AD,
BC $100\sqrt{3}$ N; CD 200 N; AB, AD
in tension

*Chapter 9*

**9.1** **1** (a) 2 kg m s$^{-1}$ due E
(b) 4 kg m s$^{-1}$ due W (c) 9 kg m s$^{-1}$
due E (d) $\sqrt{2}$ kg m s$^{-1}$ S.E.

   **2** (a) $2\sqrt{2}$, $(\mathbf{i} + \mathbf{j})/\sqrt{2}$ (b) 3, $-\mathbf{j}$
(c) $\sqrt{2}$, $(\mathbf{i} - \mathbf{j})/\sqrt{2}$
(d) 2, $(-3\mathbf{i} + 4\mathbf{j})/5$

   **3** 12 kg m s$^{-1}$ vertically upwards

   **4** 25 kg m s$^{-1}$

**9.2** **1** (a) 15 m s$^{-1}$ (b) 4 s (c) 0.1 s
(d) 160 N

   **2** (a) 40 m s$^{-1}$ (b) 40 m s$^{-1}$
(c) 50 m s$^{-1}$ (d) 3 s

   **3** 400 N

   **4** 250 N

   **5** N 30° W

   **6** $250\sqrt{5}$ N

   **7** 15 N

   **8** 0.2 s

   **9** $3\sqrt{2}$ N s S.E.

   **10** $5\sqrt{3}$ N s vertically upwards

**9.3** **1** 60 N

   **2** 4 N

   **3** 250 N; 250 N

   **4** $20\pi$ N

   **5** 500 N

**9.4** **1** $4u$

   **2** $6\frac{2}{3}$ m s$^{-1}$

**3** 0.5 kg
**4** 10.5 m s$^{-1}$
**5** 2.5 m s$^{-1}$
**6** 8$mu$/5
**7** 10 kg m s$^{-1}$; 0.2 m s$^{-1}$
**8** 5
**9** 9 m s$^{-1}$
**10** $16\frac{2}{3}$ m s$^{-1}$

**9.5** **1** (a) 0.5 (b) 0.2 (c) 3.6 m
(d) 90 cm
**2** (a) 4 m s$^{-1}$ (b) 20 cm (c) 1.6 s
(d) 5 m
**6** 1/3; 11°
**8** 4 N s
**10** $1/\sqrt{2}$; 1/4

**9.6** **1** (a) 6 m s$^{-1}$, 10 m s$^{-1}$
(b) 15 m s$^{-1}$, 18 m s$^{-1}$ (c) 7 m s$^{-1}$,
8 m s$^{-1}$ (d) 3 m s$^{-1}$, 5 m s$^{-1}$
**2** (a) 4 m s$^{-1}$; 1/4 (b) 21 m s$^{-1}$;
14 m s$^{-1}$ (c) 6 m s$^{-1}$; 1/9
(d) 6 m s$^{-1}$; 6 m s$^{-1}$
**3** (a) 1/8, 8:1 (b) 1/2, 1:3
(c) 1/4, 1:1 (d) 1/3, 2:1
**4** 4 m s$^{-1}$, 22 m s$^{-1}$
**5** 1/4
**6** 1/3
**7** 3 m s$^{-1}$, 4.5 m s$^{-1}$
**8** 7/8
**9** 3/10
**10** 40 J

## Miscellaneous exercises 9

**3** 6:1
**5** 1/5
**6** 2 N s
**7** 3 N s
**8** 1/3
**9** 1/4
**13** 2$mu$
**15** $(1 + M/m)\sqrt{(ga)}$
**17** 10 m s$^{-1}$
**20** (a) $\frac{1}{2}mu\sqrt{3}$ (b) $\frac{1}{4}mu\sqrt{3}$
**21** 100 N
**23** (a) $u$/2, 3$u$/2 (b) 5/7
**24** 1:24
**25** $u$/30; 10$a$/(9$u$), $a$/54
**26** 7$u$/20, $u$/10
**27** $M/(m + M)$
**28** 62.5 kJ, 10 kJ; 72.5 kW; 31.25 m
**30** $\tan\theta\cot\phi$
**31** $mv/\sqrt{3}$
**32** $1/\sqrt{e}$
**33** $u\sqrt{e}$
**34** (i) 1/3 (ii) 6$mu$; (i) 3$u$/2 (ii) 2$u$
**36** (i) 5$mu$ (ii) 19$u$/60, $u$/15

**37** 1/4, 3/4; $u$/8, $u$/4, 3$u$/4
**38** 10$mu^2$/$d$, 2$d$/(5$u$) (i) 1/5
(ii) 8$mu$/5 (iii) 32$mu^2$/25
**39** (i) $u$ in direction $\overrightarrow{BA}$ (ii) 4/5
(iii) 3$mu^2$ (iv) $u$, 4$u$/3 (v) 1/9
**40** (i) 10/11 m s$^{-1}$ (ii) 13.27 kN;
0.4, 4200 J

## Chapter 10

**10.1** **1** 4 rad s$^{-1}$, 2/$\pi$
**6** $\sqrt{(gh)}$
**10** $mv^2/a$
**10.2** **1** $4\sqrt{2}$ m s$^{-2}$
**2** 10$\pi$ m s$^{-2}$
**3** (a) $(2n + 1)\pi$/2 (b) $n\pi$
**4** 2$\pi a$/(e$T$)
**10.3** **1** $\sqrt{(3g)}$ m s$^{-1}$
**7** 3$mg$, $mg$
**9** $a$
**10.4** **1** 5 m, 2$\pi$ s
**2** $2\sqrt{3}$ m
**3** 5/$\pi$ m s$^{-1}$
**6** 0.18 s
**7** 50/$\pi$
**9** $5\sqrt{2}$ cm
**10** 48 m s$^{-2}$
**10.5** **1** 5 m; $\pi$ s
**2** 8 N
**3** (a) 4 N (b) $2\sqrt{3}$ N (c) 3.2 N
**4** (a) 0.5 m s$^{-1}$ (b) $(\pi + 8)$/10 s
**5** (a) 5 cm; 2$\pi$/5 s (b) 10 cm; 2$\pi$/5 s
**6** (a) $2\pi\sqrt{(l/g)}$; $l$/2 (b) 3$mgl$/2
**7** $\frac{1}{2}\sqrt{(gc)}$; $g$/2
**8** $c$; 2$mg$, 0
**9** $n$/2
**11** $T$/4
**12** $\frac{1}{2}(\pi + 1)\sqrt{(l/g)}$
**14** (a) $2\pi\sqrt{(l/g)}$, $l$/2 (b) $2\pi\sqrt{(l/g)}$,
$l$/4
**20** $(1/\pi^2)$ cm
**10.6** **1** (a) [L$^3$] (b) [ML$^2$T$^{-3}$]
(c) [T$^{-2}$]

## Miscellaneous exercises 10

**1** 4$mg$/3
**2** 3.54 a.m.
**4** 100 m s$^{-2}$
**6** 2.5 m; $\pi$ s
**8** $\mu mg$/3
**9** 12.05 p.m. and 6.55 p.m.
**10** 1/3 s
**11** $a$/4; $(1/\pi)\sqrt{(g/a)}$
**12** $2\pi\sqrt{(a/3g)}$
**13** 2/3 s
**15** $\pi$/2, 6 m s$^{-2}$

**16** $1 \text{ m s}^{-1}$

**17** $4 \text{ m s}^{-1}$

**18** 10 N, 4 N

**20** $\cos^{-1}(1/4)$

**21** $\pi\sqrt{(a/g)}$

**22** $2\pi\sqrt{(l/2g)}, l/(2\sqrt{2})$; $2\pi\sqrt{(l/g)}$, $(2 - \sqrt{2})l/4$

**23** $\sqrt{(2g/l)}$; 35:24

**24** $(k/\omega^2)(\omega t - \sin \omega t) + ut$; $u\omega + k = 0$

**25** (i) $x = 3, v = -12$ (ii) $x = 5$

**26** $2/3 \text{ m s}^{-2}$, 10 m

**27** $60°$

**28** $2mg/\sqrt{3}, mu^2/l - mg/\sqrt{3}$

**29** $mg/2, mg/2$ (i) $3l/2$ (ii) $2\pi\sqrt{(l/2g)}$

**30** $5mu^2/(9a) + 5mg/8$, $5mu^2/(9a) - 5mg/8$

**31** $(60 - \pi^2 n^2/100)$ N, $(20\sqrt{15})/\pi$ (i) $60\sqrt{5}$ N (ii) 120 N

**33** (i) $\pi$ (ii) $2\pi$ (iii) $3\pi/2$

**34** 10 m; 8 or 12 Ns in the opposite direction to the velocity

**35** 3; $l/2$

**36** $2\pi\sqrt{(l/2g)}, \sqrt{(gl/2)}$; $\sqrt{(3gl/8)}, \frac{1}{2}mg$

**37** $4\sqrt{5}$ m; $\pi$ s; $(32\sqrt{5} - 16)$ J; $\pi/3$ s

**39** (i) $5mg/3$ (ii) $5mg/12$

**40** $\sqrt{3\pi an}; M(g + 2\pi^2 an^2)$; $1/(6n)$

## Chapter 11

**11.1** **1** (a) 120 (b) 360

**2** (a) 5040 (b) 2520

**3** 720

**4** (a) 256 (b) 24

**5** 40

**6** 10 080

**7** 35

**8** 6

**9** 240

**10** 72

**11** 60

**12** 12

**13** 531 441 (a) 24 (b) 264

**14** 40 320, 1260

**15** 420, 120

**11.2** **1** (a) 28 (b) 28

**2** (a) 210 (b) 252

**3** 22 100

**4** 35

**5** 200

**6** 15

**7** 15, 20

**9** 120

**10** 650

**11.3** **1** 36, 4, 12

**2** (a) $\{6, 6\}$ (b) $\{1, 1\}, \{1, 2\}, \{2, 1\}$ (c) $\{6, 4\}, \{4, 6\}, \{5, 5\}$

**3** 12, 156, 650, 1352

**4** 3, 3, 4

**5** $\{(1, 3), (3, 1)\}$

**11.4** **5** 3/4

**6** (a) 0 (b) 1/9 (c) 1/18

**7** (a) 1/13 (b) 3/4

**8** 1/11

**9** (a) 3/13 (b) 10/13

**10** 5/12

**11** 216, 1/54

**12** 1/4

**11.5** **1** (a) 1/3 (b) 2/9 (c) 5/9 (d) 4/9

**2** 2/3

**3** 49/52

**4** (a) no (b), (c) and (d) yes

**5** (a) 5/52 (b) 2/13 (c) 6/13

**6** 1/12, 1/6, 1/4, 0

**11.6** **1** none

**7** 3

**11.7** **1** 1/2

**2** 1/8, 1/2

**3** 1/12, 1/4

**4** 1/3

**5** 1/6

**6** 1/36

**7** 1/4, 1/3

**8** 5/11

**10** 5/11

**11** 1/10

**12** 1/3

**13** 18/125

**14** (a) 1/39 (b) 0

**15** 1/3, 1/4, 2/3

**11.8** **1** (a) 1/10 (b) 3/10 (c) 3/5

**2** (a) 2/17 (b) 13/34

**3** 2/3

**4** 1/5

**5** (a) 1/27 (b) 1/9 (c) 2/9

**11.9** **1** 3/8

**2** 3/8

**3** 27/64

**4** 63/256

**6** (a) $(4/5)^4$ (b) 32/625

**7** 25/216

**8** 1/2

**10** (a) 0.53 (b) 0.0012

## Miscellaneous exercises 11

**1** 840 (a) 360 (b) 120 (c) 240

**2** (a) 4 (b) 10 (c) 40

**3** 2100

**4** (a) 56 (b) 24 (c) 32

**5** 5400

6  5/8
7  2/3, 1, 0, 1/4
9  (a) 1/256  (b) 1/64  (c) 3/32
10  16/45
11  3/10, 0, 1/15, 1/20; no
12  2/3; 11/126
13  (i) 1/21 (ii) 2/7 (iii) 4/9
(iv) 5/42 (v) 1 (vi) 4/3; 0
14  (i) 0.3 (ii) 0.3 (iii) 0.5
15  793/2048
16  (a) 5/16  (b) 5/8
17  (a) 194  (b) 182
18  (i) 1/4 (ii) 2/7
19  9!/2 (i) 72(5!) (ii) 180(5!)
20  11/12
21  (a) 1/221  (b) 32/221  (c) 13/17;
25/33
22  (a) 3/13  (b) 1/2  (c) 8/13
23  (a) 188/221  (b) 19/34; 105/188
24  59/576
25  20, 160, 3600
26  126; 56

27  $A$ and $B$
28  (i) 5/12 (ii) 8/15 (iii) 1/12
29  125/216, 75/216, 15/216, 1/216;
2p loss
30  (i) 0.02 (ii) 0.78 (iii) 0.76
(iv) 1/30
31  26; 2/3
32  (a) 3/8  (b) 2/5  (c) 1/2
33  1/3
34  36/91, 30/91, 25/91
35  (a) 81/256  (b) 1/256  (c) 3/128
(d) 3/32
36  10/21
37  (a) 1/21  (b) 1/126  (c) 1/14
(d) 5/112
38  (a) 1/9  (b) 5/9 (c) 1/2
39  32/59
40  13/16; 1/4
41  (a) 5/8  (b) 1/6  (c) 5/24
42  (a) 5/36  (b) 1/12  (c) 5/36; 2/81
43  1/3; 1/22

# Index